INTRODUCTION TO DATA ANALYSIS
AND STATISTICAL INFERENCE

Carl N. Morris

University of Texas at Austin
Austin, Texas

John E. Rolph

The Rand Corporation
Santa Monica, California

PRENTICE-HALL, INC.
Englewood Cliffs, New Jersey 07632

Library of Congress Cataloging in Publication Data

Morris, Carl N.
      Introduction to data analysis and statistical
inference.

      Bibliography: p.
      Includes index.
      1. Mathematical statistics—Data processing.
I. Rolph, John E., joint author.   II. Title.
QA276.4.M67   519.5   80-27931
ISBN 0-13-480582-8

Printed in the United States of America

10 9 8 7 6 5 4 3 2 1

Editorial/production supervision: Nancy Milnamow
Cover design; Dawn Stanley
Manufacturing buyer: Joyce Levatino and Gordon Osbourne

Prentice-Hall International, Inc., *London*
Prentice-Hall of Australia Pty. Limited, *Sydney*
Prentice-Hall of Canada, Ltd., *Toronto*
Prentice-Hall of India Private Limited, *New Delhi*
Prentice-Hall of Japan, Inc., *Tokyo*
Prentice-Hall of Southeast Asia Pte. Ltd., *Singapore*
Whitehall Books Limited, *Wellington, New Zealand*

Dedicated to the memory of
Winston Kai-wan Chow,
our good friend and colleague.

We wrote this text to meet different needs at Rand during the past decade. Principally, it grew out of a two-quarter service course given jointly by the authors in the Rand Graduate Institute of Policy Studies (RGI) since 1971. That course introduces students with varied mathematical backgrounds to statistics and data analysis so they can conduct and criticize policy studies. It also has given us a chance to assemble and present those simpler statistical and computational methods that we and other Rand data analysts have found especially useful for understanding and analyzing data on human populations.

Students in the RGI course have been graduate students in policy studies. The minimal mathematical background typically has been lower division calculus, sometimes taken a number of years earlier. RGI students take this as a terminal course, so we have covered a variety of methods, none in great mathematical depth. Other students had substantial statistical backgrounds. They still benefited from the course after the first several chapters because it is oriented toward data analysis and real data, topics often not included in elementary or intermediate courses, but always challenging even for experienced analysts.

We have required students to use the computer quite extensively. The text should be suitable for use with any standard statistical package, but in recent years we have used it exclusively with STATLIB. A description of STATLIB is given in the companion manual to this text; STATLIB: A Statistical Computing Library by William A. Brelsford and Daniel A. Relles (Prentice-Hall, 1981). We recommend that this text be used with the STATLIB manual. The computer has allowed us to rely heavily on data to introduce concepts and so substitute examples for proofs of theorems. This has motivational advantages and permits introduction of more advanced material than a pure theory course would allow.

Besides being used in the policy analysis course at RGI for which it was developed, Chapters 4 and 7-14 have been used at the University of Texas for a one-semester applied course for masters level students in statistics. The last two-thirds of the text (Chapters 6-14) can be used for a one-semester second course in undergraduate statistics at the post-calculus level.

We believe data analysts, public policy analysts, economists, social scientists, and other researchers in government and industry will find this text useful for reference and self-teaching. We have chosen the topics and methods presented in this text for several qualities: they are useful in public policy research; the methods are multivariate, as needed for multidimensional real data; they are reasonably simple (some exceptions exist); and good computational packages are available where lengthy calculations are required.

Chapter 1 gives an introduction and overview of the text. All the chapters in the book are summarized there and the organization explained. Because the book contains more material than might normally be covered in a one-semester or a two-quarter course, topics may be selected to suit various purposes. We have generally covered all but one or two chapters in a two-quarter course.

We appreciate the help of several of our colleagues. Daniel A. Relles has participated in teaching the course from earlier versions of these notes on three separate occasions. Besides giving us his helpful comments on the existing chapters, he coauthored Chapter 10 and contributed substantially to Chapter 9. In addition, Dr. Relles is a codeveloper of STATLIB, the statistical computing system mentioned above and described in the companion publication. Thus, he has been instrumental in adapting this computing system so it can be used in such a service course. We owe much to the Rand Graduate Institute students who took this course during the years 1971-1979 and offered insightful comments, vigorous complaints, and puzzled questions that led us to clarify many sections of the text. We are indebted to Cambridge University Press for permission to reprint tables from E. S. Pearson and H. O. Hartley (eds.), <u>Biometrika</u> <u>Tables</u> <u>for</u> <u>Statisticians</u>, Vol. 1, Cambridge University Press, Cambridge, England, 1954. Finally, we greatly appreciate the superb work of Martha Cooper who typed this version and many earlier ones during the years we have taught the course.

A Ford Foundation grant awarded to the Rand Graduate Institute and a grant to Rand by the (then) Department of Health, Education, and Welfare, partially supported preparation of this text. Chapter 3 is a revised version of Set #11 in <u>Statistics</u> <u>by</u> <u>Example</u>, edited by F. Mosteller et al. (Addison-Wesley, Menlo Park, 1973), pp. 119-140.

CONTENTS

FIGURES

Appendix Tables

Part I. DATA ANALYSIS

Chapter 1

INTRODUCTION

Introduction and Overview of This Text

The reader will find this text organized into four main parts: I,
Data Analysis; II, Statistical Inference for Linear Models; III, Robustness; and IV, Additional Topics. Part I classifies data and introduces
the appropriate statistics that are used throughout Part II, but it does
not distinguish between the sample (the data at hand) and the population
(all observations that theoretically could be collected). Part II
acknowledges this difference and presents the relevant sampling distributions for estimates of population parameters. By the end of Part II
the reader will have covered a wide range of methods used in statistical
applications. Validity of sampling distributions depends on meeting
certain modeling assumptions. Part III describes these assumptions, how
to verify them, and provides corrective actions when they fail. Part IV
introduces other methods in analysis and statistical design: methods
for combining sample information with nonsample information or intercom-
ponent information in Chapter 12, "Bayes and Empirical Bayes Inference,"
and methods for acquiring data from populations in Chapter 13, "Sampling
Methods," and Chapter 14, "Experimental Design: Problems in Planning
Experiments Involving Human Populations to Evaluate Social Programs."

The reader will find many statistical methods introduced here; many
others are being used by statisticians and applied scientists. We hope
to show that while there are many methods, there are fewer central concepts. We try to bring out this relationship between methods and concepts. This will be facilitated if the overview presented in the

3

remainder of this chapter is reviewed frequently during study of the book. Because this chapter will serve as an introduction and as a review, we will use some terms and methods here that will not be defined until later.

Most data analytic methods operate on a data matrix. Virtually every attempt to understand multivariate data should begin by organizing the data into a two-way array, the n rows being the "individuals" or "units of analysis" and the k columns being the measured "variables." Table 1.1 provides an example with the unit of analysis being people, but in other examples the unit of analysis might be years, countries, firms, objects, or any of thousands of other possibilities. In this example, a sample of n persons was interviewed and each was asked seven questions: his annual health expenditure for the past year, whether he was hospitalized during that period, his annual income, his age, his health status (excellent, good, fair, or poor), his state of residence, and his sex (M = male, F = female). An observation label is defined so each person has a numerical identifier.

The measurement level of each variable is classified as

a. Continuous

b. Categorical (ordered)

c. Categorical (unordered)

d. Categorical (binary).

"Continuous variables" have numerical scales taking on a range of values, as expenditure, income, and age in Table 1.1. Health status is an "ordered categorical variable" with four categories. The categories are ordered, although one may be unable to assign numerical values

Table 1.1

EXAMPLE OF A DATA MATRIX WITH n UNITS OF ANALYSIS AND k = 7 VARIABLES.
SOME VARIABLES ARE CONTINUOUS, SOME ARE ORDERED BUT NOT NUMERICAL,
SOME ARE CATEGORICAL OR DICHOTOMOUS

Observation Label (Person)	Expenditure $ (Continuous)	Hospital (Binary)	Income $ (Continuous)	Age (Continuous)	Health Status (Ordered Categorical)	State (Categorical)	Sex (Binary)
1	1400	yes	8175	23	Fair	CAL	M
2	50	no	16225	37	Exc.	NY	F
...	...	...	...	...	...	...	...
n	300	no	11100	60	Good	TEX	M

5

indicating the relative magnitudes of the categories. The hospital, state, and sex responses are "categorical," with hospital and sex called "binary" because they involve exactly two categories. The types of variables partly determine the analysis.

The variables have asymmetric roles in a statistical analysis. The response variable of central interest is called the "dependent variable." Variables affecting, causing, or explaining the dependent variable are called "independent variables." A researcher usually wishes to know how some variable--health expenditures in this case--is affected by another variable--say, income. He must decide whether or not to use the other variables available to him when investigating this relationship.

Descriptions and summary statistics for one variable, means and standard deviations being the most important of these, are described in Chapter 2 for the different measurement levels. Chapter 3 introduces the simplest example involving two variables: explaining a dependent variable from one independent variable. Simple linear regression and simple correlation are presented here. When several independent variables are analyzed together, simple regression and simple correlation give way to multiple regression and multiple and partial correlation in Chapter 4.

Part II treats issues that arise when the sample at hand must be used to make inferences about the larger population from which it was drawn. Chapter 5 introduces probability ideas and some of the main distributions that arise when the data have been sampled at random from a population. These ideas are used in Chapter 6 to make sampling and modeling assumptions and then to derive sampling distributions, confidence intervals, and test statistics for the statistics presented in Chapters 3 and 4.

When only categorical independent variables appear without any continuous independent variables being used to explain a continuous dependent variable, the Analysis of Variance (ANOVA) arises. Although ANOVA is introduced as a separate topic in most other statistics courses, Chapter 7 treats it as a special case of multiple regression by creating dummy variables from ordered and unordered variables. The term Analysis of Covariance (ANOCOVA) is reserved for an ANOVA where continuous and categorical independent variables are analyzed together.

The assumptions for inference in linear models introduced in Chapters 6 and 7 include, among other things, that the dependent variable be continuous. Chapter 8 introduces logistic regression, a model appropriate for binary dependent variables. Maximum likelihood logit estimation methods are applicable when the independent variables are categorical, continuous or a mixture of the two. A maximum likelihood logit analysis could be used with the data of Table 1.1 to estimate how the probability of hospitalization depends on the other (independent) variables. Chapter 8 also shows how discriminant analysis and minimum logit chi-square methods are sometimes appropriate alternatives to maximum likelihood logit estimation. Categorical dependent variables other than binary categories are considered in portions of Chapters 2 and 11 as contingency tables but otherwise go untreated in this text.

The methods just described are summarized in Table 1.2. This table shows that the analyst has one goal: Use the data to model the dependence of one variable (the dependent variable) on other variables and then interpret nature or predict the future based on these results. The technique used depends on the measurement level of the data.

Part III covers the validity of the interpretations and extrapolations made from the fitted model. The results must depend as much as possible on the data and as little as possible on unverified modeling assumptions. Results that depend little on modeling assumptions are called "robust." Computed residuals, first introduced in Chapter 4 to develop multiple regression, are reintroduced in Chapter 9 for checking linear model assumptions. Chapter 9 shows how linear models based on nonlinear transformations of the dependent variable often can correct

Table 1.2

TYPES OF DEPENDENT AND INDEPENDENT VARIABLES AND
METHODS APPROPRIATE TO THEIR ANALYSIS. CHAPTER
NUMBERS ARE INDICATED IN BRACKETS []

| | Independent Variables | | |
	Continuous	Mixed	Categorical
Continuous Dependent Variable	Regression [3, 4, 6]	ANOCOVA [4, 7]	ANOVA [7]
Binary Dependent Variable	1. Discriminant Analysis [8] 2. Maximum Likelihood Logit [8]	Maximum Likelihood Logit [8]	1. Minimum Logit Chi-square [8] 2. Maximum Likelihood Logit [8] 3. Contingency Table [2, 11]

the model misspecification for possible nonnormality in the dependent variable. When this fails, the analyst sometimes can use robust estimation methods, described in Chapter 10. Part III concludes in Chapter 11 with the discussion of nonparametric statistical tests and of tests appropriate for categorical data.

The final portion of this text, Part IV, treats two other important ideas in statistics. Chapter 12, "Bayes and Empirical Bayes Inference," considers problems with sparse sample information but where the data analyst has additional information available which he can combine with sample information to make better inferences. Bayes' theorem shows how sample information should be used to modify a priori information held by the data analyst. Empirical Bayes inference can improve precision of estimates when many parameters must be estimated but insufficient data exist to estimate any one of them well. As the name suggests, empirical Bayes procedures are much like Bayes procedures except that the a priori information is estimated by the data and so the data analyst is protected from faulty judgment.

The final two chapters are concerned with cost-effective acquisition of data. Chapter 13, "Sampling Methods," presents methods commonly used in sample surveys to provide unbiased samples, reduce survey costs, or increase the precision of the desired estimates. Chapter 14, "Experimental Design: Problems in Planning Experiments Involving Human Populations to Evaluate Social Programs," discusses a wider set of topics for planning studies and experiments that includes but goes well beyond the sampling issues raised in Chapter 13. The key idea in an experiment is that the experimenter identifies the experimental treatment (in this case the variants of a social program) and the experimental units (the human subjects who will be exposed to these treatments), and then ensures that the subjects are assigned to the treatments in such a way that measured differences between the treatments will reflect treatment differences, and not systematic differences between the subjects. Chapter 14 covers this and the broad set of issues that confront someone planning an experiment or observational study with special emphasis on human populations, including those things that can go wrong.

Chapter 2

SIMPLE DESCRIPTIVE STATISTICS

2.1. UNIVARIATE DATA

A first step in data analysis is structuring the data so that it appears to be something more than an incoherent jumble of numbers. Our first example of a data set is the percentage of the metropolitan area population who are federal government employees,[*] in each of the 125 U.S. cities whose metropolitan area populations were more than 250,000 in 1970. These data are *one dimensional* or *univariate* since there is only one quantity being measured for each city. For the moment we shall not be concerned with other data that might also be collected about these cities. Later, when several measurements are considered simultaneously, the data will be described as *multivariate*. The first statistical and graphical techniques we introduce will be appropriate only for univariate data.

Looking at a list of 125 percentages does not lead to rapid comprehension of the data set. Thus, a first step is to *sort* or *order* the data in ascending (or sometimes descending) order. This is done in Table 2.1. The city in this table with 10.93 percent is Washington, D.C. Such a data point is called an *outlier* since it is so far from the other points. A useful tabulation is given in Table 2.2 that shows the frequency with which cities appear in intervals of one-half of one percent. The graphical display of information in Table 2.2 given in Figure 2.1 is called a *histogram*. It shows the shape of the distribution: most cities have less than two percent federal government employees, and almost all cities have less than 5½ percent.

---

[*] Taken from E. Keeler and W. Rogers, *A Classification of Large American Urban Areas*, The Rand Corporation, R-1246-NSF, May 1973.

11

When constructing tables and histograms we must decide how large an interval to use. Our choice of width 0.5 yields 22 intervals between 0 and 11 percent, although only 12 intervals contain any observations. Using shorter width like 0.1 would make the histogram look more "jagged" and fail to show the general shape of the frequency distribution as well. One should choose as short an interval as is consistent with visually displaying the general shape of the distribution of the data.

Two more numerical ways of describing the distribution include the *sample cumulative distribution function (sample CDF)* and the *sample quantiles*. The sample CDF is the function $F(x) = \{$the number of $x_i \leq x\}/n$ where $x_1, \ldots, x_n$ are the data. To take examples from Table 2.1, $F(0.40) = 17/125 = 0.134$, $F(1.05) = 63/125 = 0.504$, $F(10.93) = 1.000$. It is not nexessary that x be an observed percentage, for example $F(0) = 0$ (no cities had zero government employees), $F(1) = 60/125 = 0.48$ (48 percent of the cities had no more than one percent government employees) and $F(50) = 1$ (all cities had fewer than 50 percent government employees). The sample CDF, which increases from 0 to 1 as x increases, gives a complete description of the ordered data and is useful in both statistical data analysis and statistical inference.

The *sample quantile* at p, $0 < p < 1$, is the value $y^*(p)$ for which the fraction p of the sample is less than or equal to y^* and the fraction 1-p is larger than y^*. For example, for the data of Table 2.1, if $p = 0.134$ ($= 17/125$), then $y^* = 0.40$ and if $p = 0.504 = 63/125$, then $y^* = 1.05$. In general the quantile $y^*(p)$ is chosen to satisfy the relation $F(y^*) = p$, so $y^*(p)$ is just the inverse of the distribution function. Other examples of quantiles include the median and quartiles, defined in the next section.

12

Table 2.1

RANKED PERCENTAGE FEDERAL GOVERNMENT EMPLOYEES IN 125 U.S. CITIES

Rank	Per- cent- age	Rank	Per- cent- age	Rank	Per- cent- age	Rank	Per- cent- age
1	.27	33	.61	65	1.08	97	1.84
2	.28	34	.66	66	1.10	98	1.89
3	.31	35	.66	67	1.10	99	1.92
4	.34	36	.67	68	1.12	100	2.02
5	.34	37	.68	69	1.15	101	2.02
6	.35	38	.69	70	1.16	102	2.05
7	.35	39	.73	71	1.17	103	2.12
8	.36	40	.74	72	1.21	104	2.14
9	.36	41	.75	73	1.23	105	2.31
10	.37	42	.75	74	1.25	106	2.42
11	.37	43	.75	75	1.26	107	2.43
12	.37	44	.75	76	1.27	108	2.45
13	.38	45	.79	77	1.28	109	2.49
14	.39	46	.84	78	1.28	110	2.72
15	.40	47	.84	79	1.29	111	2.82
16	.40	48	.86	80	1.29	112	2.94
17	.40	49	.87	81	1.31	113	2.96
18	.41	50	.88	82	1.32	114	2.98
19	.42	51	.89	83	1.37	115	3.10
20	.43	52	.90	84	1.38	116	3.22
21	.43	53	.91	85	1.40	117	3.71
22	.45	54	.92	86	1.40	118	4.14
23	.45	55	.94	87	1.42	119	4.29
24	.45	56	.95	88	1.54	120	4.76
25	.45	57	.97	89	1.55	121	5.07
26	.46	58	.98	90	1.57	122	5.08
27	.47	59	.98	91	1.63	123	5.31
28	.53	60	.98	92	1.66	124	5.41
29	.56	61	1.02	93	1.68	125	10.93
30	.57	62	1.03	94	1.76		
31	.58	63	1.05	95	1.78		
32	.59	64	1.07	96	1.79		

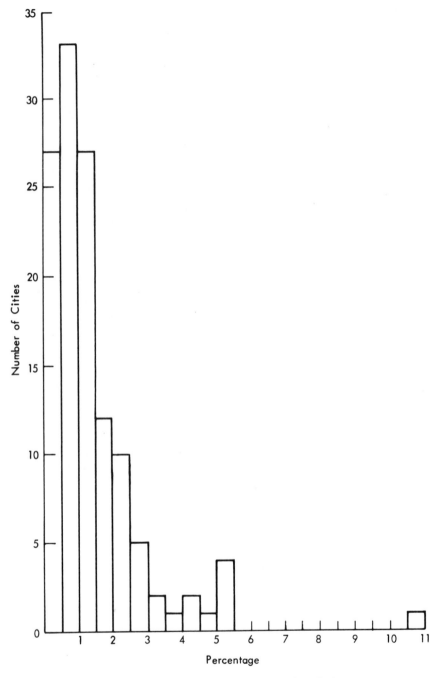

Figure 2.1 - Histogram of Percentage Federal Government
Employees in 125 U.S. Cities.

Table 2.2

GROUPED PERCENTAGE OF FEDERAL GOVERNMENT EMPLOYEES
IN 125 U.S. CITIES

At least (percent)	Less than (percent)	Number of Cities
0	0.5	27
0.5	1.0	33
1.0	1.5	27
1.5	2.0	12
2.0	2.5	10
2.5	3.0	5
3.0	3.5	2
3.5	4.0	1
4.0	4.5	2
4.5	5.0	1
5.0	5.5	4
-	-	-
-	-	-
-	-	-
10.5	11.0	1
		125

A slightly better convention that is nearly equivalent to the preceding is to let $y_1 \leq y_2 \leq \cdots \leq y_n$ be the n ordered observations (e.g., n = 125 in Table 2.1) and then define the p-th quantile $y^*(p)$ as

$$
y^*(p) = \begin{cases}
y_1 & \text{if } p < \dfrac{0.5}{n} \\
y_n & \text{if } p > 1 - \dfrac{0.5}{n} \\
y_i + (y_{i+1} - y_i)(np + 0.5 - i) & \text{with } i \text{ being} \\
& \text{the integer such that } i \leq np + 0.5 < i+1.
\end{cases}
$$

Hence the p-th quantile is the np + 0.5 observation, interpolating if necessary. These conventions guarantee that the minimum and maximum quantiles are the two extreme values of the data, and that linear interpolations be used when no one datum corresponds to the desired quantile. For simplicity, the value y_i with i the nearest integer to np + 0.5 is often used. To convert to percentages, the p-th quantile often is

called the 100 p-th *percentile*. In the examples above, if p = 17/125

then i = 17 and $y^*(17/125) = 0.40 + (0.41 - 0.40)(0.5) = 0.405$. If

p = 63/125 then i = 63, $y^*(p) = 1.05 + (1.07 - 1.05)0.5 = 1.06$. The

median (defined in the next paragraph) is $y^*(0.5)$ which is 1.05 for these

data, since i = 63 and np + 0.5 - i = 0.

2.2. MEASURES OF LOCATION

A histogram is sometimes too complicated and the analyst may prefer

to use only one or two numbers to summarize the data. It is common

practice to use one number to measure the center or a typical value of

the data and a second number to measure how far a typical value is from

this center. The *median* is an example of a one number summary of the

location of the center of a distribution. It is defined as a number

for which at least half the data points are as large and half of the

data points are as small. Hence, if there are an odd number of data

points, the median is the "middle" value while if there are an even

number of points the median is usually taken as the average of the two

"middle" values. Thus, the median is the 0.5 quantile or 50-th percentile.

From Table 2.2, the median is in the interval 1.0 to 1.5, but Table 2.1,

which lists the *ordered data*, indicates that the median is the value for

city 63 (63 = (1 + 125)/2) of 1.05 percent.

The most commonly used measure of location is the *mean* value. The

mean or *average* is calculated by summing all values and dividing by the

number of values. The mean percentage for the 125 cities is 1.45 percent,

a value substantially higher than the median. Symmetric frequency distri-

butions have identical means and medians. The difference between the

16

mean and median occurs because (i) the distribution is "skewed" to the
right (the tail on the right and not the left), and (ii) because there
is an *outlier* on the right side, namely, Washington, D.C.

If the mean is used as the central value of the data set, the
analyst should be aware of the effect of outliers. For example, removing
Washington reduces the mean from 1.45 to 1.38. Outliers sometimes occur
due to miscopied data, or a once in a lifetime event. Then it is useful
to have location measures that are relatively insensitive to them.
Although the median has this property, it wastes a lot of the information
in the data because it ignores the distances between the values. A compromise between the mean and the median which is resistant to the effect
of outliers is the *trimmed mean,* computed as follows. If there are n
data points, remove a fixed proportion p of the points from the upper
end and from the lower end of the distribution and average the remainder.
That is, delete the largest np points and the smallest np points. Then
the trimmed mean is the mean of the remaining n(1-2p) points. The *midmean* is the particular case $p = \frac{1}{4}$ that averages only the middle half
of the data.

From Table 2.1, the midmean of our data set is 1.07, computed as the
average of the numbers beginning with rank 32 (0.59) and ending with
rank 94 (1.76), thereby excluding the lowest and highest 31 observations.
For hand computation, the *trimean* is frequently preferred to the midmean.
First define the *lower quartile* as the 25th percentile and the *upper
quartile* as the 75th percentile, $y^*(0.25) = 0.5875$ and $y^*(0.75) = 1.765$,
respectively. Then the *trimean* is the average of these two quartiles
and the twice-counted median: $(0.5875 + 2(1.05) + 1.765)/4 = 1.11$.

The *mode*, defined as the most frequently occurring value of the data, occasionally is used as a measure of location. For these data there are two modes, 0.45 and 0.75, each occurring just four times. Were these values computed to more decimal places, there would be 125 modes each occurring one time. The mode of the histogram, Figure 2.1, is 0.75 percent (the center of the second interval) indicating this to be the most frequent percentage of federal employees. Thus, the histogram mode is substantially less than either the mean or median.

2.3. MEASURES OF SPREAD

Besides a measure of location, it is useful to have measures of how dispersed the data are. There are a number of measures. The difference between the upper and lower quartiles, called the *interquartile range*, is one such measure. The interquartile range, which is the width that contains the center half of the cities, is $1.765 - 0.5875 = 1.18$ from Table 2.1. While the interquartile range measures the spread of the data near the middle, the *range*, defined to be the maximum minus the minimum, measures the total spread of the distribution: $10.93 - 0.27 = 10.66$. The interquartile range clearly is more resistant to the effect of outliers than the range. Other interquantile ranges can be constructed for any $p < \frac{1}{2}$, by taking $y^*(1-p) - y^*(p)$, not just for $p = 0, \frac{1}{4}$. Then the central $1-2p$ observations are contained by an interval whose length is this range.

An alternative to defining dispersion from quantiles is to choose a center and a distance measure, and use the average distance of the data points from the center as a measure of spread. Common choices of the data centers are the median or mean while common distance measures are absolute distance or squared distance. Suppose our observations (not

necessarily ordered) are x_1, ..., x_n, with median n. Then the *mean*

absolute deviation about the median is

$$\text{MAD}(x) = \frac{1}{n} \left(|x_1-m| + |x_2-m| + \ldots + |x_n-m| \right) = \frac{1}{n} \sum_{i=1}^{n} |x_i-m| .$$

This is the average amount that the observations in the sample differ

from the median of the sample and therefore measures how dispersed the

sample is. The root mean-squared deviation about the mean $\bar{x}$ is called

the *standard deviation*

$$s_x = \sqrt{\frac{1}{n} \sum_{i=1}^{n} (x_i-\bar{x})^2} .$$

Note its similarity to the MAD. Instead of taking absolute differences

around a central point, squared distances are used, and hence the square

root must be taken at the end to get the result back on the right scale.

For our city data MAD(x) and s_x are 0.83 and 1.41, respectively. It is

always true that $\text{MAD}(x) \leq s_x$, because by squaring the data before adding,

s_x gives more weight to observations far from the center of the data.

The square of the standard deviation is called the variance, and

given by the formula

$$\text{var}(x) = \frac{1}{n} \sum_{i=1}^{n} (x_i-\bar{x})^2 .$$

The standard deviation is by far the most commonly used measure of spread

largely because of the nice mathematical properties of its square, the

variance. The reader should verify for himself that all the dispersion

and location measures given here, except the variance, preserve the units

of measurement. That is, if all the data points are multiplied by a

positive constant, the measures are multiplied by the same positive

19

constant. E.g., if $z_i = cx_i$, then

$$MAD(z) = |c|MAD(x), \quad s_z = |c|s_x.$$

An example that might apply to the city data is multiplication by 1/100, which changes percentages to proportions.

2.4. CATEGORICAL DATA

Data sets frequently are categorical rather than numerical. An example from the city data is the region of the United States in which each city is located. Suppose the cities are categorized into the four regions: Northeast, South, Midwest, and West and assigned the numbers 1, 2, 3, 4 for the four regions. Then the size and order of the numbers have no real meaning, and using these numbers to get a mean or median region makes no sense. (Although one might instead use latitude and longitude and then take their means.) Thus, with categorical data, one only can examine the frequency distribution of the data given in Table 2.3 rather than use one or two number summaries.

Table 2.3

DISTRIBUTION OF 125 U.S. CITIES BY REGION

Region	Number of Cities	Proportion
Northeast	30	0.24
South	41	0.33
Midwest	30	0.24
West	24	0.19
Total	125	1.00

Categorical data with only two categories are a special case which may be scaled arbitrarily into numerical data, perhaps most conveniently by 0 and 1. Then averages become meaningful, since the average of zeros and ones is the proportion of ones. The above restrictions therefore apply most appropriately to cases with more than two categories.

2.5. CLASSIFICATION OF DATA

Data are classified as *categorical*, *ordinal* and *numerical*. *Ordinal data* permit greater than or equal comparisons, but do not have a scale to determine "how much greater than." The English language provides many relative, but imprecise labels leading to ordinal scales. For example, a question of health status may be answered poor, fair, good or excellent; or a product sold in small, medium, large, and giant sizes. Ordinal scales may be finely graded without a numerical scale, for example the world's top 50 tennis players are ranked (or *rank ordered*) each year, but without specifying how much better one player is than another.

Most statistical methods are not appropriate for ordinal data (although some are) and the researcher then must either ignore the ordering and treat it as categorical, or scale the data and analyze it as numerical data. Some methods appropriate to ordinal data will be introduced later in nonparametric inference, but we usually will be concerned with either categorical or numerical data.

2.6. BIVARIATE AND MULTIVARIATE DATA

The most interesting statistical analyses in our interrelated world are those that investigate relationships between two or more variables, i.e., *bivariate* and *multivariate data*. The variables may again be categorical, ordinal or numerical. If all variables are categorical,

a *contingency table* organizes and displays the data systematically.
Table 2.4 is a *two-way contingency table* of the number of cities cross-
classified by region and congressional power. The congressional power
variable is coded as the value 1 for a city having either a representative or senator who is chairman or the ranking minority member of an
important committee provided that congressman's political power base includes a portion of the city. Otherwise, the variable is coded as 2.
The subtotals at the borders of the rows and columns provide the *marginal
distributions* of each variable.

Table 2.4

CONTINGENCY TABLE OF U.S. CITIES BY REGION
AND CONGRESSIONAL POWER

| | | | Region | | n_{i+} |
			West 1	Other 2	
Power Index	High	1	12	17	29
	Low	2	12	84	96
n_{+j}			24	101	125

Contingency tables may be viewed as one bivariate distribution or
as several univariate distributions. For example, each column of Table 2.4
is a separate distribution of congressional power conditional on the region.
Similarly the two rows are separate regional distributions conditional on
the power index.

Our general notation for contingency tables is the matrix of Table 2.5.
This table has r rows, c columns and marginal totals indicated by the +
subscripts. It is frequently convenient to suppress the sample size by
defining the proportions

$$p_{ij} = n_{ij}/n, \quad p_{i+} = n_{i+}/n, \quad p_{+j} = n_{+j}/n.$$

Table 2.5

THE GENERAL $r \times c$ CONTINGENCY TABLE

n_{11} $\cdots$	n_{1j} $\cdots$	n_{1c}	n_{1+}
$\vdots$	$\vdots$	$\vdots$	
n_{i1}	n_{ij}	n_{ic}	n_{i+}
$\vdots$	$\vdots$	$\vdots$	
n_{r1}	n_{rj}	n_{rc}	n_{r+}
n_{+1} $\cdots$	n_{+j} $\cdots$	n_{+c}	$n = n_{++}$

Table 2.6 contains these values computed from Table 2.4.

Table 2.6

CELL PROPORTIONS p_{ij}

Congres- sional Power	Region West 1	Other 2	p_{i+}
1	.096	.136	.232
2	.096	.672	.768
p_{+j}	.192	.808	1.000

Suppose we ask how much the West differs from the rest of the country in congressional power. Tables 2.4 and 2.6 provide the necessary information for comparison. The proportion of western region cities with high congressional power is $p_{11}/p_{+1} = n_{11}/n_{+1} = 12/24$ while $p_{12}/p_{+2} = 17/101$

is the proportion for cities outside the West. Thus, the proportion
of western cities with congressional power exceeds the proportion of
cities elsewhere, 0.500 to 0.168, a factor of three to 1. In this case,
region and power are said to be *associated*.

Were there no association between region and power, the proportion
of western cities with power would equal the proportion of other cities
with power, i.e.,

$$\frac{p_{11}}{p_{+1}} = \frac{p_{12}}{p_{+2}}$$

or equivalently

$$\frac{p_{11}}{p_{+1}} = \frac{p_{1+} - p_{11}}{1 - p_{+1}}$$

which may be solved for p_{11}:

$$p_{11} = p_{1+}p_{+1} \ .$$

That is, there is *no association* between two variables if and only if
the proportion p_{ij} in each cell is the product of two marginal proportions p_{i+} times p_{+j}.

$$p_{ij} = p_{i+}p_{+j} \qquad\qquad \text{for all } i, j.$$

$$n_{ij} = np_{i+}p_{+j} = n_{i+}n_{+j}/n$$

Note that there is association in Table 2.6 since $0.096 \neq (0.192)(0.232)$.
Nonassociation then means that the marginal column proportions are the
same as every column proportion for unassociated variables, and similarly
the marginal row proportions also are the same as every row proportion.
These relations hold for any $r \times c$ contingency table.

The expected proportions, assuming no association, are given paren-
thetically in Table 2.7. For example, 0.045 = (0.192)(0.232).

Table 2.7

CELL PROPORTIONS AND EXPECTED CELL PROPORTIONS

Congres- sional Power	Region		
	1 West	2 Other	
1	.096 (.045)	.136 (.187)	.232
2	.096 (.147	.672 (.621)	.768
	.192	.808	1.000

2.7. MEASURES OF ASSOCIATION

We have deduced relationships that must hold if two variables
defining a contingency table are unrelated or alternatively that the
two distributions for one variable are the same. It is rare in
practice that observed distributions are identical. Thus a distance
between an observed distribution and the distribution with the same
marginals but with no association is useful in measuring association.
Two such measures are considered.

By analogy to the mean absolute difference, the *mean absolute
difference measure of association* D is simply the average absolute
difference between observed and expected proportions

$$D = \frac{1}{rc} \sum_i \sum_j |p_{ij} - p_{i+}p_{+j}|$$

and has a simple interpretation. For the data of Table 2.7, D is 0.051
which is the same as every deviation, a property that always holds for

2 x 2 tables. The smallest value of D, D = 0, corresponds to no association, while the largest occurs for Table 2.8, being as diagonal as possible while having the same marginals as Table 2.7. In Table 2.8 D = 0.147. Thus the observed D is 35 percent of the maximum possible.

Table 2.8

TABLE GIVING EXTREME VALUE OF D

Congres- sional Power	Region		
	West 1	Other 2	
1	.192	.040	.232
2	0	.768	.768
	.192	.808	

A less intuitive, but much more commonly used measure than D is the *chi-square distance measure*, usually written as χ^2. This measure is actually a squared distance, and has the feature of increasing with the sample size and of giving heavier weight to unlikely cells. It is used frequently because it has a convenient sampling interpretation, as we shall see in Chapter 11, but seems less useful than D for descriptive purposes. The definition of χ^2 is

$$\chi^2 = \Sigma \Sigma \frac{(n_{ij} - np_{i+}p_{+j})^2}{np_{i+}p_{+j}}.$$

The two measures are related because it is always true that

$$D \leq \frac{1}{\sqrt{n} \ rc} \chi$$

where $X = \sqrt{\chi^2}$, and D often is close to this bound. Furthermore D and χ^2 tend to order tables similarly and for these reasons comparisons of association in contingency tables often will be the same with either measure.

For Table 2.6, $\chi^2 = 11.97$ and $\chi/\sqrt{n}$ rc $= 0.077$ which exceeds the value of D $= 0.051$.

Chapter 3

SIMPLE LINEAR REGRESSION: CURVE FITTING, CORRELATION, AND PREDICTION

The example to be discussed will serve to illustrate curve fitting and the method of least squares. No statistical model is assumed which relates the data to a larger population. The techniques developed here apply when one has many objects with at least two measurements on each. In contrast to the categorical data considered in the contingency tables of the last chapter, both measurements are numerical, and can assume several possible values.

The main points in this chapter are: that it is useful to plot bivariate data; that a formal criterion, least squares, may be used effectively for fitting a line to data; that a best line y = Mx+B can be computed; and that certain statistics determined for fitting the line give useful summary information about the relation between x and y. If there are several possible predictions then the best one is defined as that with the smallest squared prediction error, being the same as that with the largest squared correlation with the variable being predicted. The standard deviation about the prediction line measures the "typical" prediction error, and this standard deviation is a function of the square correlation. The method discussed is the simplest version of linear regression, correlation, and analysis of variance that we will encounter in later chapters.

3.1. THE PROBLEM

The final standings for the regular season of the 1970 National Hockey League (NHL) are given in Table 3.1. The first three columns of

Table 3.1

REGULAR SEASON FINAL 1970 NATIONAL HOCKEY LEAGUE STANDINGS

Eastern Division

Teams	Won	Lost	Tied	Points	Percentage*	GF	GA
Chicago	45	22	9	99	0.651	250	170
Boston	40	17	19	99	0.651	277	216
Detroit	40	21	15	95	0.625	246	199
New York	38	22	16	92	0.605	246	189
Montreal	38	22	16	92	0.605	244	201
Toronto	29	34	13	71	0.467	222	242

Western Division

	Won	Lost	Tied	Points	Percentage	GF	GA
St. Louis	37	27	12	86	0.566	224	179
Pittsburgh	26	38	12	64	0.421	182	238
Minnesota	19	35	22	60	0.395	224	257
Oakland	22	40	14	58	0.381	169	243
Philadelphia	17	35	24	58	0.381	197	225
Los Angeles	14	52	10	38	0.250	168	290

*Points divided by 152 = (wins + 0.5 ties)/76.

Table 1 give the number of wins, losses, and ties for each team. Since so many games result in ties, teams are ranked on the basis of "points," with teams receiving two points for each win, one for a tie, and none for a loss. The fourth column gives the number of points for each team. The number of points divided by 152 (twice the number of games each team played, and the maximum number of points that a team can win) is the team's percentage in the usual sense, counting each tied game as half a win and half a loss. Each team's percentage appears in the fifth column, while the goals scored for (GF) and the goals scored against (GA) each team are given in the last two columns.

It is evident from Table 3.1 that the season ended with several teams deadlocked in the standings. In 1970 the rule for breaking deadlocks was that the deadlocked team with the greater number of wins was placed higher, even though it also had the greater number of losses. Therefore, Chicago was placed above Boston and Oakland was placed over Philadelphia. Of two teams having the same number of points and wins, the team with the most goals (GF) was placed higher. Hence, New York was awarded fourth place. Only the top four teams in each division qualify for the playoffs, and therefore the Montreal Canadiens failed to do so.

The rule that breaks ties on the basis of GF and not GA is a bad one that led to an extraordinary situation in Montreal's last game. New York had finished its season with 246 goals. Montreal entered its last game with Chicago with 242 goals and needed a win or a tie or five goals to finish ahead of New York. In the last period of that game, Montreal trailed Chicago 5-2. At this point, Montreal needed at least three more goals to obtain a season total of 247 and finish ahead of New York for the season. Since they needed at least three goals to win or tie, it was immaterial to Montreal who won the game, and for Montreal, defense became irrelevant. They had to have three goals in the last period, and it made no difference how much Chicago scored. They would gladly have lost 20-3 in the last period. The Canadiens therefore removed their goalie in favor of a sixth attacker in an effort to get these goals. This plan was brilliant in theory but lacking in execution, and as it happened, Montreal did not score again and ultimately lost 10-2. But the failure of Montreal's strategy does not alter the fact that the rule ranking teams only on the basis of their offensive worth is a bad one and encouraged a team, quite reasonably, not to try to win a game. A repercussion was that had Chicago lost or tied that last game, Chicago would not have won the championship. A tie breaking system that considered both GF and GA would have forced Montreal to continue to try to win the game, but by as large a margin as possible. Then Chicago might have lost, and had they done so, Boston would have won the championship.

In this chapter, we will compare several methods of predicting a team's performance on the basis of its GF and GA. The statistical method we will use is sometimes called "linear least-squares curve fitting," and sometimes called "simple linear regression." We will see that the pre-dictors which are best in terms of our least-squares criterion are those which have the largest correlation (in absolute value) with points earned by the teams. We will see that GF is not nearly as good a predictor of a team's worth as other predictors. Finally, we will introduce some ideas connected with prediction intervals.

3.2. GRAPHICAL DISCUSSION OF THE PROBLEM

Our object is to study predictors of a team's strength which are based on GF and GA. We develop notation in order that the ideas will be clear and so our methods can be applied to other data. Let n = 12 be the number of teams, and number the teams 1 to 12 in the order that they appear in Table 3.1. We use the variable i to refer to teams, so that i = 1 for Chicago and i = 12 for Los Angeles. Let y (subscripted) represent a team's points (column 4 of Table 3.1) so that y_1 = 99, y_2 = 99, ..., y_{12} = 38, etc. Let u and v (both subscripted) represent GF and GA respectively for each team. Hence u_1 = 250, ..., u_{12} = 168, v_1 = 170, ..., v_{12} = 290.

The predictors of team strength that we shall study are of the form

$$x = f(u, v).$$

The variable x is the predictor, and u and v are GF and GA. The function f may be chosen in many ways, but we will consider only the following three choices for f. We will refer to

$$x = f(u, v) = u$$

as predictor I. This corresponds to predicting a team's strength on the basis of the goals it scores, and ignores GA. The second predictor, predictor II, is the ratio

$$x = f(u, v) = u/v$$

of GF to GA. Finally, we consider predictor III,

$$x = f(u, v) = u-v$$

being the difference between the goals scored GF for and against GA a team. This is how much the team outscored its opponents. Thus predictors

II and III depend on both offense and defense, and increase as either GF increases or GA decreases. Table 3.2 gives values of these predictors together with the team's points.

It is helpful to plot team points and the predictors on a graph. In Figures 3.1 and 3.2 below, each team's points y is plotted with predictor I in Figure 3.1 and with predictor III in Figure 3.2. The points lie near a straight line, and such a line is drawn. Obviously there are many straight lines that one might draw through the cluster of points. Each line satisfies an equation y = mx+b, with different values of m and b determining different lines, m being the slope of the line and b the y-intercept, that is, the value of y when x = 0. The interpretation of such a line is that if x were given as a value of the predictor for a team, then we would predict y = mx+b as the number of points the team won. We therefore distinguish between the predictor x and the prediction mx+b. Applying this to the 12 teams already under consideration, team i scored y_i points and would be predicted to have scored mx_i+b points where $x_i = f(u_i, v_i)$ is one of our predictors.

The error made in the prediction is then

$$e_i = y_i - (mx_i+b),$$

being the difference between the true number of points scored and the predicted number of points. Figures 3.1 and 3.2 show some of the errors e_i. The errors are positive when the dot falls above the line and negative when the dot falls below the line. If the dot coincides with the line the error is zero. Since the dots do not lie on a straight line in Figures 3.1 and 3.2, it is impossible to choose m and b so that all errors are zero. The values of m and b for each of the plotted lines are indicated

Table 3.2

TEAM'S POINTS AND THE PREDICTORS

Team	Points			
i	y_i	I	II	III
1	99	250	1.471	80
2	99	277	1.282	61
3	95	246	1.236	47
4	92	246	1.302	57
5	92	244	1.214	43
6	71	222	0.917	−20
7	86	224	1.251	45
8	64	182	0.765	−56
9	60	224	0.872	−33
10	58	169	0.695	−74
11	58	197	0.876	−28
12	38	168	0.579	−122

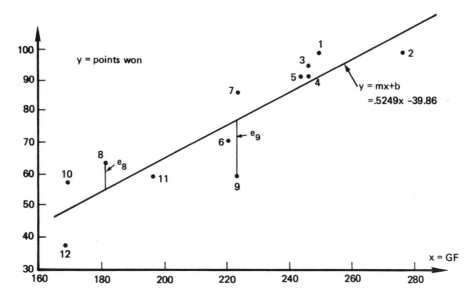

Figure 3.1 - Points won versus GF predictor for 12 teams.

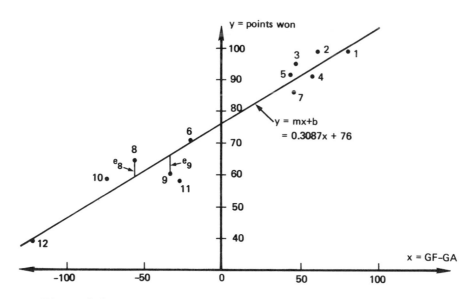

Figure 3.2 - Points won versus GF-GA indicator for 12 teams.

on the graph. Different lines would have different values of m and b and would therefore have different errors e_i. Given a choice of lines, we would prefer the one that makes the errors as small as possible.

Observe that most points in Figure 3.2 are closer to the prediction line in Figure 3.2 than are the corresponding points in Figure 3.1. For example, e_9 in Figure 3.1 is about −18, but e_9 in Figure 3.2 is about −6. That is, team 9 (Minnesota) is predicted to have 78 points in Figure 3.1, although it actually had 60, and 60−78 = −18. The error is always the vertical distance between the point and the line. In Figure 3.2, Minnesota is predicted to have 66 points, and 60−66 = −6. The fact that the points in Figure 3.2 are generally closer to the line in Figure 3.2 means that the prediction errors for predictor III are generally smaller than the prediction errors for predictor I. Thus, predictor III is a better predictor of team points than predictor I. Of course, to make this claim, we must be certain that each of the lines we have drawn to represent the relation between y and x is the best possible one through the points. For example, if a very bad line were used in Figure 3.2, the Figure 3.2 predictions based on that line would be worse than the Figure 3.1 predictions and we would think that predictor I was the better predictor.

3.3. ALGEBRAIC DESCRIPTION OF THE BEST PREDICTION LINE

We observe in the preceding section that many reasonable lines may be drawn through the points in Figure 3.1. We will define a measure of goodness of fit of a line to the points that will tell us how close the line is to the points on the average. Then we will find the closest line to the points in terms of our measure. The prediction error

$$e_i = y_i - (mx_i + b)$$

36

for teams i = 1, 2, ..., 12 will be used to define the measure. We would like to find the values of m and b that make e_i closest to zero for every team. However, it rarely will happen that some line is better than every other line for every team. To compromise, the average error will be used. A negative error of a given magnitude is just as serious as a positive error, so we use the absolute values of the e_i, $|e_i|$, since they are never negative. Mathematically, it is simpler yet to work with squared errors e_i^2 which are never negative, and which also decrease as prediction improves. The measure of goodness-of-fit to a line with slope m and intercept b is the average of squared errors

$$D^2 = \frac{1}{12}(e_1^2 + e_2^2 + e_3^2 + \ldots + e_{11}^2 + e_{12}^2)$$

$$= \frac{1}{12}[(y_1 - mx_1 - b)^2 + \ldots + (y_{12} - mx_{12} - b)^2].$$

The strings of dots mean that we include the squared errors for all 12 teams in the summation. With the usual summation notation, D also can be written for n teams as

$$D^2 = \frac{1}{n} \sum_{i=1}^{n} e_i^2 = \frac{1}{n} \sum_{i=1}^{n} (y_i - mx_i - b)^2.$$

One line is better than another if it gives a lower value of D^2. The best possible line is the one that has the lowest possible value for D^2. It turns out that the best possible line can be calculated without great difficulty. We will denote the slope of the best possible line by M and the intercept by B. This line Mx+B is called the "least squares line" because it makes the sum of squares, D^2, as small as possible. The best values M and B are calculated according to the procedure described below.

The value n is the number of pairs of data points. The mean or average of the values x_i is the sum of the values from x_1 through x_n divided by n, denoted by

$$\bar{x} = (x_1 + x_2 + \ldots + x_n)/n = \sum_{i=1}^{n} x_i/n,$$

and similarly the mean of the y-values is

$$\bar{y} = (y_1 + y_2 + \ldots + y_n)/n = \sum_{i=1}^{n} y_i/n.$$

The covariance between any pair of variables is the average of the product of their values, after each of the values has been altered by subtracting its mean. That is

$$Cov(x, y) = [(x_1 - \bar{x})(y_1 - \bar{y}) + \ldots + (x_n - \bar{x})(y_n - \bar{y})]/n$$

$$= \sum_{i=1}^{n} (x_i - \bar{x})(y_i - \bar{y})/n.$$

The covariance of a variable with itself can also be calculated and is called the variance. Thus

$$Var(x) = Cov(x, x) = \sum_{i=1}^{n} (x_i - \bar{x})^2/n$$

and

$$Var(y) = Cov(y, y) = \sum_{i=1}^{n} (y_i - \bar{y})^2/n$$

since in this case each product becomes a square. The variance is always positive or zero. It has a positive or zero square-root which is called the standard deviation. Thus, the standard deviation of x, denoted s_x is given by

$$s_x = \sqrt{Var(x)}$$

and the standard deviation of y, s_y is

$$s_y = \sqrt{Var(y)} \ .$$

The correlation coefficient between the variables x and y is denoted by r and is calculated as

$$r = Cov(x, y)/(s_x s_y).$$

The correlation coefficient r is always between -1 and 1.   We first calculate all of these quantities from our data.

    Means, variances, and standard deviations are computed from data on x and y alone.   Only the covariance cov(x, y) and the correlation coefficient r involve both x and y.   The correlation coefficient is a normalized version of the covariance, designed to account for the scale of x and y.   Thus r would not change if each value x_i or each value y_i were multiplied by a positive number, but the covariance would.   To interpret the covariance cov(x, y), if x_i and y_i are simultaneously larger than their respective means $\bar{x}$, $\bar{y}$ or simultaneously smaller, then $(x_i - \bar{x})(y_i - \bar{y}) > 0$ since either $x_i - \bar{x} > 0$ and $y_i - \bar{y} > 0$ or otherwise $x_i - \bar{x} < 0$ and $y_i - \bar{y} < 0$.   Thus cov(x, y), being a sum of such terms, would be positive if most of the terms $(x_i - \bar{x})(y_i - \bar{y})$ are positive, and otherwise cov(x, y) would be negative.   Inspection of the plots in Figures 3.1 and 3.2 show that cov(x, y) and r would be positive in both cases, for the reasons just given.   The lines drawn there also are sloped positively.   This is no accident, as the reader will see by examining the formula for M below.   Negatively sloped data and lines, which do not occur in this example, correspond to negative values of r and cov(x, y), because $(x_i - \bar{x})(y_i - \bar{y}) < 0$ in most cases.

It may seem that the best line should pass through the point $(\bar{x}, \bar{y})$, which means that the average value $\bar{y}$ of y would be the best prediction if x is equal to its average value. This is true, and consequently the best line must satisfy

$$y = \bar{y} + M(x-\bar{x})$$

so that if $x = \bar{x}$, then $y = \bar{y}$. This doesn't tell us what M is, but if the value of M is known, then we must have

$$B = \bar{y} - M\bar{x}.$$

The value of M is calculated by

$$M = Cov(x, y)/Var(x)$$

$$= r\, s_y/s_x.$$

Clearly $r > 0$ if the slope M is positive, $r < 0$ for negative slope. If $r = 0$ then x is useless for linear prediction of y and $y = \bar{y}$ is used. Having M, then B is calculated by the preceding formula. The predicted value of y for given x is then

$$p = Mx + B$$

with error at x_i of $e_i = y_i - p_i$. This prediction using these values of M and B gives the smallest possible value of D (the nonnegative square root of D^2) which we denote D_{min} for the minimum value of D. Then

$$D^2_{min} = \frac{1}{n} \sum_{i=1}^{n} e_i^2 = \frac{1}{n} \sum_{i=1}^{n} (y_i - p_i)^2 = \frac{1}{n} \sum_{i=1}^{n} (y_i - Mx_i - B)^2.$$

Another formula for calculating D_{min} is

$$D^2_{min} = s_y^2\,(1-r^2) \qquad \text{or} \qquad D_{min} = s_y\sqrt{1-r^2}.$$

We also denote

$$s_{y|x} = D_{min} = s_y \sqrt{1-r^2}$$

read "the standard deviation of y given x" as the standard deviation of y about the regression line. We see that this amount does not depend on x and that $s_{y|x} \leq s_y$. If $r = 0$, $s_{y|x} = s_y$ so that knowing x does not affect our knowledge of y. At the other extreme, $r = \pm 1$ gives $s_{y|x} = 0$ so that knowledge of x results in perfect knowledge of y. This happens only if the points (x_i, y_i) lie exactly on a a straight line. The last two formulas show that the value of r^2 determines the goodness of the predictor x of y, since s_y does not depend on x. The larger the value of r^2, the better the prediction.

We will organize the computation of these values in the tables that follow. Comparing the work done in the tables with the formulas given above facilitates understanding the formulas.

All the computation tables work in the same way. In the computation of means and covariances, the first three columns are taken from Table 3.2, and all the other columns are calculated from the first three.

For definiteness, we will discuss Table 3.3. Here GF is used as a predictor and the values of x are acquired from Table 3.2 for predictor I. The first thing to do is to sum the columns y and x. We get 912 and 2649. The average or mean is obtained by dividing this sum by n = 12. We get $\bar{y} = 76$ and $\bar{x} = 220.75$. Note that $\bar{y}$ is just the number of games that the teams played. Columns 4 and 5 are obtained by subtracting $\bar{y}$ and $\bar{x}$ from columns 2 and 3 respectively. The values $Y = y-\bar{y}$ and $X = x-\bar{x}$ are our shorthand notations for the terms needed for calculating the covariances discussed above. Column 6 is the square of column 4 as is column 8 of column 5. Column 7 is calculated as the product of the entries in columns 4 and 5. The entries in all the columns are summed, and if the calculations are correct, the sums in columns 4 and 5 will be zero. Then each sum is divided by n = 12 to get the averages. The average values that appear in columns 6, 7, and 8 are Var(y), Cov(x, y) and Var(x) respectively, as you will see if you compare the formulas above with what is being done in the table. In the last five lines of Table 3.3, s_y and s_x are calculated as the square roots of Var(x) = Cov(x, x) and of Var(y) = Cov(y, y) respectively, and then using this, r is calculated. If r is not between -1 and 1, an error has been made. Finally M and B are calculated.

Table 3.4 is really a continuation of Table 3.3, and would have been put beside Table 3.3 had there been space. The first three columns of Table 3.4 are the same as those of Table 3.3. In the fourth column, the prediction

$$p = Mx + b = 0.5249x - 39.8652$$

is calculated for each team, the values of M and B being the one obtained in Table 3.3. The errors are calculated in column 5 as the difference between the number of points won in column 2 and the predicted number of points in column 4. The errors are squared and entered in column 6. The columns are again summed, and the sums of columns 2 and 3 will be the same as for the corresponding ones of Table 3.3. If the calculations have been done correctly, the sum for column 4 will agree with that of column 2 (because $\bar{y} = M\bar{x} + B$) and the sum of the errors in column 5 will be zero. Averages are then computed by dividing the sums by n = 12. The value of D^2_{min} appears as the average under column 6. Alternatively,

D_{min} may be calculated with the formula involving s_y and r^2. The value of r^2 is also calculated at the end of Table 3.4 from the value of r in Table 3.3.

The computations for Tables 3.5, 3.6, 3.7, and 3.8 all proceed in the same way, except Tables 3.5 and 3.6 use the ratio GF/GA as a predictor and Tables 3.7 and 3.8 use the difference GF-GA. These predictors are taken from Table 3.2. Tables 3.5 and 3.7 are just like Table 3.3, and Tables 3.6 and 3.8 are like Table 3.4.

Table 3.3

COMPUTATION TABLE FOR MEANS AND COVARIANCES
USING GF AS PREDICTOR

i	y	x	$Y=y-\bar{y}$	$X=x-\bar{x}$	Y^2	XY	X^2
1	99	250	23	29.250	529	672.750	855.5625
2	99	277	23	56.250	529	1293.750	3164.0625
3	95	246	19	25.250	361	479.750	637.5625
4	92	246	16	25.250	256	404.000	637.5625
5	92	244	16	23.250	256	372.000	540.5625
6	71	222	-5	1.250	25	-6.250	1.5625
7	86	224	10	3.250	100	32.500	10.5625
8	64	182	-12	-38.750	144	465.000	1501.5625
9	60	224	-16	3.250	256	-52.000	10.5625
10	58	169	-18	-51.750	324	931.500	2678.0625
11	58	197	-18	-23.750	324	427.500	564.0625
12	38	168	-38	-52.750	1444	2004.500	2782.5625
Sum:	912	2649.000	0	.000	4548	7025.000	13384.2500
Avg:	76	220.750	0	.000	379	585.417	1115.3542

x = GF

$Y = y-\bar{y} = y-76$

$X = x-\bar{x} = x-220.750$

$Var(y) = Cov(y,y) = Avg(Y^2) = 379 \quad s_y = \sqrt{Var(y)} = 19.47$

$Var(x) = Cov(x,x) = Avg(X^2) = 1115.3542 \quad s_x = \sqrt{Var(x)} = 33.3969$

$r = Cov(x,y)/[s_x \cdot s_y] = Avg(XY)/[s_x \cdot s_y] = 0.9004$

$M = Cov(x,y)/Var(x) = Avg(XY)/Var(x) = r \cdot s_y/s_x = 0.5249$

$B = \bar{y}-M \cdot \bar{x} = Avg(y)-M \cdot Avg(x) = -39.9652$

Table 3.4

COMPUTATION TABLE FOR PREDICTIONS AND ERRORS
USING GF AS PREDICTOR

i	y	x	p=M·x+B	e=y-p	e^2
1	99	250	91.35	7.65	58.48
2	99	277	105.52	-6.52	42.56
3	95	246	89.25	5.75	33.03
4	92	246	89.25	2.75	7.55
5	92	244	88.20	3.80	14.42
6	71	222	76.66	-5.66	31.99
7	86	224	77.71	8.29	68.79
8	64	182	55.66	8.34	69.53
9	60	224	77.71	-17.71	313.50
10	58	169	48.84	9.16	83.94
11	58	197	63.53	-5.53	30.63
12	38	168	48.31	-10.31	106.36
Sum:	912	2649	912.00	.00	860.78
Avg:	76	220.750	76.00	.00	71.73

$$\text{Var}(y|x) = D^2_{\min} = \text{Avg}(e^2) = s^2_y \cdot [1-r^2] = 71.73$$

$$s_{y|x} = D_{\min} = \sqrt{\text{Avg}(e^2)} = s_y \cdot \sqrt{1-r^2} = 8.47$$

$$r^2 = .8107$$

Table 3.5

COMPUTATION TABLE FOR MEANS AND COVARIANCES
USING GF/GA AS PREDICTOR

i	y	x	$Y=y-\bar{y}$	$X=x-\bar{x}$	Y^2	XY	X^2
1	99	1.471	23	.432	529	9.942	.1868
2	99	1.282	23	.244	529	5.614	.0596
3	95	1.236	19	.198	361	3.759	.0391
4	92	1.302	16	.263	256	4.212	.0693
5	92	1.214	16	.176	256	2.809	.0308
6	71	.917	−5	−.121	25	.605	.0146
7	86	1.251	10	.213	100	2.131	.0454
8	64	.765	−12	−.274	144	3.284	.0749
9	60	.872	−16	−.107	256	2.668	.0278
10	58	.695	−18	−.343	324	6.172	.1176
11	58	.876	−18	−.163	324	2.930	.0265
12	38	.579	−38	−.459	1444	17.443	.2107
Sum:	912	12.460	0	.000	4548	61.568	.9032
Avg:	76	1.038	0	.000	379	5.131	.0753

x = GF/GA

$Y = y-\bar{y} = y-76$

$X = x-\bar{x} = x-1.038$

$\text{Var}(y) = \text{Cov}(y,y) = \text{Avg}(Y^2) = 379 \qquad s_y = \sqrt{\text{Var}(y)} = 19.47$

$\text{Var}(x) = \text{Cov}(x,x) = \text{Avg}(X^2) = .0753 \qquad s_x = \sqrt{\text{Var}(x)} = .2743$

$r = \text{Cov}(x,y)/[s_x \cdot s_y] = \text{Avg}(XY)/[s_x \cdot s_y] = 0.9606$

$M = \text{Cov}(x,y)/\text{Var}(x) = \text{Avg}(XY)/\text{Var}(x) = r \cdot s_y/s_x = 68.1694$

$B = \bar{y}-M \cdot \bar{x} = \text{Avg}(y)-M \cdot \text{Avg}(x) = 5.2169$

Table 3.6

COMPUTATION TABLE FOR PREDICTIONS AND ERRORS
USING GF/GA AS PREDICTOR

i	y	x	p=M·x+B	e=y-p	e^2
1	99	1.471	105.47	-6.47	41.81
2	99	1.282	92.64	6.36	40.48
3	95	1.236	89.49	5.51	30.40
4	92	1.302	93.95	-1.95	3.78
5	92	1.214	87.97	4.03	16.24
6	71	.917	67.75	3.25	10.55
7	86	1.251	90.52	-4.52	20.47
8	64	.765	57.35	6.65	44.27
9	60	.872	64.63	-4.63	21.47
10	58	.695	52.63	5.37	28.87
11	58	.876	64.90	-6.90	47.65
12	38	.579	44.71	-6.71	45.00
Sum:	912	12.460	912.00	.00	350.98
Avg:	76	1.038	76.00	.00	29.25

$$\mathrm{Var}(y|x) = D^2_{\min} = \mathrm{Avg}(e^2) = s_y^2 \cdot [1 - r^2] = 29.25$$

$$x_{y|x} = D_{\min} = \sqrt{\mathrm{Avg}(e^2)} = s_y \cdot \sqrt{1 - r^2} = 5.41$$

$$r^2 = .9228$$

Table 3.7

COMPUTATION TABLE FOR MEANS AND COVARIANCES
USING GF-GA AS PREDICTOR

i	y	x	$Y=y-\bar{y}$	$X=x-\bar{x}$	Y^2	XY	X^2
1	99	80	23	80	529	1840	6400
2	99	61	23	61	529	1403	3721
3	95	47	19	47	361	893	2209
4	92	57	16	57	256	912	3249
5	92	43	16	43	256	688	1849
6	71	−20	−5	−20	25	100	400
7	86	45	10	45	100	450	2025
8	64	−56	−12	−56	144	672	3136
9	60	−33	−16	−33	256	528	1089
10	58	−74	−18	−74	324	1332	5476
11	58	−28	−18	−28	324	504	784
12	38	−122	−38	−122	1444	4636	14884
Sum:	912	.000	0	.000	4548	13958.000	45222
Avg:	76	.000	0	.000	379	1163.167	3768

x = GF−GA

$Y = y-\bar{y} = y-76$

$X = x-\bar{x} = x-0$

$\text{Var}(y) = \text{Cov}(y,y) = \text{Avg}(Y^2) = 379 \quad s_y = \sqrt{\text{Var}(y)} = 19.47$

$\text{Var}(x) = \text{Cov}(x,x) = \text{Avg}(X^2) = 3768.5000 \quad s_x = \sqrt{\text{Var}(x)} = 61.3881$

$r = \text{Cov}(x,y)/[s_x \cdot s_y] = \text{Avg}(XY)/[s_x \cdot s_y] = 0.9733$

$M = \text{Cov}(x,y)/\text{Var}(x) = \text{Avg}(XY)/\text{Var}(x) = r \cdot s_y/s_x = 0.3087$

$B = \bar{y}-M \cdot \bar{x} = \text{Avg}(y)-M \cdot \text{Avg}(x) = 76.000$

Table 3.8

COMPUTATION TABLE FOR PREDICTIONS AND ERRORS
USING GF-GA AS PREDICTOR

i	y	x	$p = M \cdot x + B$	$e = y - p$	e^2
1	99	80.000	100.69	-1.69	2.86
2	99	61.000	94.83	4.17	17.41
3	95	47.000	90.51	4.49	20.19
4	92	57.000	93.59	-1.59	2.54
5	92	43.000	89.27	2.73	7.44
6	71	-20.000	69.83	1.17	1.38
7	86	45.000	89.89	-3.89	15.13
8	64	-56.000	58.72	5.28	27.93
9	60	-33.000	65.81	-5.81	33.81
10	58	-74.000	53.16	4.84	23.43
11	58	-28.000	67.36	-9.36	87.57
12	38	-122.000	38.34	-.34	.12
Sum:	912	.000	912.00	.00	239.79
Avg:	76	.000	76.00	.00	19.98

$$\text{Var}(y \mid x) = D^2_{min} = \text{Avg}(e^2) = s^2_y \cdot [1 - r^2] = 19.98$$
$$s_{y \mid x} = D_{min} = \sqrt{\text{Avg}(e^2)} = s_y \cdot \sqrt{1 - r^2} = 4.47$$
$$r^2 = .9473$$

The reader may want a mathematical proof that Mx+B, with the M and B defined in this section, actually minimizes D^2 over all possible choices of lines, that is, that the choice of m and b that minimize

$$D^2 = \frac{1}{n} \Sigma (y_i - mx_i - b)^2$$

is actually m = M and b = B. There are two approaches to proving this, one using calculus, and the other purely algebraic. Those familiar with calculus know that the method is to differentiate D^2 once with respect to m and once with respect to b, set both results equal to zero, and solve the two resulting linear equations simultaneously for m and b, to obtain M and B.

The algebraic method is actually more helpful in that it yields a value for D^2 for any line. Let $e_i = y_i - Mx_i - B$ and $B = \bar{y} - M\bar{x}$. Thus, for any line with m and b we may write

$$y_i - mx_i - b = e_i - (m-M)(x_i - \bar{x}) - (b - B + (m-M)\bar{x}),$$

so

$$D^2 = \frac{1}{n} \Sigma (y_i - mx_i - b)^2$$

$$= \frac{1}{n} \Sigma e_i^2 + (m-M)^2 \frac{1}{n} \Sigma (x_i - \bar{x})^2 + (b - B + (m-M)\bar{x})^2$$

$$= D_{min}^2 + (m-M)^2 s_x^2 + (b - B + (m-M)\bar{x})^2.$$

We have used the fact that $\Sigma e_i (x_i - \bar{x})$, $= 0$, $\Sigma (x_i - \bar{x}) = 0$, and $\Sigma e_i = 0$ to establish this identity. It is clear now that $D^2 \geq D_{min}^2$, and only by taking m = M and b = B can we make $D^2 = D_{min}^2$. Thus M and B define the least squares line.

3.4. DISCUSSION OF RESULTS: CORRELATION, PREDICTION INTERVALS

Table 3.9

CONSTANTS FOR PREDICTORS

Predictor	M	B	D_{min}^2	$s_{y|x} =$ D_{min}	r	r^2
GF	0.5249	−39.86	71.73	8.47	0.9004	0.8107
GF/GA	68.17	5.22	29.25	5.41	0.9606	0.9228
GF-GA	0.3087	76	19.98	4.47	0.9733	0.9473

Table 3.9 summarizes the most important results of Tables 3.3 through 3.8. The straight lines graphed in Figures 3.1 and 3.2 correspond to the predictors with constants M and B for GF and GF-GA in the table above. The dots in Figure 3.1 do not appear to be as close to the line as the dots in Figure 3.2. This is the case, as D_{min}^2, which is the average squared prediction error, is much larger for GF than for GF-GA. The value of $s_{y|x} = D_{min}$ may be thought of as a typical prediction error. The values of $s_{y|x}$ for GF and GF-GA indicate that knowledge of GF enables one to predict a team's points with an error typically around 8½, but knowledge of GF-GA permits predictions with typical error of 4½ points. Inspection of Tables 3.4 and 3.8 confirms this. The worst prediction in Table 3.4 involves team number 9, Minnesota, which on the basis of GF alone, would be predicted to have 18 more points than it did. Several other predictions are in the 8-10 point error category. However, Table 3.8 shows that GF-GA produces only one absolute prediction error greater than six. Team number 11, Philadelphia, would be expected to have about nine more points on the basis of its GF-GA record. The magnitude of the prediction error

51

for Minnesota drops from about 18 using GF alone to about 6 using GA-GA, because the latter measure accounts for their below average defensive skills. The ratio GF/GA also gives much better predictions than GF alone, but it is not as good as GF-GA.

The standard deviation $s_{y|x}$ about the fitted line $y = Mx+B$ decreases as the absolute value of the correlation coefficient r increases, or as r^2 increases. Therefore the predictor with the largest correction coefficient in absolute value will be the best predictor. The largest possible value of $|r|$ is one, when $r = 1$, or $r = -1$. Then D_{min} would be zero, all the points (x_i, y_i) would lie on a straight line, and prediction would be perfect. Had we only wanted to find the best predictor of the three GF, GF/GA and GF-GA, much of the computation of Tables 3.3-3.8 would have been unnecessary, for we would only have had to calculate r. Then the predictor with the highest value of r^2 would have been the best.

The fact that all three values of M in Table 3.9 are positive indicates that all three predictors are positively associated with a team's strength. That is, the predictors tend to be larger for better teams. Had we used GA as a predictor, its coefficient M would have been negative, for better teams tend to have lower GA. This must be true, since if $x = GA$ then as x increases, the prediction

$$y = Mx + B$$

should decrease. Only a negative value of M would do this.

We may be tempted to generalize from our set of twelve teams in 1970 to other years or leagues. We must do this, for example, if we were to recommend GF-GA instead of GF/GA for breaking deadlocks. We cannot construct exact prediction intervals without referring to a statistical model, which would specify how future observations would relate to the past. In Figure 3.3 we have plotted the lines

$$Mx + B \pm s_{y|x} = 0.3087x + 76 \pm 4.47$$

and

$$Mx + B \pm 2s_{y|x} = 0.3087 + 76 \pm 8.94,$$

representing the predictions of points plus and minus one or two standard deviations about the prediction line using the GF-GA predictor. From Figure 3.3 or from inspection of Table 3.8, we see that $7/12 = 58$ percent of the points lie within $Mx+B \pm s_{y|x}$ and $11/12 = 92$ percent within $Mx+B \pm s_{y|x}$. These percentages provide a rough idea of how far future values might fall from our line. These calculations would be more reliable if we had more data, perhaps drawing from other years. The statistical model that we will consider in later chapters gives more precise answers to these questions under specified assumptions. In particular, we would find that the true prediction intervals are quadratic (not linear as in our example) with the greatest accuracy near $x = \bar{x}$. The dotted line in Figure 3.8 plots a more correct version of the prediction interval.

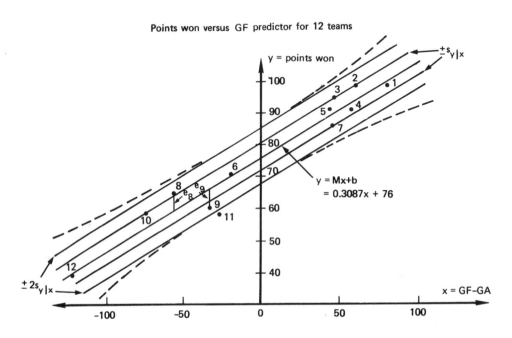

Figure 3.3 – Points won versus GF–GA indicator for
12 teams crude prediction intervals.

3.5. CONCLUSIONS AND INTERPRETATION OF RESULTS

Let us now accept GF-GA as our predictor of a team's strength. If a team's predicted points exceed the number of points that it actually earned, then one interpretation is that the team was unlucky because it scored and defended well enough to merit a better record than it had. Such a team probably lost more than its share of close games. Alternatively, teams that earned more points than their GF-GA record would predict might be termed lucky, and perhaps won more than their share of close games. For fun, we will use the predictions of Table 3.8 to create a new final standings based on predicted points instead of actual points. We have rounded all predictions off to the nearest integer.

Table 3.10 seems to indicate that Philadelphia was a better team than its fifth place finish suggests, and that Pittsburgh was not as strong as its second place finish suggests. Several other teams also altered their position.

If we use the GF-GA predictor to break the New York-Montreal deadlock, we find that for New York, GF-GA is 57 and Montreal has GF-GA of 43. Using GF-GA to break deadlocks in the standings would have, in this case, placed New York over Montreal, as GF alone does. So the deserving team reached the playoffs. If GF-GA also were used to break the Chicago-Boston and Oakland-Philadelphia deadlocks, Chicago would have maintained its first-place position, but Philadelphia would have been placed ahead of Oakland and would have reached the playoffs.

Table 3.10

FINAL 1970 NHL STANDINGS; BASED ON GF-GA PREDICTIONS

Eastern Division

Team	Predicted Points	Predicted Position	Actual Points	Actual Position
Chicago	101	1	99	1
Boston	95	2	99	2
New York	94	3	92	4
Detroit	91	4	95	3
Montreal	89	5	92	5
Toronto	70	6	71	6

Western Division

Team	Predicted Points	Predicted Position	Actual Points	Actual Position
St. Louis	90	1	86	1
Philadelphia	67	2	58	5
Minnesota	66	3	60	3
Pittsburgh	59	4	64	2
Oakland	53	5	58	4
Los Angeles	38	6	38	6

Chapter 4

MULTIPLE REGRESSION, PARTIAL AND MULTIPLE CORRELATION

INTRODUCTION

We have just treated the linear least squares problem of predicting

one variable from another. This chapter considers the more general

linear least squares problem of predicting one variable from two or

more variables.

The approach is data analytic: no inferences will be made from

the data at hand to a larger population. Inference problems in regres-

sion are deferred until Chapter 6. Instead we concentrate on develop-

ing needed formulas, measures, and concepts, building on what already

has been learned in Chapter 3. The data are chosen for this section

to facilitate hand computations, and we show how to reduce the problem

of computing multiple regressions to a series of simple (one variable)

regressions. This approach to multiple regression makes meaningful

the operations of adjusting for each independent variable, and makes

clear the logical relation between the partial regression and partial

correlation coefficients in this chapter and the simple regression

and correlation coefficients of Chapter 3. Considerable emphasis is

placed on residuals, both for their role here as the elements on which

simple regressions are based, and because they will be important

determinants for model choice later.

The reader should reproduce the hand computations in this chapter

for himself. By doing so he will become more familiar with the concepts

and operations of multiple regression. In later chapters we will assume

that all computations will be carried out by a computer, but we believe that the reader will be better prepared to interpret computer results if he is thoroughly familiar with the computing logic.

Matrix theory is discussed in Section 4.3. We have deemphasized the matrix approach to multiple regression for two reasons. First, we want multiple regression methodology to be accessible to students who are unfamiliar with matrix theory. Second, while the matrix approach is extremely powerful and concise for theory development, its abstraction tends to hide some of the concepts that are necessary for applications, e.g., partial correlation, partial regression, step-by-step adjustment, residual analysis, and stepwise regression.

Finally, some readers who have taken courses in statistics before may wonder why such a general and powerful technique as multiple regression is introduced so early. We have found that statistical procedures that do not handle several variables at once are of limited value in the social and policy sciences, and believe that regression analysis, which does account for many variables, should be introduced early enough that it may be exhaustively studied. The cost of introducing less sophisticated methods sooner is the delayed and inadequate consideration of more powerful methods that ultimately will be needed. However, the more elementary topics that often are covered in other courses first will be discussed later as special cases of the theory being developed here. Further, the reader is spared from considering the inference aspects of regression models until later chapters, and regression models will be considered from many viewpoints throughout much of the remainder of the book. Thus, although the model being presented in this chapter is complicated, only limited aspects of it will be discussed at first, and the reader will have many more opportunities throughout this book to increase his understanding of multiple regression methodology.

4.1. REGRESSION WITH TWO INDEPENDENT VARIABLES

Chapter 3 introduced linear regression using one independent variable to predict one dependent variable. Multiple regression, the subject of this chapter, is completely analogous to linear regression except more than one independent variable is used to predict the dependent variable. This section treats the simplest case of multiple regression, the case of two independent variables.

The following data are given for eight persons in Table 4.1.

Table 4.1

WAGES FOR MEN AND WOMEN

i (person)	y (hourly wages)	x_1 (years employed)	x_2 (sex)
1	2	1	0
2	4	1	0
3	4	2	0
4	6	2	0
5	2	2	2
6	4	2	2
7	4	3	2
8	6	3	2
Sum	32	16	8
Average	$\bar{y} = 4$	$\bar{x}_1 = 2$	$\bar{x}_2 = 1$

The eight individuals are all employees at the same business institution and are similar in all respects other than the number of years that they have been employed at the institution, x_1, and their sex, x_2 (men are given the code 0 and women 2 for the x_2 variable). The x_2 variable is called a "dummy variable." It simply indicates the sex of the employee, the choice of its two values being immaterial. The hourly wages in dollars are given as y. The data are fictitious, but the problem is otherwise realistic.

We are to determine how x_1 (years employed) and x_2 (sex) predict

wages. We will see that a simple regression can be misleading, but that

a multiple regression gives a valid description of the relationship

between the variables. We will also see that a multiple regression can

be decomposed into a series of simple regressions on residuals. We then

will present formulas for the regression coefficients, the partial

(conditional) correlations, conditional standard deviations, and the

multiple correlation coefficient. These quantities will be interpreted

in the context of the present problem.

If sex alone (that is, excluding x_1) is used to predict wages,

then it clearly is useless, because the four men have precisely the same

wages as the four women. However, we will see later that it is a mistake

to ignore x_2 and that sex in fact bears a strong relation to wages for

these data.

We start by plotting wages against x_1 (which will be referred to as "exp"

for experience) in Figure 4.1. The simple regression of y on x_1 alone is

$$y = 2 + x_1. \qquad (4.1)$$

This can be justified by noting that the means of y at the values of

$x_1 = 1$, 2, 3 are 3, 4, 5, respectively. Since these means lie on a

straight line, that line must be the regression line. To be sure of

this, the reader should compute the regression line from the formulas

of the preceding chapter.

But this is not the full story, for the regression line for the

four men (persons 1, 2, 3, 4) excluding the women, is

$$y_{men} = 1 + 2x_1 \qquad (4.2)$$

60

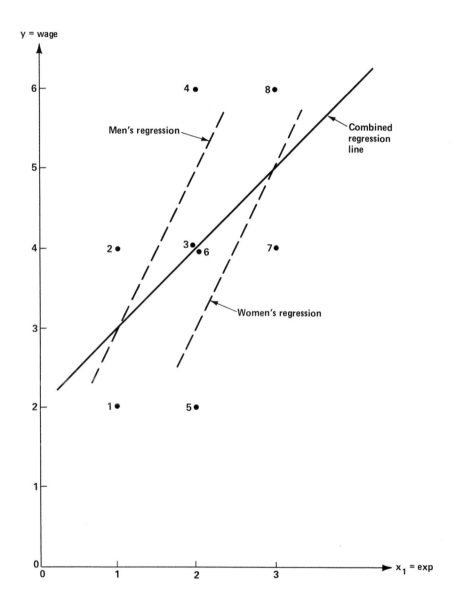

Figure 4.1 – Plot of wages versus experience. The indices refer
to the eight individuals.

and for the women (persons 5, 6, 7, 8) is

$$y_{women} = -1 + 2x_1. \qquad (4.3)$$

Again, these equations are simple to determine from the diagram, because the line is determined in each case by the mean wages at the two experience levels for each sex. Alternatively, these lines may be calculated from the simple regression formulas. The last two formulas indicate that experience has a more dramatic effect on wages than was realized from considering both sexes together in a simple regression. The reason is that women had more experience than men, but were underpaid because of their sex. The result from the simple regression $y = 2 + x_1$ is that these two effects tend to cancel making it appear that experience is less important. Correspondingly, the regression of wages on sex alone is $y = 4$ dollars per hour which does not even depend on sex.

Separating men and women gets us out of the problem here, but it has two drawbacks in general, because four regression coefficients must be estimated, whereas only three will be needed if the equation

$$y = b_0 + b_1x_1 + b_2x_2 \qquad (4.4)$$

is fitted. Furthermore, we will see that (4.4) with the best b_0, b_1, b_2 predicts as well as (4.2) and (4.3) combined, so that elimination of the extra coefficient is accomplished without loss of precision. In more general cases, x_2 may take many more than two values, so it might not even be feasible to derive (4.2) and (4.3) by splitting the data, although (4.4) still would be readily available.

As in Chapter 3, the least squares choices of b_0, b_1, b_2 are those that minimize the average squared errors, $e_i = y_i - (b_0+b_1x_{i1}+b_2x_{i2})$,

$$D^2 = \frac{1}{n} \sum_{i=1}^{n} e_i^2 = \frac{1}{n} \sum_{i=1}^{n} (y_i - b_0 - b_1 x_{i1} - b_2 x_{i2})^2. \qquad (4.5)$$

The value b_0 always is determined by the means if b_1, b_2 are known. That is, the means satisfy equation (4.4)

$$\bar{y} = b_0 + b_1 \bar{x}_1 + b_2 \bar{x}_2 \qquad (4.6)$$

so that

$$b_0 = \bar{y} - b_1 \bar{x}_1 - b_2 \bar{x}_2 . \qquad (4.7)$$

Formulas for b_1 and b_2 will be given later, but rather than give complicated formulas, we will develop them by building on our knowledge of simple linear regression and thereby reveal their meaning.

Suppose we are to determine b_1. From (4.4), b_1 determines the increase in y due to a unit increase in x_1 *holding* x_2 *fixed*. That is, if x_2 never varied in the data, then b_1 would be the simple regression coefficient between y and x_1. But for men, $b_{1,men} = 2$ and for women, $b_{1,women} = 2$ from (4.2), (4.3), so it happens that if we hold x_2 fixed at $x_2 = 0$ for men or $x_2 = 2$ for women, we get the same coefficient for experience. This suggests that $b_1 = 2$, and we shall see later that this is true. However, this argument works only for b_1, and not for b_2, in this example, so we proceed more generally as follows.

Women have more experience than men in this sample. People in the sample who have greater experience generally have higher wages. Before we can determine adequately how much sex affects wages, i.e., the correct value of b_2, we first must adjust for the differential experience. This is done as follows.

63

We have predicted y from x_1 in (4.1) as $y = 2 + x_1$ from a simple regression. The "residuals," or the errors from this are

$$\tilde{y}_i(x_1) = y_i - 2 - x_{i1} \tag{4.8}$$

for $i = 1, \ldots, 8$. This "tilde" notation always will denote residual. Even though it looks messy, it is necessary to indicate the tilde and that the residuals are computed from the fitted line using x_1 as a predictor. The errors are "uncorrelated" with x_1 since the best linear predictor of y is $2 + x_1$ (that is, the correlation coefficient between the values $\tilde{y}_i(x_1)$ and x_{i1} is zero). The errors (4.8) therefore may vary freely when x_1 is held fixed. Similarly, x_2 might be correlated with x_1. But the best linear predictor of x_2 from x_1 is

$$\hat{x}_2(x_1) = x_1 - 1 \tag{4.9}$$

which is obtained from a simple regression of x_2 (as dependent variable) on x_1 (as independent variable) or from inspecting Figure 4.2. The "hat" notation indicates predicted value, with the parenthetical x_1 specifying which variable is used to make the prediction. Note that the predicted value for individual i is $\hat{x}_{i2}(x_1)$ and the residual is denoted $\tilde{x}_{i2}(x_1) = x_{i2} - \hat{x}_{i2}(x_1)$. Therefore the residuals

$$\tilde{x}_{i2}(x_1) = x_{i2} - \hat{x}_{i2}(x_1) = x_{i2} - x_{i1} + 1 \tag{4.10}$$

represent the component of x_2 that may vary freely if x_1 is fixed. The residuals in (4.10) also are uncorrelated with x_1.

It follows that if we perform a simple regression of the components of y that may vary freely when x_1 is fixed, $y - 2 - x_1$, on the component of x_2 that may vary freely when x_1 is fixed, $x_2 - x_1 + 1$, we will obtain

64

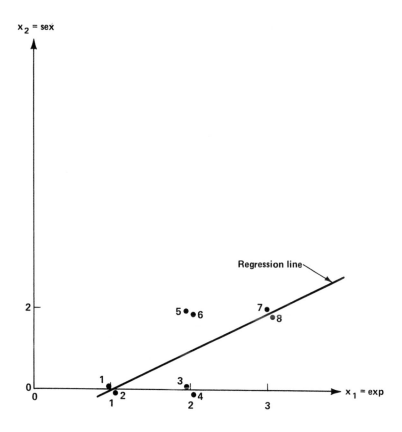

Figure 4.2 - Sex versus experience.

the multiple regression equation, and that the simple regression coefficient obtained from this operation is in fact the multiple regression coefficient b_2 that we seek.

Table 4.2

RESIDUALS OF y AND x_2 CONTROLLING FOR x_1

i	$y-2-x_1 =$ $\tilde{y}(x_1)$	$x_2-x_1+1 =$ $\tilde{x}_2(x_1)$
1	-1	0
2	1	0
3	0	-1
4	2	-1
5	-2	1
6	0	1
7	-1	0
8	1	0

These residuals are plotted in Figure 4.3. Note that residuals always have zero means. From Chapter 3, the simple regression coefficient b_2 of $\tilde{y}(x_1)$ on $\tilde{x}_2(x_1)$ is obtained with b_2 equal to the ratio of the co-variance of $\tilde{y}(x_1)$ with $\tilde{x}_2(x_1)$ divided by the variance of $\tilde{x}_2(x_1)$. Thus

$$b_2 = \Sigma \, \tilde{y}_i(x_1)\tilde{x}_{i2}(x_1)/\Sigma \, \tilde{x}_{i2}^2(x_1) \tag{4.11}$$

since the residuals sum to zero:

$$\Sigma \, \hat{\tilde{y}}_i(x_1) = 0, \; \Sigma \, \tilde{x}_{i2}(x_1) = 0. \tag{4.12}$$

We compute $\text{Cov}(\tilde{y}, \tilde{x}_2) = -4/8$, $\text{Var}(\tilde{x}_2) = 4/8$, $b_2 = -4/4 = -1$ from Table 4.2. This agrees with the slope of the line in Figure 4.3. Note that we occasionally abbreviate $\tilde{y}(x_1)$ and $\tilde{x}_2(x_1)$ by $\tilde{y}$ and $\tilde{x}_2$.

The correlation between these two sets of residuals $\tilde{y}(x_1)$ and $\tilde{x}_2(x_1)$ will be denoted as $r_{y2|1}$, the subscripts denoting that the

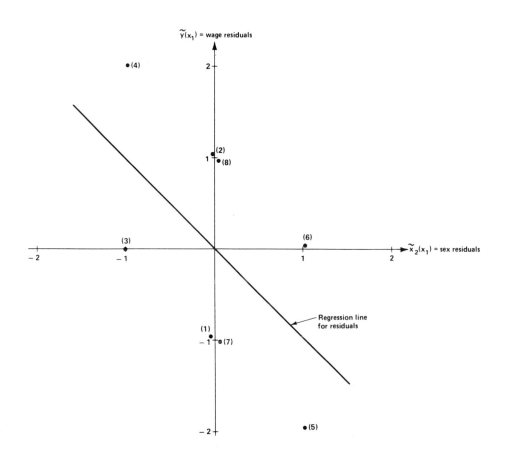

Figure 4.3 − Wage and sex residuals.

correlation between y and x_2 is computed only after removing x_1 through the process of constructing residuals. All these subscripts are necessary to keep the meaning absolutely clear. From Chapter 3 we have

$$r_{y2|1} = \frac{\text{Cov}(\tilde{y}, \tilde{x}_2)}{\sqrt{\text{Var}(\tilde{y})\text{Var}(\tilde{x}_2)}} . \tag{4.13}$$

This is called the "partial correlation coefficient between y and x_2 holding x_1 fixed," but it is calculated as a simple correlation of residuals. It will not ordinarily be equal to the simple correlation r_{y2} between y and x_2. In this example, $r_{y2} = 0$ from Table 4.1, but $\text{Cov}(\tilde{y}, \tilde{x}_2) = -4/8$, $\text{Var}(\tilde{y}) = 12/8$, and $\text{Var}(\tilde{x}_2) = 4/8$ from Table 4.2. Thus

$$r_{y2|1} = -1/\sqrt{3} = -0.577.$$

In this example the correlation between sex and wages is negative, after making an adjustment for the imbalance in experience, even though the simple correlation $r_{y2} = 0$. We conclude that although these women earn the same as men, they deserve more on the basis of their greater experience, and therefore are underpaid (assuming there are no other factors other than experience that distinguish the two sexes and are related to wages).

We have completed three simple regressions: y on x_1, x_2 on x_1, and then after obtaining residuals, $\tilde{y}(x_1)$ on $\tilde{x}_2(x_1)$, and we shall see that we have done enough to produce the full multiple regression (4.4) of y on both x_1 and x_2. As in Figure 4.3, $\tilde{y}$ is predicted from $\tilde{x}_2$ by the least squares line

$$\tilde{y}(x_1) = b_2 \, \tilde{x}_2(x_1). \tag{4.14}$$

Formulas for $\tilde{y}$ and $\tilde{x}_2$ are available from (4.8) and (4.10), and placing them in the preceding with $b_2 = -1$ yields

$$(y-2-x_1) = (-1)(x_2-x_1+1)$$

or

$$y = 1+2x_1-x_2. \tag{4.15}$$

This is the multiple regression equation that we have sought. These values $b_0 = 1$, $b_1 = 2$, $b_2 = -1$ are those that satisfy the least squares criterion of (4.5). Note that (4.6) and (4.7) are satisfied since $\bar{y} = 4$, $\bar{x}_1 = 2$, $\bar{x}_2 = 1$ from Table 4.1. Equation (4.15) is represented geometrically by a plane in three dimensional space (x_1, x_2, y) going through the means $(\bar{x}_1, \bar{x}_2, \bar{y})$. Formula (4.15) is exactly equivalent to combining (4.2) and (4.3) which were derived by separating the data into two groups, because when $x_2 = 0$ for men (4.15) reduces to (4.2), and when $x_2 = 2$ for women (4.15) reduces to (4.3).

There is another way to think of the process of obtaining the effect of x_2 on y after first adjusting for x_1. Simple regression of y on x_2 does not yield the correct answer since we wish to determine the increase in y as x_2 varies while x_1 remains fixed. Without adjustment, x_1 varies with x_2 so that the effect of x_2 on y from a simple regression is the combined result of the effect of x_2 directly on y and the indirect effect of x_2 on y that occurs because x_2 affects x_1 which in turn affects y. The indirect effect is not wanted.

We therefore can think of first adjusting x_1 to some fixed level, say x_1^*, before evaluating the effect of x_2 only. This would be done by adjusting all values of y_i, x_{i2} to what they would be at $x_{i1} = x_1^*$ (we shall see that the choice of x_1^* is immaterial). Regressing y on x_1

69

yields coefficients a_0 and a_1 and

$$\hat{y} = a_0 + a_1 x_1 \; .$$

Thus if an individual with y_i and x_{i1} had his x_{i1} adjusted to x_1^*, his value of y_i should be adjusted to

$$y_i^* = y_i + a_1 (x_1^* - x_{i1})$$

$$= y_i - (a_0 + a_1 x_{i1}) + (a_0 + a_1 x_1^*) \qquad (4.16)$$

$$= \tilde{y}_i (x_1) + (a_0 + a_1 x_1^*) \; .$$

Similarly, a simple regression of x_2 on x_1 yields coefficients c_0 and c_1 and

$$\hat{x}_2 = c_0 + c_1 x_1 \; .$$

Thus adjustment of x_{i1} to x_1^* for individual i adjusts x_{i2} to

$$x_{i2}^* = x_{i2} + c_1 (x_1^* - x_{i1})$$

$$= x_{i2} - (c_0 + c_1 x_{i1}) + (c_0 + c_1 x_1^*) \qquad (4.17)$$

$$= \tilde{x}_{i2} (x_1) + (c_0 + c_1 x_1^*) \; .$$

Now we can assume that everyone has the same value of x_1, so x_1 is effectively constant and can be ignored in studying the simple relation between the y_i^* and x_{i2}^* data.

When we do this to the data of Table 4.1, taking $x_1^* = 2$ years of employment, we change Table 4.3 to Table 4.4 (here $a_0 = 2$, $a_1 = 1$, from (4.1) and $c_0 = -1$, $c_1 = 1$ from (4.9)). Thus all individuals have the same number of years employed.

Table 4.3

WAGES FOR MEN AND WOMEN, ADJUSTING DATA
TO TWO YEARS OF EMPLOYMENT

i (person)	y_i^* adjusted wages	x_1^* years employed	x_{i2}^* adjusted sex
1	3	2	1
2	5	2	1
3	4	2	0
4	6	2	0
5	2	2	2
6	4	2	2
7	3	2	1
8	5	2	1

Note that the y_i^* and x_{12}^* columns of Table 4.4 are obtained by adding 4 and 1, respectively, to the $\tilde{y}$ and $\tilde{x}_2$ columns of Table 4.2, 4 and 1 being the predicted wages and sex for two years of experience ($4 = a_0 + 2a_1$, $1 = c_0 + 2c_1$). This is true in general because from (4.16) and (4.17), the adjusted values y_i^* and x_{i2}^* are the residual values from the x_1 prediction plus the predicted value at x_1^*. Of course persons 1, 2, 7, and 8 have their sex adjusted to 1, when only 0 and 2 are possible, but this just means that sex cannot be predicted accurately from two years of experience.

The reader can check that the simple regression of y_i^* on x_{12}^* yields the predictor

$$y^* = 5 - x_2^*$$

which agrees with Equation (4.15) when $x_1 = x_1^* = 2$ years of experience. We would have obtained the same coefficient of x_2^* had we chosen x_1^* to be any value. Thus the pure effect of sex on wages is determined.

The graphical statement of the proceeding makes the process almost obvious. Figures 4.1 and 4.2 are reproduced below in Figure 4.4. In

71

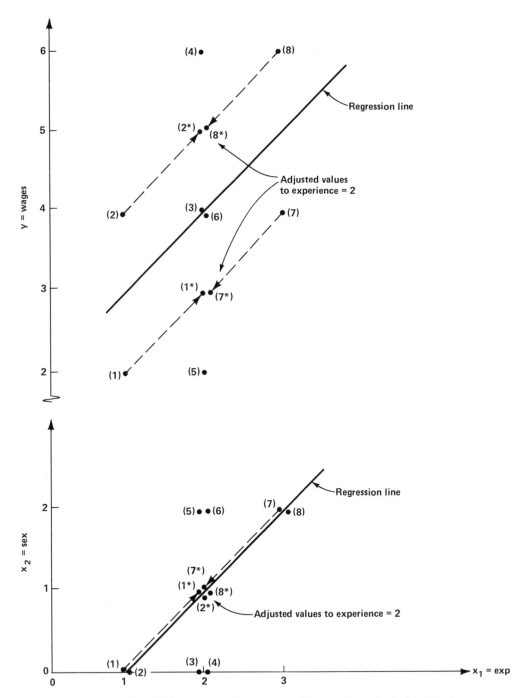

Figure 4.4 – Adjustment of wages and sex of each individual
to the two years of experience.

each case those individuals not having two years of experience have their wages and sex adjusted to that value by moving along the dotted line parallel to the regression line. The resulting values, at $x_1 = 2$, are those appearing in Table 4.3. The adjusted values are all plotted at $x_1 = 2$; no adjustment is needed for those who already have two years of experience.

The correlation between y_i^* and x_{i2}^* in Table 4.3 may now be computed with x_{i1} held constant, and therefore is the partial correlation $r_{y2|1}$. Furthermore, the simple correlation between y_i^* and x_{i2}^* is from (4.16) and (4.17) equal to the simple correlation between $\tilde{y}_i(x_1)$ and $\tilde{x}_{i2}(x_1)$, since the constants $a_0 + a_1 x_1^*$ and $c_0 + c_1 x_1^*$ do not affect the correlation (the reader should verify that correlations are unchanged by addition of constants to the data). Thus, writing corr for the correlation:

$$corr(y^*, x_2^*) = corr(\tilde{y}(x_1), \tilde{x}_2(x_1))$$

$$= r_{y2|1}$$

from (4.13). This shows that the correlation computed from the adjusted data, as in Table 4.3, is exactly the desired partial correlation.

The standard deviation of y about the regression surface $b_0 + b_1 x_1 + b_2 x_2$ is the square root of (4.5) and is denoted by $s_{y|x_1,x_2}$. The vertical bar reads "given," and this is often called the "residual standard deviation." This quantity, by analogy to Chapter 3, is given by

$$s_{y|x_1,x_2}^2 = Var(\tilde{y}(x_1))(1 - r_{y2|1}^2) .$$

It also is the square root of the variance of the y_i^* about x_{i2}^*, since it is the variance of y at a fixed point x_1 and x_2. Since

$$Var(\tilde{y}(x_1)) = s_y^2(1 - r_{y1}^2)$$

73

from Chapter 3, it follows that

$$s_{y|x_1,x_2} = s_y \sqrt{(1-r_{y1}^2)} \sqrt{(1-r_{y2|1}^2)} \ .$$

We denote

$$1 - R_{y,(1,2)}^2 = (1 - r_{y1}^2)(1 - r_{y2|1}^2) \tag{4.18}$$

and therefore can write

$$s_{y|x_1,x_2} = s_y \sqrt{1 - R_{y,(1,2)}^2} \ . \tag{4.19}$$

$R_{y,(1,2)}^2$ is called the "multiple correlation coefficient for predicting y from x_1, x_2." Then $R_{y,(1,2)}^2$ measures how accurately y may be predicted using both x_1 and x_2. From (4.18), $R_{y,(1,2)}^2$ is determined from r_{y1}^2, the worth of x_1 for predicting y and from $r_{y2|1}^2$, the additional worth of using x_2 when x_1 is already used to predict y.

For our data $r_{y2|1}^2 = 1/3$, and we may calculate

$$s_y^2 = 2, \ s_{x_1}^2 = 0.5 \ ,$$

$$Cov(y,x_1) = 0.5 \text{ so } r_{y1} = 0.5 \ .$$

Therefore $s_{y|x_1}^2 = 2(1 - (0.5)^2) = 1.5$ and $s_{y|x_1,x_2}^2 = (1.5)(1 - 1/3) = 1$. Even though sex is *marginally* uncorrelated with wages, conditional on experience, it has a partial correlation of $-1/\sqrt{3} = -0.577$, and it reduces the variance an additional 33 percent. Correlations and partial correlations may even have opposite signs.

74

We also can check that $s^2_{y|x_1,x_2} = 1$ by noting that the average

Table 4.4

PREDICTED VALUES AND RESIDUALS

i	y	$\hat{y} = 1 + 2x_1 - x_2$	residual $y - \hat{y}$
1	2	3	-1
2	4	3	1
3	4	5	-1
4	6	5	1
5	2	3	-1
6	4	3	1
7	4	5	-1
8	6	5	1

squared residual, formula (4.5), is 1 and by definition is $s^2_{y|x_1,x_2}$.

The preceding also enables us to give general formulas for the regression coefficients. Recall that in a simple regression

$$b_1 = r_{y1} \frac{s_y}{s_{x_1}} .$$

Then since x_2 must be held constant in a multiple regression

$$b_1 = r_{y1|2} \frac{s_{y|x_2}}{s_{x_1|x_2}} . \qquad (4.20)$$

Equivalently

$$b_2 = r_{y2|1} \frac{s_{y|x_1}}{s_{x_2|x_1}} . \qquad (4.21)$$

75

For calculations, it is simpler to use the following computation table. First subtract means from the y, x_1, x_2 variables of Table (4.1), and let $Y = y - \bar{y}$, $X_1 = x_1 - \bar{x}_1$, $X_2 = x_2 - \bar{x}_2$. We compute Table 4.5.

Table 4.5

COMPUTATION TABLE

Y	X_1	X_2	Y^2	YX_1	YX_2	X_1^2	X_1X_2	X_2^2
-2	-1	-1	4	2	2	1	1	1
0	-1	-1	0	0	0	1	1	1
0	0	-1	0	0	0	0	0	1
2	0	-1	4	0	-2	0	0	1
-2	0	1	4	0	-2	0	0	1
0	0	1	0	0	0	0	0	1
0	1	1	0	0	0	1	1	1
2	1	1	4	2	2	1	1	1
Sum 0	0	0	16	4	0	4	4	8
Average 0	0	0	2	.5	0	.5	.5	1

Variance-Covariance Matrix:

$$
\begin{array}{c|ccc}
 & Y & X_1 & X_2 \\
\hline
Y & 2 & .5 & 0 \\
X_1 & .5 & .5 & .5 \\
X_2 & 0 & .5 & 1 \\
\end{array}
=
\begin{pmatrix}
s_y^2 & c_{y1} & c_{y2} \\
c_{y1} & s_1^2 & c_{12} \\
c_{y2} & c_{12} & s_2^2 \\
\end{pmatrix}.
$$

Let $d = s_1^2 s_2^2 - c_{12}^2$. Then

$$b_1 = \frac{c_{y1}s_2^2 - c_{y2}c_{12}}{d} \tag{4.22}$$

$$b_2 = \frac{c_{y2}s_1^2 - c_{y1}c_{12}}{d} \tag{4.23}$$

$$b_0 = \bar{y} - b_1\bar{x}_1 - b_2\bar{x}_2 \tag{4.24}$$

$$R^2_{y,(x_1,x_2)} = \frac{b_1 c_{y1} + b_2 c_{y2}}{s^2_y} \qquad (4.25)$$

$$r_{y1} = c_{y1}/s_y s_1, \quad r_{y2} = c_{y2}/s_y s_2, \quad r_{12} = c_{12}/s_1 s_2 \qquad (4.26)$$

$$r^2_{y1|2} = \frac{R^2_{y,(x_1,x_2)} - r^2_{y2}}{1 - r^2_{y2}} \qquad (4.27)$$

$$r^2_{y2|1} = \frac{R^2_{y,(x_1,x_2)} - r^2_{y1}}{1 - r^2_{y1}} \qquad (4.28)$$

and alternatively

$$r_{y1|2} = \frac{r_{y1} - r_{y2} r_{12}}{\sqrt{(1-r^2_{y2})(1-r^2_{12})}} \qquad (4.29)$$

$$r_{y2|1} = \frac{r_{y2} - r_{y1} r_{12}}{\sqrt{(1-r^2_{y1})(1-r^2_{12})}} \qquad (4.30)$$

$$s^2_{y|x_1} = s^2_y (1 - r^2_{y1}) \qquad (4.31)$$

$$s^2_{y|x_2} = s^2_y (1 - r^2_{y2}) \qquad (4.32)$$

$$s^2_{y|x_1,x_2} = s^2_y (1 - R^2_{y,(x_1,x_2)}). \qquad (4.33)$$

These formulas can be justified by using the analysis of residuals taking one variable at a time.

Summary of Calculations

As a convenient reference we summarize the above series of simple regressions that yield the multiple regression.

The Problem: Find b_0, b_1, and b_2 for the equation

$$\hat{y} = b_o + b_1 x_1 + b_2 x_2$$

Step 1. Regress y on x_1 to obtain

$$\tilde{y}_i(x_1) \text{ residuals}$$

$$\hat{y}_i(x_1) \text{ prediction equation}$$

Step 2. Regress x_2 on x_1 to obtain

$$\tilde{x}_{i2}(x_1) \text{ residuals}$$

$$\hat{x}_2(x_1) \text{ prediction equation}$$

Step 3. Regress the y residuals (from Step 1) to obtain

$$b_2 = \frac{Cov(\tilde{y}, \tilde{x}_2)}{Var(\tilde{x}_2)} = \frac{\Sigma\, y_i(x_1)\tilde{x}_{i2}(x_1)}{\Sigma\, \tilde{x}_{i2}^2(x_1)}$$

$$\tilde{y}(x_1) = b_2\tilde{x}_2(x_1) \text{ prediction equation}$$

Step 4. Expand the prediction equation in Step 3 to obtain b_0 and b_1 and thus the full regression line --

Using the fact that $\tilde{y}(x_1) = y - \bar{y}(x_1)$ and $\tilde{x}_2(x_1) = x_2 - \hat{x}_2(x_1)$, obtain

$$y - \hat{y}(x_1) = b_2(x_2 - \hat{x}_2(x_1))$$

or

$$\hat{y}(x_1, x_2) = b_0 + b_1 x_1 + b_2 x_2$$

78

4.2. REGRESSION WITH SEVERAL INDEPENDENT VARIABLES

Now suppose a third variable, x_3 = "training" is also available to distinguish between the employees. We code $x_3 = 0$ if the employee has not had this particular kind of training and $x_3 = 2$ if the employee has been trained. We extend Table 4.1 to include training, which is given in the last column.

Table 4.6

WAGES AND CHARACTERISTICS FOR MEN AND WOMEN

i	Wages y	Years x_1	Woman x_2	Trained x_3
1	2	1	0	0
2	4	1	0	2
3	4	2	0	0
4	6	2	0	0
5	2	2	2	2
6	4	2	2	2
7	4	3	2	0
8	6	3	2	2

The multiple regression of y on x_1, x_2, and x_3 all together is obtained by: (1) predicting y from x_1, x_2; (2) predicting x_3 from x_1, x_2; and (3) regressing the y residuals on the x_3 residuals by a simple regression. The residuals of y on x_1 and x_2 were derived before as

$$\tilde{y} = y - 1 - 2x_1 + x_2 \qquad (4.34)$$

and the values are given in the last column of Table 4.4. (We could write

79

$\tilde{y}(x_1, x_2)$ for completeness.) To predict x_3 from both x_1 and x_2, we proceed as in the previous section. First regress x_3 on x_1. It is easy to determine that x_3 is uncorrelated with x_1 so $x_3 = 1$ is the prediction and x_3-1 is the formula for the residuals. We regressed x_2 on x_1 in formula (4.9), the residuals being given in (4.10) as $x_2-x_1 + 1$, the numerical values appearing in the last column of Table 4.2.

Table 4.7

RESIDUALS OF x_3 ON x_1 AND OF x_2 ON x_1

i	$x_3 - 1$	$x_2 - x_1 + 1$
1	-1	0
2	1	0
3	-1	-1
4	-1	-1
5	1	1
6	1	1
7	-1	0
8	1	0

The regression coefficient of $x_3 - 1$ on $x_2 - x_1 + 1$ is computed as the sum of the products of the two columns divided by the sum of squares of the second column of Table 4.7 (why?). It is therefore $4/4 = 1$. Hence

$$x_3 - 1 = 1(x_2 - x_1 + 1) \tag{4.35}$$

or

$$\hat{x}_3 = 2 - x_1 + x_2 \tag{4.36}$$

(We could write $\hat{x}_3(x_1, x_2)$ for completeness.) The residuals of (4.36) are therefore

80

$$\tilde{x}_3 = x_3 - 2 + x_1 - x_2 \qquad (4.37)$$

and the residuals (4.34) and (4.37) are placed in Table 4.8.

Table 4.8

RESIDUALS OF y AND x_3 ON x, AND x_2

i	$\tilde{y}$	$\tilde{x}_3$
1	-1	-1
2	1	1
3	-1	0
4	1	0
5	-1	0
6	1	0
7	-1	-1
8	1	1

The quantities in Table 4.8 are plotted in Fig. 4.5.

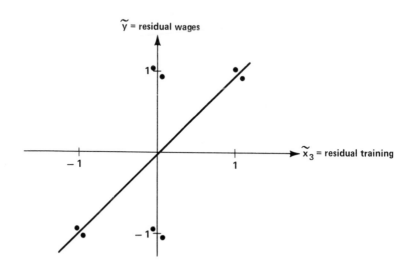

Fig. 4.5 - Wage and Training Residuals.

Clearly $\tilde{x}_3$, residual training, explains some of the wage residuals, and therefore x_3 will be useful for predicting wages. From Fig. 4.5 the prediction of $\tilde{y}$ from $\tilde{x}_3$ is $\tilde{x}_3$ itself, or

$$\tilde{y} = \tilde{x}_3 . \tag{4.38}$$

Of course, this could also be verified by a numerical calculation. Then from (4.34) and (4.37), (4.38) gives

$$y - 1 - 2x_1 + x_2 = x_3 - 2 + x_1 - x_2$$

or

$$\hat{y} = -1 + 3x_1 - 2x_2 + x_3 . \tag{4.39}$$

Formula (4.39) is the multiple regression of wages on experience, sex, and training. The effects have changed again: training is worth one dollar per hour, and since women had better training than men, the coefficient of x_2 is smaller when training is omitted than it is in (4.39). The coefficient of x_2 (sex) has changed from 0 in a simple regression, to -1 with experience as the sole covariate, to -2 using both experience and training. With more data, we would believe the -2 value is best.

The partial correlation between wages and training, holding experience and sex constant, is denoted

$$r_{y,3|1,2} \tag{4.40}$$

and is computed as the simple correlation between $\tilde{y}$ and $\tilde{x}_3$. From Table 4.8 this is

$$r_{y,3|1,2} = \frac{4}{\sqrt{8 \times 4}} = .707 \ .$$

By analogy to (4.18), the squared multiple correlation coefficient $R^2_{y,(1,2,3)}$ increases from $R^2_{y,(1,2)} = 0.5$ according to

$$1 - R^2_{y,(1,2,3)} = (1 - R^2_{y,(1,2)})(1 - r^2_{y,3|1,2}) \qquad (4.41)$$

$$= (1 - .5)(1 - .5)$$

or

$$R^2_{y,(1,2,3)} = 0.75.$$

This means that 75 percent of the variance in y is explained by x_1, x_2, x_3 together,

$$s_{y|x_1,x_2,x_3} = s_y \sqrt{1 - R^2_{y,(1,2,3)}} \ . \qquad (4.42)$$

By definition, $s_{y|x_1,x_2,x_3}$ is the standard deviation of y around the fitted regression line $b_0 + b_1 x_1 + b_2 x_2$, i.e.,

$$s^2_{y|x_1,x_2,x_3} = \frac{1}{n} \Sigma \ (y_i - b_0 - b_1 x_{i1} - b_2 x_{i2} - b_3 x_{i3})^2$$

where b_0, b_1, b_2, b_3 have been chosen to minimize this sum of squares. It can be shown that the values of b_0, b_1, b_2, b_3 we have derived in this section minimize this sum of squares. The multiple correlation in (4.43) can be solved to obtain

$$1 - R^2_{y,(1,2,3)} = \frac{s^2_{y|x_1,x_2,x_3}}{s^2_y} \qquad (4.43)$$

Table 4.9

PREDICTED VALUES AND RESIDUALS

i	y	$\hat{y} =$ $-1+3x_1-2x_2+x_3$	$\tilde{y} = y - \hat{y}$
1	2	2	0
2	4	4	0
3	4	5	-1
4	6	5	1
5	2	3	-1
6	4	3	1
7	4	4	0
8	6	6	0

which is the ratio of the variance of y around the regression surface, or equivalently the residual variance, to the total variance of y. Thus R_y^2 is the fractional reduction in variance of y provided by using the variables x_1, x_2, x_3 to explain y. It therefore is often called "the proportion of variance explained."

We also note that $R_{y,(1,2,3)}^2$, or the multiple correlation coefficient between y and any set of variables differs from an ordinary or partial correlation because rather than being a correlation between two variables, it is a correlation between one variable y and several other variables. What is meant by this is that the best linear combination of these other variables for predicting y is obtained. Then the simple correlation between y and this linear combination is obtained. This best linear combination is precisely the least squares regression equation, and

$$R_{y,(x_1,x_2,x_3)}^2 = \text{corr}^2(y, \hat{y})$$

where $\hat{y} = b_0+b_1x_1+b_2x_2+b_3x_3$. The least squares linear combination not only minimizes the sum of squared errors, but it can be shown that it maximizes the correlation between y and any linear combination of the explanatory variables given in the last columns of Table 4.6.

Finally, we plot residual wages versus predicted wages. We hope to see that there are (1) no outliers (no residuals are much greater than the others), and (2) that the standard deviation of the residuals does not vary systematically as $\hat{y}$ increases. Figure 4.6 confirms this hope.

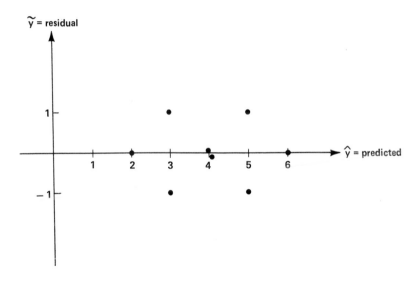

Fig. 4.6 - Plot of residual wages versus predicted wages.

4.3. MULTIPLE REGRESSION: MATRIX NOTATION

Authors' note: Readers unfamiliar with matrix theory will have difficulty with this section. We still believe that they will find enough useful information to justify at least one reading before continuing on to Section 4.4.

The techniques for computing and interpreting multiple regression coefficients, multiple and partial correlation coefficients, and residual standard deviations may be extended indefinitely to situations with more than three independent variables. This method permits computation of a regression on k-independent variables to be simplified to two regressions on k-1 independent variables, followed by a simple regression of these residuals on one another. These regressions in turn may be further simplified until ultimately a k-dimensional regression is equivalent to $k(k+1)/2$ simple regressions. Efficient computer programs perform stepwise reductions equivalent to these. The student is urged to become very familiar with this stepwise method especially because it makes so clear the interpretation that each multiple regression coefficient is a simple regression coefficient, holding the remaining independent variables fixed. Similarly, partial correlation coefficients determine the strength of residual relationships, i.e., holding other independent variables fixed, and thereby measure the increased explanatory power provided by inclusion of an additional explanatory (independent) variable.

In spite of this, it frequently is useful to express the regression model and results in "matrix notation." This is done as follows. For each $i = 1, \ldots, n$ we have

$$y_i = b_0 + b_1 x_{i1} + b_2 x_{i2} + \ldots + b_k x_{ik} + e_i$$

with e_i the error if y_i is predicted by the particular combination of $(x_{i1}, \ldots, x_{ik})$ using $b_0, b_1, \ldots, b_k$. This may be written

$$\begin{pmatrix} y_1 \\ y_2 \\ \vdots \\ y_n \end{pmatrix} = \begin{pmatrix} 1 & x_{11} & \cdots & x_{1k} \\ 1 & x_{21} & & x_{2k} \\ \vdots & \vdots & & \vdots \\ 1 & x_{n1} & & x_{nk} \end{pmatrix} \begin{pmatrix} b_0 \\ b_1 \\ \vdots \\ b_k \end{pmatrix} + \begin{pmatrix} e_1 \\ e_2 \\ \vdots \\ e_n \end{pmatrix}$$

(4.44)

or

$$\underset{\sim}{y} = \underset{\sim}{X} \underset{\sim}{b} + \underset{\sim}{e}$$

in matrix and vector notation, with y the column vector of values of the dependent variable and $\underset{\sim}{X}$ the n row by k+1 column matrix of values of independent variables. The first column of X is a vector of ones to account for the "constant term" b_0.

The vector $\underset{\sim}{b}$ of regression coefficients is to be varied and then chosen to obtain the best fit, that is, the one which makes the residuals $(e_1, e_2, \ldots, e_n)$ smallest. The least squares criterion for "smallest" is that $\underset{\sim}{b}$ should be chosen to minimize

$$\sum_{i=1}^{n} e_i^2 = \underset{\sim}{e}'\underset{\sim}{e}.$$

(4.45)

Here $\underset{\sim}{e}'\underset{\sim}{e}$ is squared length (norm) of the vector $\underset{\sim}{e}$. Any choice of $\underset{\sim}{b}$ which satisfies

$$\underset{\sim}{X}'(\underset{\sim}{y} - \underset{\sim}{X}\,\underset{\sim}{b}) = \underset{\sim}{0}$$

(4.46)

minimizes the sum of squared residuals. If $\underset{\sim}{X}$ is of "full rank" ($n \geq k+1$ is a necessary requirement for this, but also required is that no column of $\underset{\sim}{X}$ should be a linear combination of the other columns) then

$$\underset{\sim}{b} = (\underset{\sim}{X}'\underset{\sim}{X})^{-1}\underset{\sim}{X}'\underset{\sim}{y}$$

(4.47)

is the unique solution. The "full rank" condition guarantees that $\underset{\sim}{X}'\underset{\sim}{X}$ has an inverse, which is denoted $(\underset{\sim}{X}'\underset{\sim}{X})^{-1}$ above. In almost all problems, $\underset{\sim}{X}$ is constructed so that it is of full rank. Equations (4.46) are called the *Normal Equations*. The normal equations say that b is any solution causing the residuals $\underset{\sim}{e} = \underset{\sim}{y} - \underset{\sim}{X}\underset{\sim}{b}$ to be uncorrelated with (orthogonal to, perpendicular to) all the independent variables ($\sum_i e_i x_{ij} = 0$, $j = 1, 2,$ ..., k and $\sum_i e_i = 0$).

The first column vector of units in $\underset{\sim}{X}$ plays no special role in the solution as it is presented in the preceding paragraph, but rather may be thought of as any other column in X. Hence, if a fit with $b_0 = 0$ is wanted then the column of units is omitted from $\underset{\sim}{X}$ and (4.47) still is the correct solution. In fact, since $\sum_i e_i = 0$ corresponds to fitting a constant term, then if one centers all the variables at their means before finding the least squares solution, fitting

$$y_i - \bar{y} = b_1(x_{i1} - \bar{x}_1) + b_2(x_{i2} - \bar{x}_2) + \ldots + b_k(x_{ik} - \bar{x}_k) \qquad (4.48)$$

($\bar{y}$ and $\bar{x}_j$ being means of the dependent variable and the j-th independent variable), then the least squares solution $b' = (b_1, \ldots, b_k)$ is

$$b = C_{\underset{\sim}{XX}}^{-1} C_{\underset{\sim}{Xy}} \qquad (4.49)$$

with $C_{\underset{\sim}{XX}}$ the k × k matrix of covariances of the independent variables and $C_{\underset{\sim}{Xy}}$ the k × 1 vector of covariances of y with the independent variables. This is just equivalent to the form $(\underset{\sim}{X}'\underset{\sim}{X})^{-1}\underset{\sim}{X}'\underset{\sim}{y}$, but with $\underset{\sim}{X}$ taken to be the matrix of elements of $x_{ij} - \bar{x}_j$. It follows from (4.48) that

$$b_0 = \bar{y} - b_1\bar{x}_1 - \ldots - b_k\bar{x}_k . \qquad (4.50)$$

It also follows that (4.49) is the solution to the problem of maximizing the simple correlation between y and $\hat{y} = \bar{y} + \sum_j b_j(x_j - \bar{x}_j)$.

The fitting of multiple regression equations involves only variances and covariances of the observations, and their means in (4.50).

Expressions for other quantities also may be given in matrix notation, for example

$$1 - R^2_{y,X} = \frac{\Sigma \, e^2_i / n}{s^2_y} \tag{4.51}$$

$$1 - R^2_{y,X} = \frac{\underset{\sim}{y}'(\underset{\sim}{y} - X\underset{\sim}{b})}{ns^2_y} \tag{4.52}$$

if $\underset{\sim}{b}$ is given by (4.47). This simplifies to

$$R^2_{y,X} = \frac{\underset{\sim}{b}' C_{XX} \underset{\sim}{b}}{s^2_y} \tag{4.53}$$

if $\underset{\sim}{b}$ is given by (4.49). That is, only the last k components of $\underset{\sim}{b}$ are retained in (4.53).

The residual standard deviation and variance, as usual, are given by

$$s_{y|X} = s_y \sqrt{1 - R^2_{y,X}} \, . \tag{4.54}$$

The matrix presentation given above is very sketchy, and is presented primarily for those students whose mathematical background is sufficient that it be understood. It is presented here because the statistical literature frequently uses this notation so that the student needs some familiarity with it. Many texts in the bibliography present a fuller treatment of these ideas. We recommend Neter and Wasserman [Chapters 6 and 7] for a fuller explanation. Some additional material about this model also is presented in Section 1 of Chapter 7.

Matrices are used in statistics in two ways, as a mathematical tool and as a descriptive device for data sets. Understanding the former demands a mathematical background that this book does not require, since the methods of the earlier sections present alternatives which have the additional virtue of enhancing the interpretation of solutions. But matrices are extremely useful as a description for data sets, and the analyst who can readily organize his data into a rectangular matrix, the rows being the individuals for replication, the columns the variables, will find the tasks of understanding his data, formulating a model, and of choosing computer programs much simplified. If the analyst understands well the relationships and methods to be used, then the technical problems of matrix inversion and manipulation are managed by the computer for any particular application. The analyst therefore is freed from this burden, unless he is required to extend or adapt theory to unusual situations.

4.4. DIAGNOSTIC PLOTS, TRANSFORMATIONS OF VARIABLES, EXAMPLES

The yearly wages in thousands of dollars of a sample of 16 men is given as y in Table 4.11. We are interested in relating this quantity to x_1, their years in the labor force and x_2, their race (coded 0 for nonwhites and 1 for whites).   We shall take the obvious first step of regressing y on x_1 and x_2.   The simplest procedure for doing this is to use a computer to obtain

$$y = 4.260 + .668 \ x_1 + 4.025 \ x_2 \qquad\qquad (4.55)$$

with a prediction error $s_{y|x_1,x_2} = 1.649$ thousand dollars.

Note that x_1 and x_2 are uncorrelated (this is obvious from Table 4.11 without doing any calculations). This is a useful fact. In general, if x_1 and x_2 are uncorrelated, then the simple regression of y on x_1 alone will produce the same regression coefficient b_1 as will the multiple regression of y on x_1 and x_2 together. Can you prove this? In this case, $b_1 = 0.668$ from either regression. There is another simple fact, that when x_1 and x_2 are uncorrelated,

$$R^2_{y|x_1,x_2} = r^2_{y1} + r^2_{y2} \ . \qquad\qquad (4.56)$$

Table 4.11

WAGES AND CHARACTERISTICS OF 16 MEN IN THE LABOR FORCE

Worker i	Annual Wages in Thousands of Dollars y	Years in Labor Force x_1	Race x_2	Interaction $x_3 = x_1 x_2$	log y
1	5.455	0	0	0	1.696
2	6.600	0	0	0	1.887
3	6.630	4	0	0	1.892
4	8.022	4	0	0	2.082
5	8.059	8	0	0	2.087
6	9.751	8	0	0	2.277
7	9.796	12	0	0	2.282
8	11.853	12	0	0	2.473
9	6.545	0	1	0	1.879
10	7.920	0	1	0	2.069
11	8.955	4	1	4	2.192
12	10.835	4	1	4	2.393
13	12.250	8	1	8	2.506
14	14.823	8	1	8	2.696
15	16.759	12	1	12	2.819
16	20.279	12	1	12	3.010

That is, the proportion of the variance explained is just the sum of the proportions explained by x_1 and x_2 separately. For these data $R^2_{y,(x_1,x_2)} = 0.8270$, $r_{y1} = 0.7543$ and $r_{y2} = 0.5079$. These numbers satisfy (4.56).

From (4.28), we can always write

$$R^2_{y|x_1,x_2} = r^2_{y1} + r^2_{y2|1} (1 - r^2_{y1}) \qquad (4.57)$$

so that in general the proportion of explained variance from x_1, x_2 is the proportion of explained variance from x_1 alone plus the proportion of residual variance explained by x_2 times the unexplained variance from x_1. Formula (4.57) simplifies to (4.56) when $r_{12} = 0$ by using (4.30).

The plot of residual wages on predicted wages in Fig. 4.7 suggests that a nonlinearity has been unaccounted for, and that Equation (4.55) is

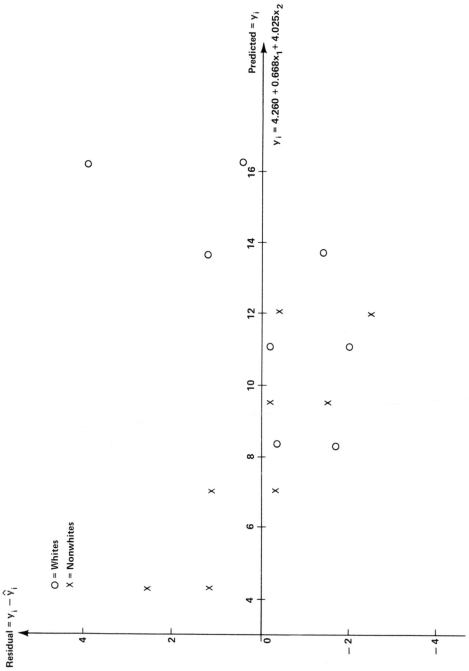

Figure 4.7 – Plot of Residual Wages on Predicted Wages.

misspecified. Figure 4.8 makes this point more clearly and suggests what is wrong. The lines given by (4.55) predicting the dependence of y on x_1 have the same slope, $668 per year for both races, but the plotted data clearly suggest that whites receive more for their experience than nonwhite. Thus, an "interaction" term is needed, meaning that an additional variable should be constructed showing that the value of experience depends on race. The simplest approach is to fit

$$y = b_0 + b_1 x_1 + b_2 x_2 + b_3 x_3 \qquad (4.58)$$

with $x_3 = x_1 x_2$ as the interaction term. Then for nonwhites, $x_2 = x_3 = 0$, so the equation $y = b_0 + b_1 x_1$ is fit, while for whites, $x_2 = 1$ and $x_3 = x_1$, so the equation $y = b_0 + (b_1 + b_3)x_1 + b_2$ is fit. Consequently, b_0 is the mean wage for a nonwhite with no experience and b_1 is the value of a year's experience. Therefore b_3 is the additional value of a year's experience for a white over a nonwhite and b_2 the additional amount that a white receives with no experience. While these results could be obtained by two simple regressions in this case, we shall proceed with a multiple regression because in more general or more complicated cases where other explanatory variables are included, or where x_2 is not a dummy variable, breaking up the sample would be infeasible. However, the method being developed works satisfactorily in those cases.

The values of x_3 are in Table 4.11. When y is regressed on x_1, x_2, x_3, the equation

$$y = 5.875 + .3992\ x_1 + .7954\ x_2 + .5383\ x_1 x_2 \qquad (4.59)$$

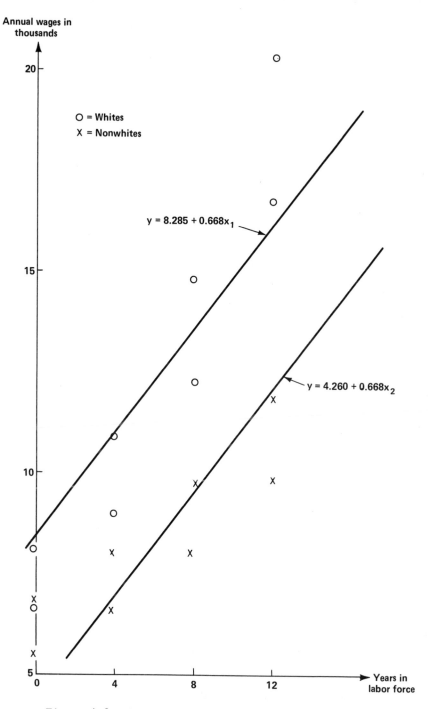

Figure 4.8 – Wage Predictions Without Interactions.

results.   This shows that experience is worth about $399 per year to nonwhites and $937 to whites (compare $668 obtained for both groups before).   The lines

$$y = 5.875 + .399 \ x_1$$

for nonwhites and

$$y = 6.670 + .937 \ x_1$$

for whites are plotted in Fig. 4.9.

While the specification is much improved by using (4.59), a disturbing nonlinearity remains.   Figure 4.9 suggests that wages are nonlinear with x_1 being larger than predicted at the extreme values of x_1 and less than predicted in the intermediate range.   Furthermore, the variation of wages appears to increase with x_1.   While this feature is not very disturbing with our sample of 16, it would be if it persisted in larger samples.

We may believe that wages increase not by a constant amount each year, but by a constant percentage.   Such a model would reflect the nonlinear increasing growth as x_1 increases and would also lead to larger absolute dispersion as x_1 increases.   Mathematically, this implies that the dependence of wages on experience is of the form

$$y = AB^{x_1} \tag{4.60}$$

with A and B being constants.   For example, if A = 5 and B = 1.10, then

$$y = 5(1.10)^{x_1}$$

would represent a 10 percent increase in wages each year from a starting

96

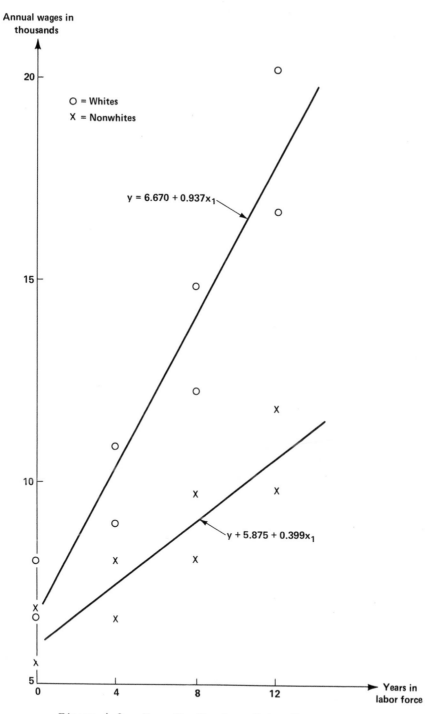

Figure 4.9 — Wage Predictions Using Interaction.

wage of 5 thousand dollars for no experience. While (4.60) is not
linear, so that linear regression methods cannot be applied to it
directly, we can take logarithms of both sides to get

$$\log y = \log A + x_1 \log B. \qquad (4.61)$$

Then if log y is regressed as

$$\log y = b_0 + b_1 x_1, \qquad (4.62)$$

it follows by comparing (4.61) and (4.62) that $b_0 = \log A$ and $b_1 = \log B$, or conversely, $A = \exp(b_0)$ and $B = \exp(b_1)$. (We are using the "natural" logarithm here.) Consequently, A and B may be estimated by regressing log (y) on x_1, thus obtaining b_0, b_1.

The values of log y are given in Table 4.11 and are plotted in Fig. 4.10. Clearly the data for each race is more linear and is more homoschedastic (has more constant variance) than in Figs. 4.8 and 4.9.

If again log (y) is regressed on x_1, x_2,

$$\log y = 1.7031 + .0636 \ x_1 + .3597 \ x_2 \qquad (4.63)$$

results. The standard error of estimate is $s_{\log y | x_1, x_2} = .129$ indicating about a 13 percent error in predicting wages.

Exponentiating (4.63) gives

$$y = 5.491 \ (1.066)^{x_1} \ (1.433)^{x_2} . \qquad (4.64)$$

In this formulation, nonwhites with no experience get \$5491 and whites without experience ($x_1 = 0$, $x_2 = 1$) get 43 percent more than nonwhites. Experience is worth 6.6 percent per year.

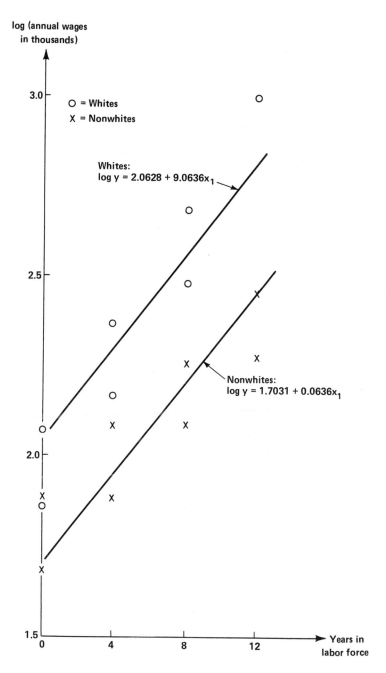

log (annual wages in thousands)

3.0

O = Whites
X = Nonwhites

Whites:
log y = 2.0628 + 9.0636x_1

2.5

Nonwhites:
log y = 1.7031 + 0.0636x_1

2.0

1.5

0   4   8   12   Years in labor force

Figure 4.10 – Prediction of log(wage) without interactions.

However, the plot of log y on x_1 for each race in Fig. 4.10 again shows that an interaction term is needed since each race has its own slope. Consequently, log y should be regressed on x_1, x_2 and x_3 = $x_1 x_2$, yielding

$$\log y = 1.7918 + .0488\ x_1 + .1823\ x_2 + .0296\ x_1 x_2 \tag{4.65}$$

which differs meaningfully from (4.63). Exponentiating both sides yields

$$y = 6(1.05)^{x_1} (1.20)^{x_2} (1.03)^{x_1 x_2} \tag{4.66}$$

with a standard error of $s_{\log y | x_1, x_2, x_3} = 0.11$ or about 11 percent of wages. As Fig. 4.11 shows, the residuals are constant and the model (4.66) seems acceptable. From (4.66), we have the following interpretation: nonwhites (x_2 = 0) without experience (x_1 = 0) have beginning wages of $6000 per year and thereafter increase at 5 percent per year. Whites (x_2 = 1) get 20 percent more than nonwhites initially (x_1 = 0) and increase 3 percent faster than nonwhites per year of experience. This represents a total increase of about 8 percent per year for whites. After six years (the average experience for the sample) whites are making 43 percent more than nonwhites, which accounts approximately for the coefficient in (4.64).

The curve (4.66) is plotted against years in labor force for both groups in Fig. 4.12. The curve does not bisect the points in Fig. 4.12. Instead of allowing for equal absolute errors, the percentage errors are equal. The residuals do not add to zero in this coordinate system, although the percentage residuals (errors) do.

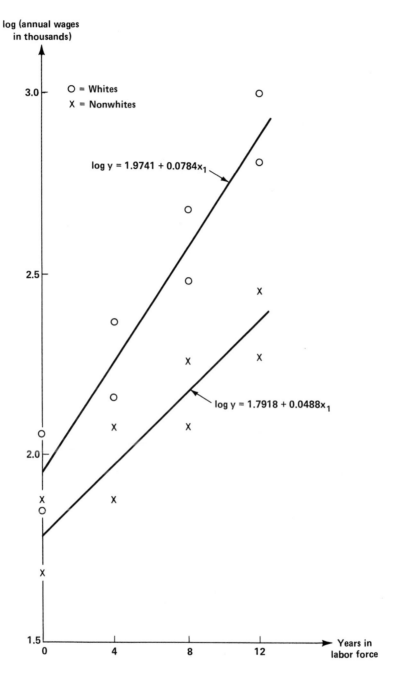

Figure 4.11 — Prediction of log(wage) using interaction.

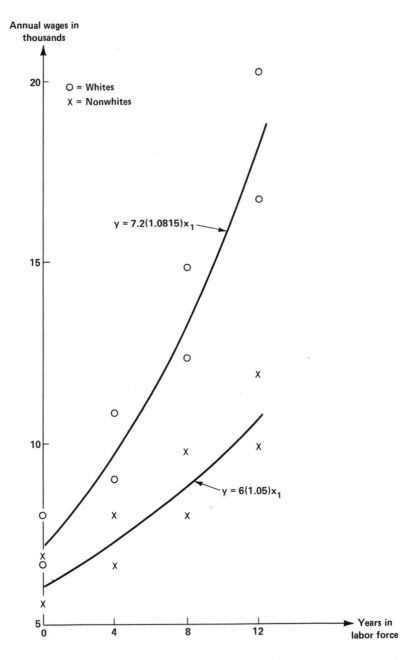

Figure 4.12 – Predicted Wages for Final Specification: Logarithmic
Regression Including an Interaction Term.

102

Part II. STATISTICAL INFERENCE FOR LINEAR MODELS

Chapter 5

PROBABILITY AND STATISTICAL DISTRIBUTIONS

5.1. SAMPLING IDEAS

In preceding chapters, regression models and other models were used
for the sole purpose of describing data. Often, the data are known to
be a *sample* from a larger *population*. In such cases the primary interest
usually is in the population. This requires making inferences from the
sample to the population. Such inferences depend on knowing how the
sample is related to the population. For example, the simplest assumption is that there is a large or infinite collection of "individuals"
or "objects" and the data at hand are a "random sample" from this population. Most of the theory of statistical inference is based on this
assumption. There are many models other than random sampling which include: biased sampling (which sometimes may be corrected by adjustments);
cluster sampling (positive dependence between sample elements); time
series (which usually requires "stationarity" assumptions); sampling
from finite populations without replacement (which requires finite sampling corrections); and "representative" and stratified sampling, which
usually occurs by deliberate design. If data are acquired without reference to a population, then one must define the population from which
the data may be viewed as a random sample, and all inferences should be
restricted to that population. In such circumstances the assumption
of "randomness" may be based on ignorance and if subsequent information
(a reduction in ignorance) indicates that the sample does not represent
the population initially assumed, then new interpretations must be
made. Only simple random sampling will be treated in this chapter, the
more complicated alternatives being left for later.

We assume that a clearly defined population of individuals is given. "Individuals" is a broad term used to refer to the units that comprise the sample. Individuals can be people, aircraft parts, cities, coin flips, times at bat, dates, etc. Associated with each individual is one or more measurements. If a full census of individuals in the population could be taken, the population characteristics or "parameters" could be determined from that census. A full census frequently is infeasible for a number of reasons. Examples are infinite populations and cost or time constraints for finite populations. So in most cases, a sample must be taken from the population. The population parameters then can be *estimated* by computing the equivalent quantities in the sample. Different samples from the same population will produce different estimates of the same population parameter. One objective of *statistical inference* is to estimate the population parameters of interest from the actual sample and to determine how different this estimate might have been had other samples been obtained. To do this, we calculate the sample estimator, its standard deviation, and the associated "confidence intervals" for the unknown parameter.

Most inference methods are based on the assumption that the sample in hand is a random sample from the specified population (sometimes called the sample space). A random sample (i.e., a "simple random sample") means,

1. All individuals in the population have an equal probability of being in the sample.

2. Individuals are chosen *independently*.

Independence means that the probability that a given individual is selected does not depend on which individuals are selected. (This will be made more precise later.) Cluster sampling provides an example where equal probability holds but not independence. For example, suppose the population in a small town is 1,000 homes and two homes are to be chosen at random (possibly allowing the same home to be drawn twice). A cluster sampling method (devised to reduce traveling costs) might specify that the population is to be broken into the 500 homes on the west side and the 500 homes on the east side. A coin is to be flipped, and if heads, two homes are to be chosen randomly on the west side; if tails, two homes are to be chosen on the east side. Then every home has two chances in 1,000 to be chosen so the equal probability assumption holds. However, independence does not hold, since if the first home chosen is on the west side, then the probability that a given home is chosen as the second home is one chance in 500 if it is on the west side, with no chance if it is on the east.

A second problem appears in the cluster sampling example. The population in the example is finite and independence can occur only if *sampling with replacement* is undertaken (allowing the same house to be sampled twice). If, after the first west-side home was selected, the second home had to be chosen from the remaining 499, then the method is referred to as *sampling without replacement*. Independence is destroyed in sampling without replacement because the probability that a given home is selected as the second home depends on whether or not it was selected as the first. However, if a finite population has many individuals relative to the size of the sample, then sampling with and without replacement yields approximately the same results, because it is unlikely that duplications will occur when sampling with replacement. Consequently, the independence assumption holds approximately for sampling without replacement from large finite populations. For infinite populations, sampling with and without replacement are equivalent since not replacing the individual selected on the first choice does not change the probabilities for subsequent selections.

5.2. A SAMPLE OF ONE FROM A FINITE POPULATION

In the earlier chapters we studied methods of summarizing data sets. Each individual or unit in the data set had one or more characteristics (such as wages, sex, training, years experience, region) which we tabulated. Suppose exactly one individual is to be selected at random (everyone is equally likely to be selected) from the data set. Formally, we call the data set the *sample space S* for this experiment. Each individual is a *sample point* and an *event* is defined as a collection of sample points. The *probability of an event A* is defined as

$$P(A) = \frac{\text{number of sample points in A}}{\text{number of sample points in S}} = \frac{\text{No. in A}}{n} \qquad (5.1)$$

As an example of a sample space, consider the data set for the eight individuals in Table 5.1, consisting of their hourly wages, number of years employed, and sex. Think of picking one of the eight individuals at random from this sample space.

Table 5.1

WAGES FOR MEN AND WOMEN

i (person)	y (hourly wages)	x_1 (yrs. employed)	x_2 (sex)
1	2	1	0
2	4	1	0
3	4	2	0
4	6	2	0
5	2	2	2
6	4	2	2
7	4	3	2
8	6	3	2
Sum	32	16	8
Average	$\bar{y} = 4$	$\bar{x}_1 = 2$	$\bar{x}_2 = 1$

If A is the event that the individual selected has an hourly wage of 2, A contains 2 sample points: person 1 and person 5. Thus $P(A) = 2/8 = 1/4$. Similarly, if B is the event of being employed 1 year then $P(B) = 2/8 = 1/4$. The *conditional probability of the event A given the event B* is defined as

$$P(A|B) \;=\; \frac{P(A \text{ and } B)}{P(B)} \;=\; \left(\frac{\text{Number of sample points in } both \text{ A } and \text{ B } / \text{ n}}{\text{Number of sample points } / \text{ n} \text{ in B}} \right) \qquad (5.2)$$

Since there is one person with a wage of 2 who has been employed 2 years

$$P(A|B) \;=\; \frac{1/8}{2/8} \;=\; 1/2.$$

Formula (5.2) also can be written $P(A \text{ and } B) = P(B)P(A|B)$ and can be extended to more than two events as

$$P(A \text{ and } B \text{ and } C....) \;=\; P(A)P(B|A)P(C|A \text{ and } B)...$$

A rule that assigns a number or a value to each sample point (or outcome of a random experiment) is called a *random variable.* For finite sample spaces, the *probability distribution* of a random variable is given by a list of the values of the random variable together with their corresponding probabilities. The random variable Y (hourly wages) given in Table 5.1 has the distribution:

Value of Y	Probability
2	0.25
4	0.50
6	0.25

Other random variables that can be defined on this sample space are X_1 (years employed), X_2 (sex), as well as $Y-X_1$, X_2/Y, X_1^2, and so on. The probability distributions of these random variables are specified by calculating the probabilities for each value. Table 5.2 gives the values of some variables.

Table 5.2

VALUES OF VARIABLES

Person	Y	X_1	X_2	$Y-X_1$	X_2/Y	X_1^2
1	2	1	0	1	0	1
2	4	1	0	3	0	1
3	4	2	0	2	0	4
4	6	2	0	4	0	4
5	2	2	2	0	1	4
6	4	2	2	2	.5	4
7	4	3	2	1	.5	9
8	6	3	2	3	.33	9

From this table we can calculate directly the probability distributions of these random variables, since each person has probability 1/8 of selection. For example,

Value of X_1	probability
1	0.25
2	0.50
3	0.25

Value of X_2/Y	probability
0	0.50
0.33	0.125
0.50	0.25
1	0.125

The right-hand column of probabilities is called the *density* or *frequency function* of the random variable.

The *mean* or *expected value* or *expectation* of a random variable X is denoted by E(X) and is defined as

$$E(X) = \Sigma x P(X = x) \qquad\qquad (5.3)$$
$$\text{x}$$

Summation over values x of X.

We will follow the usual convention of using capital letters (X) to denote the random variable and lower case letters (x) to denote possible values the random variable may take. Thus for hourly wages,

$$E(Y) = 2P(Y{=}2) + 4P(Y{=}4) + yP(Y{=}6) = 2 \cdot .25 + 4 \cdot .5 + 6 \cdot .25 = 4.$$

Alternatively, the expectation of Y is computed by multiplying the value of Y for each individual i by the probability of that individual being selected and adding these products.

$$E(Y) = \Sigma y(i)P(i)$$
$$\text{i}$$

Summation over individuals i.

Here $y(i)$ denotes the value of Y for individual i. Since in a population of size n each individual has probability $1/n$, this yields $E(Y) = \frac{1}{n} \Sigma y(i)$. That is, $E(Y)$ is the *average value of Y.* Here $E(Y) = \frac{1}{8} (2{+}2{+}4{+}4{+}4{+}6{+}6) = 4.$

Some properties of the mean which follow from the definition (5.3) are

1. The expected value of a sum of random variables is the sum of the expected values:

$$E(X_1 + X_2 + \ldots) = E(X_1) + E(X_2) + \ldots$$

2. The expected value of any constant times a random variable is that constant times the expected value

$$E(cX) = cE(X)$$

3. The expected value of a constant is that constant

$$E(c) = c$$

4. And combining the above, for any linear combinations,

$$E(a_1X_1 + a_2X_2 + \ldots + a_nX_n + c) = a_1E(X_1) + a_2E(X_1) + \ldots + a_nE(X_n) + c .$$

The *variance* of a random variable X (usually denoted $\sigma^2(X)$) is defined as the expectation of the random variable $(X-\mu)^2$ where $\mu = E(X)$. Since the variance is the expected value of the random variable $(X-\mu)^2$ it also can be calculated as the average value of that random variable. That is,

$$\sigma^2(X) = E(X - \mu)^2 = \sum_x (x - \mu)^2 P(X = x) \qquad (5.4)$$

where the summation is over values x of X, or

$$\sigma^2(X) = \sum_i (x(i) - \mu)^2 P(i) \qquad (5.5)$$

where the summation is over individuals i. For the example given in Table 5.2, since $E(Y) = \mu_Y = 4$ Equation (5.4) gives

$$\sigma^2(Y) = (2-4)^2(.25) + (4-4)^2(.5) + (6-4)^2(.25) = 1 + 0 + 1 = 2,$$

where Equation (5.5) uses the average value of $(Y-\mu_Y)^2$ to get

$$\sigma^2(Y) = \frac{1}{8} (4+4+0+0+0+0+4+4) = \frac{16}{8} = 2 .$$

This latter expression is equivalent to the definition of the squared sample standard deviation s_y^2 given in Chapter 2. An alternative formula for the variance that is often easier for computation is $\sigma^2(X) = E(X^2) - \mu^2$. A useful relationship between the mean and variance of any random variable X is that the value of the constant c which makes the expectation of $(X-c)^2$ smallest is c = E(X). For this choice of c, the expected value of $(X-c)^2$ is $\sigma^2(X)$.

The nonnegative square root of the variance of a random variable X is called the *standard deviation of X* as denoted by $\sigma(X)$. Thus $\sigma^2(Y)$ = 2 above means that the average of the squared distance from y to μ is 2. Hence the square root of this squared distance, the "root-mean squared distance," $\sigma(Y)$ is 1.414.

112

Some easily checked properties of $\sigma^2(X)$ and $\sigma(X)$ are, for constants c, a, b:

1. $\sigma^2(X+c) = \sigma^2(X)$

    $\sigma(X+c) = \sigma(X)$

2. $\sigma^2(cX) = c^2\sigma^2(X)$

    $\sigma(cX) = |c|\sigma(X)$

3. In general $\sigma^2(a+bX) = b^2\sigma^2(X)$.

    $\sigma(a+bX) = |b|\sigma(X)$.

Returning to our example given in Table 5.2, the *joint probability distribution* of two random variables may be defined by a list of possible pairs of values of the two random variables together with these corresponding probabilities. As in the one variable case these probabilities are called the *joint density* or *frequency function*. Thus from Table 5.2, the joint distribution of Y and X_1 is given by

Values of (Y,X_1)	Probability
(2,1)	.125
(2,2)	.125
(4,1)	.125
(4,2)	.25
(4,3)	.125
(6,2)	.125
(6,3)	.125

The joint probability distribution of k variables is defined analogously by a list of "k-tuples" (or k-vectors) of values together with their corresponding probabilities.

The *covariance of two random variables* X and Y is defined by

$$\text{Cov}(X,Y) = E(X-\mu_X)(Y-\mu_Y) \tag{5.6}$$

where $\mu_X = E(X)$ and $\mu_Y = E(Y)$.

All of the concepts defined thus far are identical to their defini-
tions in Chapter 3 if we think of randomly selecting exactly one individual from a finite population. The covariance of Y and X_1 can be computed directly from the joint distribution as

$$\text{Cov}(Y, X_1) = (2-4)(1-2)(.125) + (2-4)(2-2)\ (.125) + (4-4)(1-2)(.125)$$
$$+ (4-4)(2-2)(.25) + (4-4)(3-2)(.125) + (6-4)(2-2)(.125)$$
$$+ (6-4)(3-2)(.125) = .125(2+2) = .5$$

Some useful properties of covariance, for constants a, b, a_i, b_j, are

1. $\text{Cov}(X, Y) = EXY - (EX)(EY)$
2. $\text{Cov}(X, Y) = \text{Cov}(Y, X)$
3. $\text{Cov}(X, X) = \text{Var}(X)$
4. $\text{Cov}(aX, bY) = ab\ \text{Cov}(X, Y)$
5. $\text{Cov}(X, a) = 0$
6. $\text{Cov}(X+a, Y+b) = \text{Cov}(X, Y)$
7. $\text{Cov}(X_1+X_2, Y_1+Y_2) = \text{Cov}(X_1, Y_1) + \text{Cov}(X_1, Y_2) + \text{Cov}(X_2, Y_1)$
$$+ \text{Cov}(X_2, Y_2)$$
8. $\text{Cov}(a_0 + \sum_{i=1}^{m} a_i X_i b_0 + \sum_{j=1}^{n} b_j Y_j) = \sum_{i=1}^{m} \sum_{j=1}^{n} a_i b_j\ \text{Cov}(X_i, Y_j)$.

The *correlation coefficient,* $\rho(X, Y)$, between X and Y is defined as

$$\rho(X, Y) = \frac{\text{Cov}(X, Y)}{\sigma(X)\sigma(Y)} \tag{5.7}$$

(where $\frac{0}{0}$ is interpreted as 0). Note that $\rho(X, X) = 1$, $\rho(aX+b, cX+d) =$
$\text{sign}(ac)$, and $\rho(aX+b, cY+d) = \text{sign}(ac)\rho(X, Y)$.

All the concepts on regression and correlation for data sets presented in Chapters 3 and 4 have a probability interpretation. Suppose
one of the eight individuals will be selected at random from Table 5.2.
You are required to predict his (or her) hourly wage, and you lose
the square of your error or prediction. Then your best prediction is
the mean of Y, or 4, and your expected loss using this best prediction
is

$$\sigma^2(Y) = E(Y - 4)^2 = 2 \ .$$

114

If you know you will be told the number of years the selected individual has been employed, your *best linear predictor* is the regression of Y on X_1 or $2 + X_1$ (Eq. 4.1) and your expected loss is

$$E[Y - (2+X_1)]^2 = \sigma^2(Y)[1 - \rho^2(Y, X_1)] = 1.5 .$$

Thus $\rho^2(Y, X_1) = .25$ is the *"worth of X_1"* as a linear predictor of Y since $\rho^2(Y, X_1)$ is the proportion your expected loss is reduced from $\sigma^2(Y)$ by knowing X_1. Similarly, the squared multiple correlation coefficient = 0.25, written $R^2_{y,(1,2)}$ in Eq. 4.19, is the proportion that your expected loss is reduced from $\sigma^2(Y)$ by knowing both X_1 *and* X_2 (and predicting linearly).

5.3. SOME MORE ON PROBABILITY DISTRIBUTIONS

In the preceeding section, we sampled one unit from a finite population with all units having equal probability of selection. For finite populations, the probabilities of selection need not be equal for all elements. More generally the population may be either countably or uncountably infinite. The relationships presented in section 2 do not change in any substantive way, but the calculations become mathematically more complicated. When the population is finite or countably infinite, the only change is that unit i has probability P(i) of being selected and the expectation of a random variable X is given by

$$E(X) = \Sigma\ x(i)\ P(i)$$

where the summation is over i's in the population, or

$$E(X) = \Sigma_x\ xP(X = x)$$

where the summation is over values x of X.

For our purposes, the uncountably infinite case will be synonomous with "absolutely continuous" distributions so that the expectations of random variables will be given by the integral:

$$E(X) = \int_{-\infty}^{\infty} x f_X(x) dx$$

where $f_X(x)$ if still called the *density of X* and is a nonnegative
function with

$$\int_{-\infty}^{\infty} f_X(x) dx = 1 \ .$$

All the concepts defined earlier carry over with summation signs replaced by integral signs. For example, one of the most commonly
occurring densities is

$$f_X(x) = \frac{1}{\sqrt{2\pi}} e^{-x^2/2}$$

which is called the *standard normal density*. You can check that if a
random variable X has a standard normal distribution, then

$$E(X) = \int_{-\infty}^{\infty} x \frac{1}{\sqrt{2\pi}} e^{-x^2/2} dx = 0$$

and

$$\sigma^2(X) = \int_{-\infty}^{\infty} (x - 0)^2 \frac{1}{\sqrt{2\pi}} e^{-x^2/2} dx = 1 \ .$$

In this text we rarely will have to integrate densities explicitly.
Students who have some knowledge of integral calculus should refer to
a good statistics and probability text such as Hogg and Craig.

5.4. RANDOM SAMPLING AND THE BINOMIAL DISTRIBUTION

Random Sampling. *A random sample of size n from a probability*
distribution having density f is defined as a sequence of n *independent*
and *identically distributed* (i.i.d.) random variables X_1, X_2, ..., X_n

whose densities are all f. Since most of our statistical inference procedures will assume a random sample, we now define the concept of *independence* more precisely.

Earlier, the conditional probability of an event A given an event B was defined as $P(A|B) = \dfrac{P(A \text{ and } B)}{P(B)}$. Since a random variable taking on a particular value is an event, we define the *conditional probability distribution* of the (discrete) random variable Y given the value x of the random variable X by its density $P(Y=y|X=x)$.[*] For example, the probability distribution of Y given $X_1 = 1$ in Table 5.2 is given by $P(Y=2|X_1=1) = \frac{1}{2}$ and $P(Y=4|X_1=1) = \frac{1}{2}$. We say that *two random variables X and Y are independent* if all the conditional probability distributions are the same as the unconditional probability distributions, i.e., $P(A|B) = P(A)$ or $P(A \text{ and } B) = P(A)P(B)$. More generally *the random variables X_1, X_2, X_3, ..., are independent* if the distribution of each one, given the values of all the others, is always the same as its unconditional distribution (that is, does not depend on the values of the other variables). Thus, random sampling is an important application of independence.

One useful property of n independent random variables X_1, ..., X_n is

$$\sigma^2(X_1 + \dots + X_n) = \sigma^2(X_1) + \dots + \sigma^2(X_n) . \qquad (5.8)$$

That is, the variance of the sum is the sum of the variances. To see that this is true, you should first check that *if two random variables are independent, they are uncorrelated* (the converse is *not* true).[**] This follows from the fundamental fact that the expectation of the product of any number of independent variables is the product of expectations:

---

[*] The definition is analogous for continuous random variables.

[**] Zero correlation is sometimes called "linear independence." Zero correlation between *all* functions of the random variable implies full independence.

117

$$EX_1X_2 \ldots X_n = (EX_1)(EX_2) \ldots (EX_n) , \qquad (5.9)$$

if $X_1, \ldots, X_n$ are independent, since for example,

$$EX_1X_2 = \Sigma \; \Sigma \; x_1 x_2 P(X_1 = x_1, \; X_2 = x_2)$$

$$= \Sigma \; \Sigma \; x_1 x_2 P(X_1 = x_1) P(X_2 = x_2)$$

$$= (\Sigma \; x_1 P(X_1 = x_1))(\Sigma \; x_2 P(X_2 = x_2))$$

$$= (EX_1)(EX_2) .$$

Then $\text{Cov}(X_1, X_2) = EX_1X_2 - EX_1 EX_2 = 0$. In general, if all pairs of variables in a sum of random variables are uncorrelated, then (5.8) holds. This follows from taking two uncorrelated random variables X_1 and X_2 and observing

$$\sigma^2(X_1 + X_2) = \text{Cov}(X_1 + X_2, \; X_1 + X_2) = \text{Cov}(X_1, \; X_1) + \text{Cov}\;(X_1, X_2) + \text{Cov}(X_2, \; X_1)$$

$$+ \; \text{Cov}(X_2, \; X_2) = \sigma^2(X_1) + \sigma^2(X_2) + 2\;\text{Cov}(X_1, \; X_2) .$$

Thus $\sigma^2(X_1 + X_2) = \sigma^2(X_1) + \sigma^2(X_2)$ if $\text{Cov}(X_1, \; X_2) = 0$ which is true if X_1 and X_2 are uncorrelated.

The number most commonly calculated from a sample is the mean or average. If the sample is random, the mean $\overline{X} = \dfrac{1}{n} (X_1 + \ldots + X_n)$ also has a probability distribution. We now give the expectation and variance of $\overline{X}$.

1. If $X_1, \ldots, X_n$ is a random sample from a distribution with mean μ, then the mean of the sample mean $\overline{X}$ is the same as the mean of the distribution:

$$E(\overline{X}) = E(X_1) = \ldots = E(X_n) = \mu . \qquad (5.10)$$

2. The variance of $\overline{X}$ is σ^2/n, where σ^2 is the variance of the distribution of X_i,

$$\sigma^2(\overline{X}) = \frac{\sigma^2}{n}$$

and thus the standard deviation of $\overline{X}$ is

$$\sigma(\overline{X}) = \frac{\sigma}{\sqrt{n}} \, . \tag{5.12}$$

Eq. (5.10) follows directly from the definition of expected value while (5.11) is mainly a consequence of (5.8).

5.5 THE BINOMIAL DISTRIBUTION AND THE NORMAL APPROXIMATION

Suppose all the individuals of a population can be classified into one of two categories. For convenience, we will refer to these categories as "success" and "failure." The parameter of interest is p, the proportion of individuals in the population who are "successes." A random sample of size n is taken from this population, and we denote by X the number of individuals in the sample who are classified "success." We shall see that X has a probability distribution that is known as the "binomial distribution." We will also see that for n large enough, this distribution may be approximated by the "normal distribution." An understanding of these distributions is important, because they provide the basis for inference from a binomial sample to a binomial population. In Chapter 6, inference for regression models and other models will be considered, but we will not go into them in as much detail as we do here. However, while the distributions there are more complicated to work with, the conceptual issues are the same as those in making inferences for the simpler binomial setting.

The *probability distribution* of a binomial random variable X describes, before the sample is drawn, the probability of getting the possible values of X. If there are n observations, X takes integer values between 0 and n. We shall see how to compute probabilities for X from published tables.

A sample from a binomial population with a specified parameter p may be taken by using a table of pseudo random numbers (see Table 1). For example, to obtain a sample of size n = 10 from a population with p = 0.6, read across the top row of Table 1 and classify all numbers less than 0.60000 as a "success." This gives SFFSSSSSSF, interpreting each block of five as one number with a decimal in front. That is, we read the first numbers as 0.53479, 0.81115, ..., etc. The value X = 7 is obtained.

The binomial distribution is the distribution of the number of successes from a sample of size n. One usually justifies the use of the binomial distribution by verifying the assumptions of random sampling. Our notation for saying that X has a binomial distribution with n and p as parameters is

$$X \sim \text{Bin}(n, p) .$$

There is one binomial distribution for each integer $n \geq 1$ and each p, $0 \leq p \leq 1$.

The binomial density is

$$b(x; n, p) = P(X = x|n, p) = \binom{n}{x} p^x (1-p)^{n-x},$$ (5.13)

being the probability of exactly x successes in n trials if p is the probability of a success on any one trial.

The "choice function" $\binom{n}{x}$ is the number of distinct subsets of size x that can be chosen from a set of size n. In this case, it is the number of distinct arrangements of x successes in a sequence of n trials. The formula for the choice function is

$$\binom{n}{x} = \frac{n!}{x!\,(n-x)!} = \frac{n(n-1)(n-2)\ldots(4)(3)(2)(1)}{x(x-1)\ldots(3)(2)(1)(n-x)\ldots(3)(2)(1)}.$$ (5.14)

Values of the choice function also can be obtained from Pascal's triangle where each entry is the sum of the two entries immediately above.

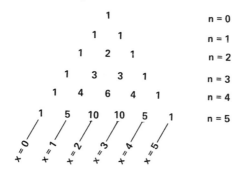

Figure 5.1 - Pascal's triangle.

Pascal's triangle gives the correct formula for $\binom{n}{x}$ because

$$\binom{n}{x} = \binom{n-1}{x-1} + \binom{n-1}{x}$$

This happens since x successes occur in n trials in exactly two ways: the first trial is a success and then x-1 successes occur in the last n-1 trials in any of $\binom{n-1}{x-1}$ ways; or the first trial is a failure and then x successes occur in the last n-1 trials in any of $\binom{n-1}{x}$ ways.

The crucial assumption for the binomial distribution is that trials are independent. The independence assumption permits the computation of the probability of any string of successes and failures to be computed as the product of the success and failure probabilities. For example, if there is a success on the first trial, then a failure, then a success, abbreviated SFS, then P(SFS) = P(S)P(F)P(S). Since p = P(S) and P(F) = 1-P(S) = 1-p, then P(SFS) = p(1-p)p = $p^2(1-p)$. It is convenient to denote P(failure) = q = 1-p as the failure probability when working with the binomial distribution. Then P(SFS) = p^2q. There are also two successes in three trials from the outcomes FSS or SSF. But note P(FSS) = P(F)P(S)P(S) = qpp = p^2q and P(SSF) = P(S)P(S)P(F) = ppq = p^2q. Hence, the probability of any sequence of successes and failures depends only on the number of successes and the number of failures, but not on the order of their occurrence. This gives

P(x successes and n-x failures in a specified order) = $p^x q^{n-x}$.   (5.15)

The probability of x successes and n-x failures added up over all orderings, (the event of exactly x successes) is given by (5.13) since $\binom{n}{x}$ is the number of distinct orderings, each occurring with probability $p^x q^{n-x}$ as given by (5.15).

122

Taking the example above, two successes (x = 2) in three trials (n = 3) can happen as FSS(qpp), SFS(pqp) and SSF(ppq). These probabilities add to $3p^2q$ and $\binom{n}{x}p^x q^{n-x} = \binom{3}{2}p^2 q^{3-2} = 3p^2q$, which checks. You should verify $\binom{3}{2} = 3$ from (5.14) and Figure 5.1.

The probability of two successes in four trials is obtained from SSFF(ppqq), SFSF(pqpq), SFFS(pqqp), FSSF(qppq), FSFS(qpqp), FFSS(qqpp). These probabilities add to $6p^2q^2$ and you can check that this is b(2; 4, p) = $\binom{4}{2}p^2q^2$.

The binomial density for n = 10, p = 0.25 is graphed in Figure 5.2. The graph increases from x = 0 to a maximum and then declines.

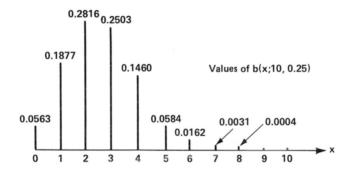

Figure 5.2 - Graph of binomial density for n = 10, p = 0.25.

For large n, binomial probabilities are more difficult to compute. To take the examples x = 0, 2, 8 from Figure 5.2, we compute

$$b(0;\ 10,\ \tfrac{1}{4}) = \binom{10}{0}(\tfrac{1}{4})^0(\tfrac{3}{4})^{10} = \frac{3^{10}}{4^{10}} = 0.0563$$

$$b(2;\ 10,\ \tfrac{1}{4}) = \binom{10}{2}(\tfrac{1}{4})^2(\tfrac{3}{4})^8 = 45 \cdot \frac{3^8}{4^{10}} = 0.2816$$

$$b(8;\ 10,\ \tfrac{1}{4}) = \binom{10}{8}(\tfrac{1}{4})^8(\tfrac{3}{4})^2 = 45 \cdot \frac{3^2}{4^{10}} = 0.000386$$

Since the computations become difficult, it is convenient to table the values.

Given in Table 3 are values for the binomial distributions with
n = 1, 2, ..., 25, p = .05, .10, .15, ..., .45, .50. The values
tabled are

Prob (no more than x successes in n trials with success
probability p)

$$= \sum_{i=0}^{x} b(i; n, p).$$

Since the tables do not cover all parameter values for the binomial
distribution, we must learn to deduce what is needed from the values
given in Table 3.

Example 1: Suppose $X \sim \text{Bin}(10, 0.25)$, i.e., n = 10.

(a) $P(X = 3) = P(X = 0, 1, 2, 3) - P(X = 0, 1, 2)$

$$= .7759 - .5256 = .2503.$$

(b) $P(X \geq 3) = 1 - P(X = 0, 1, 2) = 1 - .5256 = .4744.$

(c) $P(2 \leq X \leq 5) = P(X = 0, 1, 2, 3, 4, 5) - P(X = 0, 1)$

$$= .9803 - .2440 = .7363.$$

(d) $P(2 \leq X \leq 5 \text{ or } X = 8) = P(2 \leq X \leq 5) + P(X = 8)$

$$= .7363 + (1.000 - .9996) = .7367.$$

Example 2: $X \sim \text{Bin}(10, 0.75)$, that is n = 10, p = 0.75.

(e) Find $P(X \leq 3)$. Let Y = number of failures. Then $Y \sim$
Bin(10, .25). Then $P(X \leq 3) = P(Y \geq 7) = 1 - .9965 = .0035.$

Example 3: $X \sim B(30, 0.5).$

(f) Find $P(X \leq 15)$. This problem cannot be solved with the
tables. We must use more sophisticated tables, do the calculations ourselves by using the formulae, or use the *normal
approximation* to the binomial, to be covered shortly.

124

The mean and standard deviation of a binomial population are very important quantities, and have simple formulae as given below. Thus, if $X \sim \text{Bin}(n, p)$ then the expection or mean of X is

$$E(X) = \mu_X = \Sigma x b(x; n, p) = np. \tag{5.16}$$

The variance of X is

$$\sigma^2(X) = E(X - \mu_X)^2 = \Sigma(x - np)^2 b(x; n, p) = npq. \tag{5.17}$$

The standard deviation, is

$$\sigma_X = \sqrt{npq} . \tag{5.18}$$

The mean of X in (5.16) and the variance of X in (5.17) can be verified by carrying through considerable algebra to simplify the summations. An easier method is to define the independent random variables $I_1, \ldots, I_n$ to be 0 or 1 with $P(I_i = 1) = p$, and observe that

$$X = \sum_{i=1}^{n} I_i .$$

Now

$$E(I_i) = 0(1-p) + 1 \cdot p = p$$

and

$$\sigma^2(I_i) = (0-p)^2 (1-p) + (1-p)^2 p = p(1-p) = pq.$$

Thus

$$E(X) = nE(I_i) = np$$

and by (5.8),

$$\sigma^2(X) = \sum_{i=1}^{n} \sigma^2(I_i) = npq$$

thus proving (5.16) and (5.17).

Define the sample proportion of successes $\hat{p} = X/n$ to be the number of successes divided by the number of trials. Since the mean of $\hat{p}$ is p (no matter what p is) we say that $\hat{p}$ is an "unbiased" estimator of p. That is,

$$E(\hat{p}) = E(X/n) = \frac{1}{n} E(X) = \frac{np}{n} = p \text{ for all } p.$$

125

The variance of $\hat{p}$ is

$$\sigma^2(\hat{p}) = \sigma^2 \left(\frac{X}{n}\right) = \frac{1}{n^2} \sigma^2(X) = \frac{npq}{n^2} = \frac{pq}{n}$$

and the standard deviation is

$$\sigma_{\hat{p}} = \sqrt{pq/n} \ .$$

We see that the mean number of successes X for a sample of size n is just n times the success probability for any one trial. Note that σ_X increases as $\sqrt{n}$. But $\sigma_{\hat{p}}$ decreases as $1/\sqrt{n}$, which means that $\hat{p}$ will get close to p as n increases.

The Normal Distribution

The two most common and most important sampling distributions in statistics are the binomial distribution and the normal distribution. The standard normal distribution (also called the Gaussian or bell-shaped distribution) is a continuous distribution taking an uncountably infinite number of possible values. Its density function

$$\phi(z) = \frac{1}{\sqrt{2\pi}} \exp(-z^2/2) = \frac{0.39894}{(1.6487)z^2}$$
$$= 0.39894/10^{(z/2.146)^2} \tag{5.19}$$

is plotted in Figure 5.3. Since it is a density, $\phi(z)$ has area 1.

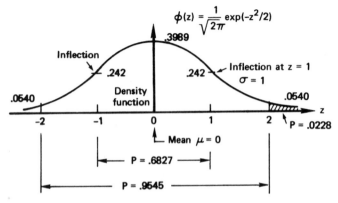

Fig. 5.3 - The standard normal density function.

The density is symmetric about 0 and its inflection points lie ±1 standard deviation from the mean. Approximately 2/3 of the time a random selection from this distribution will be between ±1 standard deviations of the mean and 19 times out of 20 within two standard deviations.

The distribution given by (5.19) has mean zero and standard deviation one, which is why it is called the "*standard* normal distribution." Normal distributions with other means and standard deviations are described below.

The standard normal *distribution function* $F(z) = P(Z \le z)$ is the area under the density curve ϕ to the left of z. It increases from 0 at $-\infty$ to 0.5 at 0 to 1 at ∞. The values of $F(z)$ are given in Table 2.

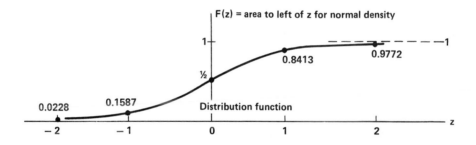

Figure 5.4 - The standard normal cumulative
distribution function.

Probability calculations for standard normal distributions are carried out using the fact that $F(z) = P(Z \le z)$ is given in Table 2 for $z \ge 0$. When $z < 0$, the symmetry of the normal distribution must be used to get probability values.

$P(Z < 1) = .8413$, $P(-1 < Z < 1) = 2[P(Z < 1) - .5] = .6826$

$P(Z < 2) = .9772$, $P(-2 < Z < 2) = 2[P(Z < 2) - .5] = .9544$

$P(Z < 3) = .9987$, $P(-3 < Z < 3) = 2[P(Z < 3) - .5] = .9974$.

We next describe the general normal distribution, allowing for an arbitrary mean and standard deviation. We say that X has a normal distribution with mean μ and standard deviation σ written $X \sim N(\mu,\sigma^2)$ if $Z = (X-\mu)/\sigma$ has the standard normal distribution. Now,

$$P((X - \mu)/\sigma \leq z) = F(z). \quad F(z) = \int_{-\infty}^{z} \frac{1}{\sqrt{2\pi}} e^{-t^2/2} \, dt.$$

In this case, we can write $X = \sigma Z + \mu$, Z being standard normal, and hence deduce that μ is the mean of X and σ is the standard deviation of X. There are many normal distributions, one for each mean and variance. Table 2 is sufficient to find the probabilities for all of them! Here is an example:

$X \sim N(2,4)$. Find $P(X > 0)$. Since $\mu = 2$ and $\sigma = 2$, $Z = \frac{X-2}{2}$. Now $X > 0$ if $Z > -1$. Therefore, $P(X > 0) = P(Z > -1) = P(Z < 1) = 0.8413$.

One important use of the normal distribution is that when n is large, the binomial distribution is approximately normal with the binomial mean and variance. That is, if

$X \sim Bin(n,p)$, and $\hat{p} = X/n$ then approximately:

$X \sim N(np, npq)$ and $\hat{p} \sim N(p, pq/n)$.

This approximation is perfect as n goes to infinity, and usually is fairly good if the binomial variance $npq \geq 4$. (This theorem is called the "central limit theorem for the binomial distribution" when the emphasis is on letting n go to infinity.)

128

Example: Let X ∿ Bin(16, .5).   Then npq = 4.

   (a) Find $P(X = 8)$ and (b) Find $P(|X - E(X)| \leq 2)$.   Solutions:

$\mu = 8$, $\sigma = 2$.   (a) $P(X = 8) = P(7\tfrac{1}{2} \leq X \leq 8\tfrac{1}{2})$.   Let $Z = \dfrac{X-8}{2}$.   With the normal approximation ($P(7\tfrac{1}{2} \leq X \leq 8\tfrac{1}{2}) = P(-\tfrac{1}{4} \leq Z < \tfrac{1}{4}) = .1974$.   Note that Table 3 gives 0.1964.

   (b) $P(|X - E(X)| \leq 2) = P(|X - 8| \leq 2.5)$

$$= P(5.5 \leq X \leq 10.5) = P(-\frac{2.5}{2} \leq Z \leq \frac{2.5}{2})$$

$$= 2[P(Z \leq 1.25) - .5] = .7888.$$

Note that Table 3 gives

$$P(|X - EX| \leq 2) = P(X = 6, 7, 8, 9, 10) = .8949 - .1051 = .7898.$$

The above examples where the error in the approximation is .001 support the rule of thumb that the normal approximation to the binomial is sufficiently good if npq $\geq$ 4.   Of course, "good" may require better approximations in some applications.

Chapter 6

STATISTICAL INFERENCE

The last chapter covered probability. There we were given parameter values for the population and calculated how likely various outcomes were in the sample. In statistical inference, we reverse this process. Given a particular value for the sample, we make statements about the characteristics of the population that produced this value.

6.1. INFERENCE FOR A BINOMIAL POPULATION

A male individual chosen at random from the U.S. population in 1970 had probability $p = 0.6143$ of living beyond the age of 70.[*] A sample of $n = 50$ black males chosen at random from the U.S. population resulted in $X = 20$ men who lived beyond 70. We would like to answer the following questions:

1. *How do we estimate the true probability p that a black male will live beyond age 70?*

2. *How accurate is our estimate?*

3. *Can we give an interval that is "likely" to contain p?*

4. *How do we decide whether the data are consistent with the hypothesis that black males have the same probability of living to age 70 as do white males?*

The methods for answering these questions are contained in the theory of point estimation, confidence intervals and hypothesis testing. We use the binomial distribution example to describe these general methods.

---

[*]This probability was obtained from life expectancy tables in the U.S. Almanac.

Point Estimation (answers questions 1 and 2)

The usual method of estimating a population proportion p is to use the sample proportion $\hat{p}$ = X/n which, for our sample, is 0.40. We already have shown that this estimate is *unbiased* $(E(\hat{p})$ = p for every value of p). Since p is unknown, it is reassuring to know that, on the average, $\hat{p}$ will be neither consistently higher nor lower than p. This property of unbiasness follows from the assumption that X~Bin(n, p) and hence from the fact that a random sample was drawn from the population of black males. The question of the accuracy of the estimate $\hat{p}$ may be answered in terms of its average distance to the quantity being estimated, p. Formally this is $E|\hat{p} - p|$. However, this quantity is difficult to calculate and is no more useful than the alternative of using the square root of the average squared distance $E(\hat{p} - p)^2$. Since p is the mean of $\hat{p}$, this square root is just the standard deviation of $\hat{p}$. Since this measures the error of estimation, the term *standard error* of the estimate is used. To summarize:

An unbiased estimate for the sample proportion p is $\hat{p}$ and the standard error of $\hat{p}$ is $\sqrt{pq/n}$.

Of course, if p is unknown, then the standard error is unknown. There are two simple approaches that may be used to overcome this.

1. If n is reasonably large, $\hat{p}$ will be reasonably close to p and therefore the true standard error may be approximated by $\sqrt{\hat{p}\hat{q}/n}$.

2. (Conservative approach.) A graph of $\sqrt{pq} = \sqrt{p(1-p)}$ in Figure 6.1 shows that the largest value of $\sqrt{pq}$ is 0.5 and that this value occurs at p = 0.5. Therefore the standard error, no matter what p is, can never be larger than the standard error for p = 0.5, which is $0.5/\sqrt{n}$. Figure 6.1 also shows that for a fairly wide

range of p values surrounding 0.5, the standard error is fairly close
to its maximum value of $0.5/\sqrt{n}$. Thus the conservative approach does
not lose much when p is near 0.5. This graph also shows that the
estimate $\hat{p}$ is more accurate when p is near 0 or 1 than when p is
near 0.5.

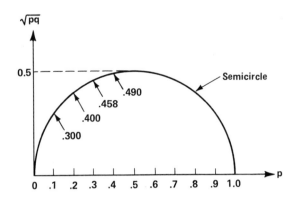

Figure 6.1 – The binomial standard deviation
as a function of the mean.

For our black male sample, $\hat{p}$ = 0.40, so that the two approaches give
standard errors of 0.069 and 0.071, respectively.

Confidence Intervals for p (answers question 3)

Assume that the normal approximation to the binomial distribution
applies (recall that npq $\geq$ 4 is usually adequate). Then $Z = (\hat{p}-p)/\sqrt{pq/n}$
has approximately a standard normal distribution. Then P(-1 < Z < 1) =
0.6827, P(-2 < Z < 2) = 0.9545, P(-2.58 < Z < 2.58) = 0.990 and
P(-3 < Z < 3) = 0.9973. Other examples can be calculated from tables
of the normal distribution. These statements say that $\hat{p}$ will differ
from p by no more than one standard deviation about 68 percent of the
time, $\hat{p}$ will differ from p by no more than two standard deviations

132

about 95 percent of the time, $\hat{p}$ will differ from p by no more than 2.58 standard deviations 99 percent of the time, and $\hat{p}$ will differ from p by no more than 3 standard deviations 99.7 percent of the time. But $\hat{p}$ differs from p by the same amount that p differs from $\hat{p}$ and so we can make the following "confidence statements":

The population proportion p is within 1, 2, 2.58, 3 (respectively) standard deviations of $\hat{p}$ with confidence 0.68, 0.95, 0.99, 0.997 (respectively) if the normal approximation is valid. In practice the standard deviation in the denominator of Z is taken to be $\sqrt{\hat{p}\hat{q}/n}$, or more conservatively, $0.5/\sqrt{n}$ rather than $\sqrt{pq/n}$ as given above.

For our black male data, $\hat{p} = 0.40$ and one standard deviation (standard error) is 0.069. Rounding the standard deviation to 0.07 gives confidence intervals of (0.33, 0.47) (0.26, 0.54), (0.22, 0.58), and (0.19, 0.61), respectively, with confidences of 0.68, 0.95, 0.99, 0.997.

Confidence intervals for a population parameter should be viewed as the interval of parameters that are consistent with the sample obtained. The sample estimate of the population parameter plus or minus two standard deviations usually provides an adequate (95 percent) confidence interval. This interval length accounts for errors due to the randomness of the sample in a correctly specified model. It does not account for possible errors due to possible incorrectness in the model or errors due to a biased sample. Model specification errors and biased samples can occur frequently, especially when complicated analyses are used. There is substantial evidence that standard confidence intervals, which account only for the randomness of the sample, but not for other errors, usually have smaller confidence than their nominal value would suggest, even in favorable circumstances. That is, a 95 percent confidence interval actually contains the true parameter value in fewer than 95 percent of the analyses that compute such intervals.

Hypothesis Testing (answers question 4)

Now $X \sim \text{Bin}(n = 50, p)$ has been observed, with p being the unknown probability that a black male lives beyond 70. The hypothesis to be tested is that $p = 0.6143$, that is, the probability that a black male in the U.S. population lives beyond 70 is the same as that for non-black males. Under this "*null hypothesis*,"

$$ Z = \frac{\hat{p} - p}{\sqrt{pq/n}} = \frac{\hat{p} - 0.6143}{\sqrt{\frac{(0.6143)(0.3857)}{50}}} = \frac{\hat{p} - 0.6143}{0.0688} $$

has a standard normal distribution. If the hypothesis is false ($p \neq 0.6143$) then the mean of Z is not zero and it will not have a standard normal distribution. Under the null hypothesis, we expect $|Z| \leq 2$, since this would happen 95 percent of the time. If we use the rule: Accept the hypothesis for values of $|Z| \leq 2$ and reject if $|Z| > 2$, then the probability of rejecting the null hypothesis if it in fact is true is about 0.05 (precisely, it is 0.0456). The probability of this rule (or "test") rejecting the null hypothesis when it is true is called the "size" or "significance level" of the test and is usually denoted by α. Commonly used values of α are 0.05 and 0.01. Tests with these α values require the estimate to differ from the hypothesized value by at least 2 or 2 1/3 standard deviations, respectively, in order for the hypothesis to be rejected. In general, the particular situation should dictate the choice of α. In our example $Z = (0.40 - 0.6143)/\sqrt{(0.6)(0.4)/50} = -3.11$. That is, the estimate 0.40 differs from the hypothesized value 0.6143 by more than 3 standard deviations. The hypothesis that p = 0.6143 is therefore rejected for any conventional value of α and is thus inconsistent with the data.

The probability at which the data are significant (the *significance probability*) is determined for this problem by computing

$$1 - \alpha^* = P(|Z| \le 3.11)$$

from the table of the standard normal distribution. Thus, $\alpha^* = 1 - 0.998 = 0.002$. Therefore, a value as extreme as the given data would occur only one time in 500 if the hypothesis were true. One would reject the hypothesis if $\alpha = 0.005$ (for example) but not if $\alpha = 0.001$, since $0.005 > \alpha^*$, but $0.001 < \alpha^*$.

A helpful way to think of hypothesis testing is to think of the analogous confidence intervals. Since a confidence interval is the interval of population parameters that are consistent with the data, the set of parameters excluded from the interval are those values that are not consistent with the data. The size of the test is simply one minus the confidence level. The hypothesis that the population parameter takes a particular value is rejected for a particular value of α if and only if the parameter is not included in the level 1-α confidence interval. The reason for rejection is that the hypothesized value differs from the estimated value by an amount too large (as measured by the confidence interval) to be explained by the randomness of the sample. If the test rejects the hypothesis, we conclude that the true parameter differs from the hypothesized value (and probably takes one of the values in the confidence interval).

By analogy with confidence intervals, we see that hypothesis tests tend not to be as conservative as claimed in practice. That is, only errors due to the randomness of the sample are accounted for, while possible errors due to misspecification of the model and biased sampling are not.

An additional "bias" brings more rejected hypotheses to our attention of accepted ones. It stems from the fact that researchers tend to look for cases where a hypothesis is rejected. In analyzing one body of data, it is common practice to perform many tests, searching for a statistically significant result. Clearly, computing 100 independent test statistics, using significance at the $\alpha = 0.05$ level will on average produce 5 *apparently* significant cases (rejections of the hypothesis), even if all 100 population parameters are precisely equal to their hypothesized values. Yet attention may be focused only on these "significant" outcomes, the others being ignored. Since only "significant results" tend to be published, the researcher may report only on these five cases without reminding his reader of the 95 nonsignificant cases. Moreover, this is magnified by many authors publishing only of their statistically significant research results. Readers of many articles therefore may get a very distorted picture of "truth." Until statistically significant results have been corroborated several times, or at least by uncommonly small significance levels, they should be viewed only as hypotheses.

There is another reason to prefer a confidence interval to a hypothesis test. Suppose, for example, the null hypothesis is p = 0.50 and we wish to be sure of rejecting this hypothesis if the population parameter is 0.55 or greater. If the true value is p = 0.52 and n = 10,000, then a value of $\hat{p} = 0.52$ yields a 95 percent confidence interval for p of

(0.51, 0.53). (Why?) The hypothesis p = 0.50 is therefore rejected, but the value p = 0.55 that is of interest does not obtain. A smaller value of n, say n = 2,000, would have produced the desired conclusion of accepting the hypothesis that p = 0.50, i.e., that p is "close enough" to 0.50. Using a confidence interval gives the complete picture of the situation for n = 10,000. That is, the reasonable values for p are in the interval (0.51, 0.53). One is free to determine later if this interval of values differs from 0.50 in an "important" way. The hypothesis test permits the conclusion that p ≠ 0.50, but says nothing about how far p may be from 0.50. One may prefer to think of p = 0.52 as not differing from 0.50 in a way that matters.

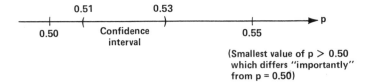

Fig. 6.2 – The problem of "too much data" being
resolved by a confidence interval.

While hypothesis tests can result in the rejection of a hypothesized value, they never can result in the *definite* acceptance of any value, no matter how much data are collected. To see this, note that the entire confidence interval consists of infinitely many possible true population values. No matter how large the sample size n (provided the entire population is not sampled) a confidence interval will have positive length and therefore an infinity of true values cannot be ruled out. However, parameter values beyond a fixed distance from a hypothesized value can be ruled out by using a sufficiently large sample.

137

Example: Ballot Position

The following example answers a fifth question that was not posed earlier: *How large must a sample be to give a fixed "precision"?*

It is claimed that his position on the ballot affects a candidate's vote. Suppose n elections having only two candidates are analyzed, and let p be the probability that the candidate listed in first position wins (we assume, among other things, that elections with incumbents are excluded, for incumbents are usually listed first). The hypothesis that ballot position does not affect the outcome is the hypothesis that p = $\frac{1}{2}$. However close the true value of p is to $\frac{1}{2}$, if it is not equal to $\frac{1}{2}$, then it is possible to choose n large enough so that a test will reject the hypothesis that p = $\frac{1}{2}$ and thus prove the existence of ballot position effect. As pointed out above, it is not possible to choose n large enough to prove ballot position has no effect at all since *all* values of p other than $\frac{1}{2}$ cannot be eliminated.

On the other hand, suppose it is claimed that the effect of ballot position is strong enough that the candidate listed first would win on average in 52 percent of the cases, an advantage sufficiently large to be of concern. This claim (hypothesis) can be tested and either accepted or rejected. We shall see that the size of the sample is important.

Suppose first that n = 500 elections are studied. Then if p = $\frac{1}{2}$, $\sqrt{pq/n}$ = 0.0224 and a 95 percent confidence interval will be $\hat{p}$ ± 0.0448. Even if $\hat{p}$ = 0.50, the values most consistent with the hypothesis of no positional effect (the confidence interval for p) would be (0.455, 0.545)

138

so the possibility that p = 0.52 could not be excluded. Therefore,
analysis of 500 cases is insufficient to shed any light on this issue.[*]

In order to be quite certain of deciding between 0.50 and 0.52, it
is necessary to choose n so large that the two values are at least four
standard errors (of $\hat{p}$) apart. This assertion follows from the calculation that for an n of this size a 95 percent confidence interval will
just contain p = 0.50, if $\hat{p}$ = 0.509 (just under 2 standard deviations)
in which case $\hat{p}$ will differ from 0.52 by over 2 standard deviations
(see Figure 6.3).

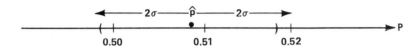

Fig. 6.3 – The "four standard deviation rule."

Similarly, if the true p is 0.52 a 95 percent confidence interval around
$\hat{p}$ = 0.511 (just under two standard deviations) excludes p = 0.50. Thus,
the four standard deviation rule implies that the confidence interval is
likely to contain the true p of interest and will also exclude the
particular false value of p of interest.

The preceding reasoning leads us to choose n so that

$$4 \sqrt{pq/n} = p_1 - p_0$$

where p_1 = 0.52, p_0 = 0.50 in this example.

Setting p = q = ½, the preceding formula implies that

———————
[*]In a 1972 Los Angeles court case, Pines vs. Layton, the defense
analyzed 500 cases and drew the conclusion that first position was no
advantage. They didn't use a hypothesis test, but had they done so,
they would have been unable to reject p = 0.50 in favor of p = 0.52,
even if p = 0.52 were the true value.

$$n = \frac{4}{(p_1 - p_0)^2}$$

is required in order to have a "good chance" to discriminate between the hypotheses p_0 and p_1. For $p_1 = 0.52$, $p_0 = 0.50$, it follows that at least n = 10,000 elections must be analyzed. Analysis of ballot position with a sample this large has never been attempted.

Formally speaking, the four standard deviation rule is equivalent to constructing a two-sided size 0.05 test of the null hypothesis $p - p_0$ which has power 0.975 at the alternative p_1. See Hogg and Craig for an excellent treatment of power of tests. Although we will not cover such matters as formally in this text, Table 6.1 below gives the exact values of the "four" in the four standard deviation rule for varying choices of precision. Readers not already familiar with the terms "power," "one-sided test," "two-sided test," etc., should skip this table. The first two columns give the size and power of the test while the third column gives the number of standard deviations the alternative must be from the null hypothesis in order for the test to reject the alternative. For the binomial case, the required sample size n is given by

$$n = \frac{e^2 pq}{(p_1 - p_0)^2}$$

where e is the tabled entry and p is the value of p_0 or p_1 nearest 1/2. Example: If $p_1 = 0.52$, $p_0 = 0.50$, $p = p_0 = 1-q$, size = $\alpha = 0.05$, power = $\beta = 0.95$, then n = $(3.290)^2 (.5)(.5)/[.52-.50]^2$ = 6765, which is substantially smaller than 10,000 given by the four-standard deviation rule above. The four-standard deviation rule actually gives power

140

$\beta = .95$ to a one-sided normal test with size $\alpha = .01$, i.e., $2.326 + 1.645 = 3.971$ is about 4. Three standard deviations requires settling for $\alpha = .05$, $\beta = .90$, so $n = (2.927)^2(.5)(.5)/(.02)^2 = 5355$.

Table 6.1

SEPARATION OF NULL AND ALTERNATIVES REQUIRED
FOR SPECIFIED SIZE AND POWER TESTS

Size = α	1-Power = $1-\beta$	Number of Standard Deviations Needed (One-Tailed Test)
		$z_\alpha + z_{1-\beta}$
0.05	0.10	1.645 + 1.282 = 2.927
0.05	0.05	1.645 + 1.645 = 3.290
0.01	0.05	2.326 + 1.645 = 3.971
0.01	0.01	2.326 + 2.326 = 4.652
		Two-Tailed: $z_{\alpha/2} + z_{1-\beta}$
0.05	0.05	1.960 + 1.645 = 3.605
0.05	0.025	1.960 + 1.960 = 3.920
0.01	0.025	2.576 + 1.960 = 4.536
0.01	0.005	2.576 + 2.576 = 5.152

141

6.2. INFERENCE ABOUT MEANS OF NORMAL POPULATIONS

In the last section we used the normal distribution to approximate the binomial distribution in order to make the appropriate calculations for binomial inference. In this section, we discuss inferences for some particularly simple situations using the normal distribution. These are estimation of, confidence intervals for, and hypothesis tests about the means of normal distribution. More complex settings for statistical inference using the normal distribution are described in Section 6.3 (regression) and in Chapter 7 (Analysis of Variance).

Inferences About a Single Mean

Suppose X_1, ..., X_n constitute a random sample from a normal distribution with mean μ and variance σ^2. For the moment assume that the information in the sample about μ is contained in the sample average[*]

$$\overline{X} = \frac{1}{n} \sum_{i=1}^{n} X_i .$$

If our interest is in estimating μ, we have already seen in Chapter 5 that the expectation and variance of $\overline{X}$ are given by:

$$E(\overline{X}) = \mu \text{ and } \sigma^2(\overline{X}) = \sigma^2/n .$$

That is, $\overline{X}$ is an *unbiased estimator* of μ whose standard deviation is $\sigma/\sqrt{n}$. It can also be shown that among all unbiased estimators of μ, $\overline{X}$ has the smallest variance. We say that X is the *minimum variance unbiased estimator* (MVUE) of μ. As in the binomial case, $\overline{X}$ can be used to construct confidence intervals for μ by observing that

---

[*]Technically, we say that $\overline{X}$ is a sufficient statistic for μ. See Hogg and Craig for a more precise definition and complete discussion of sufficiency.

$$Z = \frac{\overline{X} - \mu}{(\sigma/\sqrt{n}} \qquad (6.1)$$

has a standard normal distribution. Hence from Section 6.1, we know

that the population mean μ is within 1, 1.96, 2.58, and 3 standard devi-

ations of $\overline{X}$ with confidence of 0.68, 0.95, 0.99, and 0.997 (respectively).

As in the binomial situation, tests of hypothesis about the popula-

tion mean μ can be constructed analogously to confidence intervals using

Equation (6.1). To test the hypothesis that μ is equal to some pre-

specified value μ_0, we reject the null hypothesis that $\mu = \mu_0$ if

$$\left| \frac{X - \mu_0}{\sigma/\sqrt{n}} \right| > 1.96 \ .$$

That is, reject if $\overline{X}$ is more than 1.96 standard deviations away from μ_0.

This test has size $\alpha = .05$. This is the same as rejecting the null

hypothesis if μ_0 falls outside a 95 percent confidence interval for μ.

The value 1.96 comes from the standard normal table corresponding to

the probability of the absolute value of a standard normal random vari-

able exceeding it being $\alpha = 0.05$.[*] This table can be used to find the

appropriate cutoff point for other values of α, e.g., if $\alpha = 0.01$, the

cutoff point is 2.33.

More commonly, σ^2 is unknown. The natural and correct inclination

in this case is to replace σ^2 in our formulas with its estimated value

calculated from the sample. Two issues arise: (i) how to estimate σ^2

and (ii) how to change our inferences to account for using the estimate

of σ^2 instead of σ^2 itself.

Were μ known, the natural estimate of σ^2 would be the average of

the values of $(X_i - \mu)^2$:

---

[*] We frequently use the value 2 to approximate 1.96, as we did in
Section 6.1.

$$\frac{1}{n} \sum_{i=1}^{n} (X_i - \mu)^2 \ .$$

Since μ is unknown, we must estimate it from the data and thus lose one "degree of freedom" from the n degrees of freedom provided by the n observations. The appropriate estimate for σ^2 is:

$$\hat{\sigma}^2 = \frac{1}{n-1} \sum_{i=1}^{n} (X_i - \overline{X})^2 \ .$$

It is straightforward to show that $\hat{\sigma}$ is an unbiased estimate of σ^2; that is $E(\hat{\sigma}^2) = \sigma^2$. Note that this formula for $\hat{\sigma}^2$ differs from the value of s^2 given in Chapter 2 in that n-1 rather than n is in the denominator. We will make such "sampling corrections" in all our work on inference. The -1 is needed because one degree of freedom is lost in the sum of squares to estimating μ by $\overline{X}$.

Inferences Using $\hat{\sigma}^2$, the Unbiased Estimator of the Variance

Is the distribution of

$$t = \frac{\overline{X} - \mu}{\hat{\sigma}/\sqrt{n}}$$

still standard normal? The answer is no, since $\hat{\sigma}$ is a random variable whose value is calculated from the data. The distribution of the above quantity t is called a "student's t-distribution with n-1 degrees of freedom." Table 4 of the t-distribution is used in the same way as the normal distribution tables for confidence intervals and hypothesis testing, as described above. For large values of n, $\hat{\sigma}$ is a good estimate for σ. Thus, you can see by inspecting Table 4 that the t-distribution with a large number of degrees of freedom (say above 30) is essentially the same as a standard normal distribution. For a small number of degrees of freedom, the "tail" of the t-distribution is thicker than that of the normal distribution.

144

Confidence intervals for and tests of hypotheses about μ using $\hat{\sigma}^2$ are constructed the same way as when σ^2 is known except that (i) $\hat{\sigma}^2$ replaces σ^2 and (ii) the appropriate cutoff value is obtained from the t-distribution tables rather than the standard normal distribution tables. The t-distribution will reappear repeatedly when inference in regression and analysis of variance is discussed.

Inferences About the Difference Between Two Means

We cover one other commonly occurring inference problem in this section. Suppose there are two samples from two different populations and we wish either (i) to estimate the difference between the two means or (ii) test whether the two means are equal. Formally, let $X_1, \ldots, X_n$ be a random sample from a $N(\mu_X, \sigma_X^2)$ distribution and let $Y_1, \ldots, Y_m$ be a random sample from a $N(\mu_Y, \sigma_Y^2)$ distribution. For the moment assume that both σ_X^2 and σ_Y^2 are known. The difference of the sample averages $\bar{X} - \bar{Y}$ is the minimum variance unbiased estimator of the difference $\Delta = \mu_X - \mu_Y$ and the standard deviation of $\bar{X} - \bar{Y}$ is $\sigma_{\bar{X} - \bar{Y}} = \sqrt{\sigma_X^2/n + \sigma_Y^2/m}$. Thus $\bar{X} - \bar{Y} \sim N(\Delta, \sigma_X^2/n + \sigma_Y^2/m)$. Confidence intervals for and tests about Δ are then based on

$$ Z = \frac{\bar{X} - \bar{Y} - \Delta}{\sqrt{\sigma_X^2/n + \sigma_Y^2/m}} \tag{6.2} $$

in the usual way.

The problem of making inferences on Δ when σ_X^2 and σ_Y^2 are completely unknown is called the "Behrens-Fisher" problem. It's complete solution is beyond the scope of this text, but we will discuss two special cases: (i) n and m are large or (ii) σ_X^2 and σ_Y^2 are assumed equal.

145

(i) For large n and m (say both above 30) we can use the estimators of σ_X^2 and σ_Y^2:

$$\hat{\sigma}_X^2 = \frac{1}{n-1} \sum_{i=1}^{n} (X_i - \overline{X})^2 ,$$

$$\hat{\sigma}_Y^2 = \frac{1}{m-1} \sum_{i=1}^{m} (Y_i - \overline{Y})^2 ,$$

in Equation (6.2) and use the standard normal tables to construct tests and confidence intervals in the usual way to get an approximately correct solution.

(ii) If we assume $\sigma_X^2 = \sigma_Y^2 = \sigma^2$ where σ^2 is unknown, we see that $\text{Var}(\overline{X}-\overline{Y}) = (\frac{1}{n} + \frac{1}{m})\sigma^2$. We estimate σ^2 by

$$\hat{\sigma}_2^2 = \frac{1}{m+n-2} \left[\sum_{i=1}^{n} (X_i - \overline{X})^2 + \sum_{i=1}^{m} (Y_i - \overline{Y})^2 \right] , \qquad (6.4)$$

and it can be shown that

$$t = \frac{\overline{X} - \overline{Y} - \Delta}{\sqrt{(\frac{1}{n} + \frac{1}{m})\hat{\sigma}^2}} \qquad (6.5)$$

has a t-distribution with n+m-2 degrees of freedom. Two "degrees of freedom" out of the m+n available are lost since both μ_X and μ_Y are being estimated by $\overline{X}$ and $\overline{Y}$ respectively in (6.4). When the value $\Delta = 0$ is substituted in (6.5) for testing the hypothesis that the difference between the two means is zero, we call this the *two sample t-test*. We shall see in Chapter 7 that this is a special case of the one-way layout in the analysis of variance.

6.3. INFERENCE IN SIMPLE AND MULTIPLE REGRESSION

In our work in Chapters 3 and 4 on regression and correlation we calculated regression coefficients, correlations, variances, and standard deviations from data on *all* the individuals of the population about which inferences were made. No sampling was assumed and we concentrated on the interpretation of the calculated quantities, which were population parameters there. In this section we address the problem of making inferences about these regression parameters, from a random sample of individuals from the population.

For example, suppose that we have a population of individuals with two measurements on each individual. If we think of one measurement as the dependent variable y and the other measurement as a single independent variable x_1, we could calculate the regression coefficients, b_0 and b_1, the correlation coefficient r_{y1} and the variance $s^2_{y|x_1}$ if both measurements were available for all the individuals in the population. Now suppose that we have a random sample of 10 individuals from the population. We can calculate b_0, b_1, r_{y1}, and $s_{y|x_1}$ for our sample, but if we took another independent random sample of 10 individuals from the population and calculated b_0, b_1, r_y, and $s_{y|x_1}$ for that sample, the resulting values would be different. Just as we used the sample proportion to make inferences about the parameter p for a binomial population and the sample mean to estimate a normal population mean, we will use the sample values of b_0, b_1, r_{y1}, and $s_{y|x_1}$ to make inferences about the population values of these quantities.

In order to make inferences about a larger population from which the data were drawn, assumptions about the structure of the larger population and the data's relationship to it are needed. The simplest

147

example is the value of the dependent variable Y in the sample being represented as a linear function of the value of the independent variable x plus an error or residual. We will adopt the usual convention of using Greek letters for the population values of parameters and the Roman letters for the corresponding sample values. Thus in our population, an individual's dependent variable value Y is

$$Y = \beta_0 + \beta_1 x + \varepsilon \qquad (6.6)$$

where β_0 and β_1 are the regression coefficients calculated from the whole population, x is his independent variable value and ε is the error or residual.

We now give the other necessary assumptions and show how to make statistical inferences in the context of the above model. In later chapters we address the problem of making inferences when these assumptions may not be true. Suppose we take a random sample of n observations and get the values:

$$y_1 = \beta_0 + \beta_1 x_1 + e_1$$
$$\vdots$$
$$y_n = \beta_0 + \beta_1 x_n + e_n.$$

The x's are assumed fixed and $y_1, \ldots, y_n$ are the observed values of Y in the sample. These values e_i are the result of observing ε independently each time from a normal distribution with mean 0 and variance σ^2. The value y_i is determined by adding the sampled value e_i to $\beta_0 + \beta_1 x_i$. The unknown population parameters are β_0, β_1, and σ^2. Notice that $E(Y_i) = \beta_0 + \beta_1 x_i$ and the variance of Y_i given x_i is $\sigma^2(\varepsilon) = \sigma^2$.

The Density of the Y's for Fixed x_i's

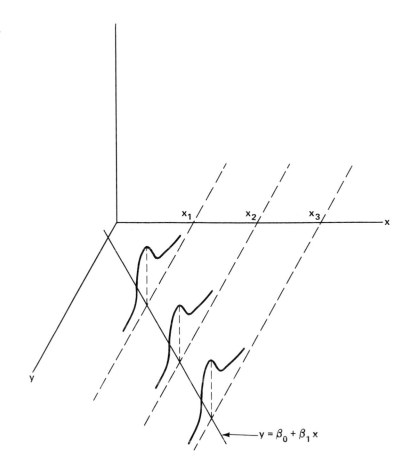

The assumption that errors have normal distributions dates back to eighteenth century investigations into the nature of experimental errors. It was observed then that the distribution of discrepancies between repeated measurements of the same physical quantity could be approximated closely by a normal curve. The mathematical properties of this distribution were first studied by Laplace, deMoivre, and Gauss. In the physical sciences it is commonly called the Gaussian distribution.

An intuitive explanation of why errors often follow a normal distribution comes from the Central Limit Theorem. We invoked a special case of this theorem in Chapter 5 when we used the normal approximation to the binomial distribution. If errors can be represented as the sum of a large number of independent random variables, all being about the same magnitude, this theorem asserts that the distribution of the errors will be approximately normal. A random variable with a binomial distribution is the sum of the number of successes in the n trials, so if n is large, the theorem allows us to approximate the binomial curve with a normal curve. Since unexplained errors are frequently the sum of contributions due to a number of different factors, it follows that their distribution should be approximately normal.

The above development has made the assumption that the explanatory variables, the x's, are fixed. In many situations individuals are sampled randomly from a population and a number of characteristics are recorded for each individual. For example, in Chapter 4 the population of eight people whose hourly wages, experience, training, and sex were recorded might have been the result of drawing a random sample from a larger population. In this case, the x's (experience, training, and sex) as well as the y's (hourly wages) can be regarded as observed values of random variables X, all having a joint probability distribution. In this situation, we think of Equation (6.6) as being the conditional probability distribution of Y given the value x of X: Thus, the x's can be regarded as fixed since we condition on their observed values.[*] This argument applies to several explanatory variables as well as to a single one.

---

[*] The interested reader should see Allan R. Sampson, "A Tale of Two Regressions," *Journal of the American Statistical Association*, Vol. 69, No. 347, 1974, pp. 682–689, for a discussion of how the "fixed x's" model can be obtained by conditioning from the "random x's" model.

Inferences About β_1

For the moment, let us assume that σ^2 is known. The slope of the least squares line calculated from a sample of size n is:

$$b_1 = \frac{\frac{1}{n} \sum_{i=1}^{n} (x_i - \bar{x}) y_i}{\frac{1}{n} \sum_{i=1}^{n} (x_i - \bar{x})^2} .$$

Since under our model, the x's are fixed (not random) and known, b_1 is a random variable depending on $(y_1, \ldots, y_n)$. Easy calculations show that the expectation and variable of b_1 are given by:

$$E(b_1) = \beta_1, \quad Var(b_1) = \frac{\sigma^2}{\sum_{i=1}^{n} (x_i - \bar{x})^2} = \frac{\sigma^2/n}{s_x^2} . \qquad (6.7)$$

There is an interesting consequence of Equation (6.7) for the variance of b_1. If we are allowed to select the x_i values at which to take observations Y_i and wished to minimize $Var(b_1)$, we would attempt to make $\sum(x_i - \bar{x})^2$ as large as possible. If there are not constraints on possible values of x_i, we could make $Var(b_1)$ arbitrarily small by choosing some x_i's with very large positive values and some x_i's with very large negative values. In practice, this means that half the x's should be placed at each of the possible extreme values. (This result is intuitively analogous to choosing two points as far apart as possible when drawing a straight line by connecting the two points to get the most "accurate" line.) This result depends critically on the linear regression model (6.6) being the correct one. When this assumption is not precisely true as is frequently the case, this result may be quite wrong.[*] For this reason most statisticians include interior

points as well as extreme points when they have a choice in the values

of the x's.

Since b_1 is a linear combination of normal random variables (the

Y_i's), it also has a normal distribution. Thus,

$$z = \frac{b_1 - \beta_1}{\sigma} \sqrt{\sum_{i=1}^{n} (x_i - \bar{x})^2}$$

is the value of a standard (mean = 0, variance = 1) normal random vari-

able. Hence from Section 6.2, we have 0.95 confidence that the true

slope β_1 is between

$$b_1 - 1.96 \, \frac{\sigma}{\sqrt{\Sigma(x_i - \bar{x})^2}}$$

and

$$b_1 + 1.96 \, \frac{\sigma}{\sqrt{\Sigma(x_i - \bar{x})^2}} \, .$$

In a similar manner, hypotheses about the value of β_1 can be tested.

Estimating σ^2

Usually σ^2 is unknown. As we did in Section 6.2, we replace σ^2

by its estimated value calculated from the sample, as follows. Were

β_0 and β_1 known, the natural estimate of σ^2 would be the average value

of e_i^2:

$$\frac{1}{n} \sum_{i=1}^{n} (y_i - \beta_0 - \beta_1 x_i)^2 \, .$$

---

[*] See G. Box and N. Draper, "A Basis for the Selection of a Response
Surface Design," *Journal of the American Statistical Association*, Vol. 54,
No. 287, 1959, pp. 622-654, for a discussion of the proper choice of the
x's when (6.6) is in error.

Since β_0 and β_1 are unknown, we must estimate these two parameters from the data. Just as we lost one "degree of freedom" when estimating a normal mean μ, we lose two "degrees of freedom" from the n degrees of freedom provided by the n observations in estimating β_o and β_1. The minimum variance unbiased estimate for σ^2 is:

$$\hat{\sigma} = \frac{1}{n-2} \sum_{i=1}^{n} (y_i - b_0 - b_1 x_i)^2 = \frac{1}{n-2} \sum_{i=1}^{n} \hat{e}_i^2 .$$

It is straightforward to show that $\hat{\sigma}$ is an unbiased estimate of σ^2. Note that this formula for $\hat{\sigma}^2$ differs from the value of $s_{y|x}^2$ given in Chapter 2 in that n-2 rather than n is in the denominator. This is another case of the "sampling correction" we will make in all our work on inference.

Inferences Using $\hat{\sigma}^2$, the Unbiased Estimate of the Residual Variance

The distribution of

$$t = \frac{b_1 - \beta_1}{\hat{\sigma}} \sqrt{\sum (x_i - x)^2} \tag{6.8}$$

is a Student's t-distribution with n-2 degrees of freedom. Equation (6.8) can be used for confidence intervals and hypothesis testing about β_1 in a manner analogous to the inferences about μ given in Section 6.2.

Testing whether the x's are really related to the Y's means testing whether $\beta_1 = 0$. If the hypothesis $\beta_1 = 0$ is true, the quantity

$$t = \frac{b_1 - 0}{\hat{\sigma}} \sqrt{\sum (x_i - \bar{x})^2}$$

has a t-distribution with n-2 degrees of freedom. Thus the observed value of t in Table 4 gives the significance probability.

153

Note that any hypothesized value for β_1 can be tested in this way. For example, if the null hypothesis is $\beta_1 = 3$, the test statistic is

$$t = \frac{b_1 - 3}{\hat{\sigma}} \sqrt{\Sigma(x_i - \bar{x})^2}$$

with the null hypothesis being rejected for those t that are sufficiently large in absolute value.

In Chapters 3 and 4, we measured the strength of the relationship between y and x by the correlation coefficient r_{yx}. There is no relationship between these variables if and only if $\beta_1 = 0$. We can rewrite

$$\hat{\sigma}^2 = \frac{n}{n-2} s_y^2 (1 - r_{yx}^2), \quad s_x^2 = \frac{1}{n} \Sigma(x_i - \bar{x})^2 \, ,$$

and $b_1 = r_{yx} s_y / s_x$. Then it follows that when testing whether $\beta_1 = 0$, we use

$$t = \sqrt{n-2} \, \frac{r_{yx}}{\sqrt{1 - r_{yx}^2}}$$

so that

$$r_{yx} = \frac{t}{\sqrt{n-2+t^2}} \, .$$

This shows that the distribution of the sample correlation coefficient can be obtained from tables of the t_{n-2} distribution (when the population correlation coefficient $\rho_{xy} = 0$) and that tests of the hypothesis $\beta_1 = 0$ and tests of the independence of y and x in a simple linear regression, based on the sample correlation coefficient, are the same.

Confidence intervals for β_1 are calculated in the usual way. For example, if the sample size n = 10, then

$$\frac{(b_1 - \beta_1)}{\hat{\sigma}} \sqrt{\Sigma(x_i - \bar{x})^2}$$

has a t-distribution with n−2 = 8 degrees of freedom. Thus from the t-table,

$$\text{Prob}\left(-2.306 \leq \frac{(b_1 - \beta_1)}{\hat{\sigma}} \sqrt{\Sigma(x_i - \bar{x})^2} \leq 2.306\right) = 0.95.$$

So that

$$b_1 \pm \frac{2.306\,\hat{\sigma}}{\sqrt{\Sigma(x_i - \bar{x})^2}} \quad \text{or} \quad b_1 \pm \frac{2.306\,\hat{\sigma}}{\sqrt{n}\,s_x}$$

is a 95 percent confidence interval for β_1.

The Prediction Problem

Suppose we wish to estimate the value Y_0 for some specified value of x, say x_0. We saw in Chapter 3 that the regression equation for predicting $\beta_0 + \beta_1 x_0 = E(Y|x_0)$ can be written

$$\hat{Y}_0 = \bar{Y} + b_1(x_0 - \bar{x}). \tag{6.9}$$

Now $\bar{Y}$ and b_1 are the only random variables in (6.9) and it is easy to show that they are uncorrelated. But $\hat{Y}_0$ is an unbiased estimator of $\beta_0 + \beta_1 x_0$ and the variance of $\hat{Y}_0$ is given by

$$\text{Var}(\hat{Y}_0) = \text{Var}(\bar{Y}) + (x_0 - \bar{x})^2\,\text{Var}(b_1)$$

$$= \frac{\sigma^2}{n} + \frac{(x_0 - \bar{x})^2\,\sigma^2}{\Sigma(x_1 - \bar{x})^2}.$$

Thus, for σ^2 unknown, the standard error of the prediction $\hat{Y}_0$ is estimated by substituting the estimate $\hat{\sigma}^2$ for σ^2 to get

$$\hat{\sigma}(\hat{Y}_0) = \hat{\sigma}\sqrt{\frac{1}{n} + \frac{(x_0 - \bar{x})^2}{\Sigma(x_i - \bar{x})^2}} = \frac{\hat{\sigma}}{\sqrt{n}}\sqrt{1 + \left(\frac{x_0 - \bar{x}}{s_x}\right)^2}. \tag{6.10}$$

Frequently we are interested in assessing how close $\hat{Y}_0$ is likely to be to Y_0 itself. That is, how well does $\hat{Y}_0$ perform in predicting Y_0. The most commonly used measure of accuracy in predictions is the *mean square error*--the expected value of the square of the difference between the prediction and the observation: $E(\hat{Y}_0 - Y_0)^2$. The mean square error (MSE) of $\hat{Y}_0$ can be calculated as follows.

$$
\begin{aligned}
MSE(\hat{Y}_0, Y_0) &= E(\hat{Y}_0 - Y_0)^2 \\
&= E[(b_0 + b_1 x_0) - (\beta_0 + \beta_1 x_0 + \varepsilon)]^2 \\
&= E[(b_0 + b_1 x_0 - \beta_0 - \beta_1 x_0) - \varepsilon]^2 \\
&= E[b_0 + b x_0 - (\beta_0 + \beta_1 x_1)]^2 \\
&\quad - 2E[\varepsilon(b_0 + b_1 x_0 - \beta_0 - \beta_1 x_0)] + E(\varepsilon^2) \ .
\end{aligned}
$$

The first term is $Var(\hat{Y}_0)$, the last term is σ^2 by definition, while the middle cross product term is zero since ε is associated with a future observation and so is independent of b_0 and b_1. Thus, from Equation (6.10), the prediction error is:

$$
\begin{aligned}
MSE(\hat{Y}_0 - Y_0) &= \sigma^2 \left[\frac{1}{n} + \frac{(x_0 - \bar{x})^2}{\Sigma(x_i - \bar{x})^2} \right] + \sigma^2 \\
&= \sigma^2 \left[1 + \frac{1}{n} + \frac{1}{n} \left(\frac{x_0 - \bar{x}}{s_x} \right)^2 \right]
\end{aligned}
$$

(6.11)

This is the formula used in Figure 3.3 of Chapter 3 to draw the confidence intervals for the predictions there. Note that the further x_0 is from the mean of x, the larger the error. Confidence intervals are calculated by referring to the t-table and using Equation (6.11).

Multiple Regression: Inferences About β_0, β_1, ..., β_k

Consider the multiple regression model:

$$Y_i = \beta_0 + \beta_1 x_{i1} + \beta_2 x_{i2} + \ldots + \beta_k x_{ik} + \varepsilon_i \qquad i = 1, \ldots, n. \qquad (6.12)$$

We have a random sample of n individuals and with each measurement of Y_i we have k other measurements $(x_{i1}, \ldots, x_{ik})$. In this case we continue to assume normally distributed errors with expectation 0 and unknown variance σ^2. The sample regression coefficients are computed as given in Chapter 4. They are unbiased estimates of the population regression coefficients.

To make an inference about a single regression coefficient, say β_i, let $\hat{\sigma}(b_i)$ be the sample estimate of the standard deviation of b_i (defined below) then

$$t = \frac{b_i - \beta_i}{\hat{\sigma}(b_i)}$$

has a Student's t-distribution with n − (k+1) = n−k−1 degrees of freedom since the k+1 parameters β_0, β_1, ..., β_k have been estimated from the data. Thus to test the hypothesis $\beta_i = 0$ we use $b_i/(\hat{\sigma}(b_i))$ as the test statistic and refer to the t-table with n−k−1 degrees of freedom for cutoff values. A 95 percent confidence interval for β_i is $(b_i - t_{n-k-1,.95}\hat{\sigma}(b_i), \; b_i + t_{n-k-1,.95}\hat{\sigma}(b_i))$ where $t_{n-k-1,.95}$ is the point in the t-distribution with n−k−1 degrees of freedom with 0.95 between $-t_{n-k-1,.95}$ and $t_{n-k-1,.95}$.

The value $\hat{\sigma}(b_i)$ is printed out in the output of most regression computer programs. For concreteness we give the formula for $\hat{\sigma}(b_i)$ for i = 3. Other values of i have analogous formulas. In the spirit of Equation (6.7), the standard deviation of b_3 in the notation of Chapter 4 is

$$\sigma(b_3) = \frac{\sigma}{\sqrt{n} \ s_{x_3|x_1,x_2,x_4 \cdots x_k}} \ .$$

Thus, the estimated standard deviation from the sample is

$$\hat{\sigma}(b_3) = \frac{s_{y|x_1 x_2 \cdots x_k}}{\sqrt{n} \ s_{x_3|x_1,x_2,x_4, \ \cdots, \ x_k}} \cdot \sqrt{\frac{n}{n-k-1}} \ .$$

The "sample correction factor" $\sqrt{n/(n-k-1)}$ results from the $k+1$ degrees of freedom that are lost in estimating the $k+1$ parameters $\beta_0, \ \beta_1, \ \cdots, \ \beta_k$. Thus for $\hat{y}_i = b_0 + \sum_j b_j x_{ij}$,

$$\hat{\sigma}^2 = \frac{1}{n-k-1} \Sigma(y_i - \hat{y}_i)^2 = \frac{n}{n-k-1} \ s^2_{y|x_1, \ \cdots, \ x_k} \ ,$$

is the unbiased estimate of σ^2, the variance of Y_i about the true regression surface $\beta_0 + \Sigma^k_{j=1} \beta_j x_{ij}$.

In Chapter 4, the partial correlation coefficient $r_{y,x_3|A}$ (where $A \equiv (x_1, \ x_2, \ x_4, \ x_5, \ \cdots, \ x_k)$ is used to shorten notation) was shown to measure the strength of the relation between y and x_3 when x_1, x_2, x_4, $\cdots$, x_k are held fixed. If $r_{yx_3|A} = 0$ then (in the sample), x_3 has no relation to y is the presence of the other variables, so $b_3 = 0$. But large absolute values of $r_{y3|A}$ (x_3 is simplified to "3" here) indicate that x_3 is important and suggest that $\beta_3 \neq 0$. This may be seen more clearly as follows: Compare to Eq. (4.21). Since $b_3 = r_{y3|A} s_{y|A}/s_{3|A}$, $\hat{\sigma}(b_3) = s_{y|3,A}/s_{3|A}\sqrt{n-k-1}$, and $s_{y|3,A} = s_{y|A}\sqrt{1-r^2_{y3|A}}$, it follows that for testing whether $\beta_3 = 0$,

$$t = \frac{b_3}{\hat{\sigma}(b_3)} = \sqrt{n-k-1} \ \frac{r_{y3|A}}{\sqrt{1-r^2_{y3|A}}}$$

or equivalently

$$r_{y3|A} = \frac{t}{\sqrt{n-k-1 + t^2}} \; .$$

This shows that the distribution of the partial correlation coefficient can be obtained from the t distribution with n-k-1 degrees of freedom. It also shows that large values of $|r_{y3|A}|$ and large values of $|t|$ measure the same thing (i.e., that x_3 is useful to predict y in the presence of $x_1 x_2 x_4 \ldots x_k$, or equivalently that $\beta_3 \neq 0$).

The F-distribution

The mathematically inclined reader may want to know that the t_{n-k-1} distribution can be written as the ratio $t = \frac{Z}{\sqrt{U/(n-k-1)}}$ where Z has a standard normal distribution and U is independent of Z, being written $U = X_1^2 + \ldots + X_{n-k-1}^2$ where the X_i's have independent standard normal distributions. The random variable U is said to have a "chi-square (χ^2) distribution with n-k-1 degrees of freedom" and t is thus the quotient of a standard normal and the square root of an independent chi-square variable divided by its degrees of freedom. We can write

$$t^2 = \frac{Z^2/1}{U/(n-k-1)} \; ,$$

as the ratio of two independent chi-square (χ^2) variables divided by their degrees of freedom, respectively. Such a ratio is said to have an *F-distribution* with 1 and n-k-1 degrees of freedom, and is usually written $F_{1,n-k-1}$. In general any ratio of the form $(V/m/(U/k)$ has an $F_{m,k}$ distribution if V and U are independent chi-square variables with m and k degrees of freedom, respectively. We will use the F-distribution below.

159

The F-tables (Table 6) give the degrees of freedom k and n-k-1 as ν_1 and ν_2 across the top of each table and down the left side of each table, respectively. A separate table is given for each of 8 significance levels covering the range of commonly used values: 0.5, 0.25, 0.1, 0.05, 0.025, 0.01, 0.005, 0.001. For example, suppose that n = 30 and k = 3 and that the value of the calculated F-statistic is 3.06. Taking ν_1 = 3 and ν_2 = 26, the tables show 3.06 to be significant at the α = 0.05 level and not significant at the α = 0.025 level and below.

Inference on the Squared Multiple Correlation Coefficient; Testing that $\beta_1 = \beta_2 = \ldots = \beta_k = 0$ Simultaneously

Let ρ^2 be the population squared multiple correlation coefficient. The corresponding value calculated from the sample is R^2 and is an estimate of ρ^2. More completely, $R^2_{y|x_1, \ldots, x_k}$ is an estimate of $\rho^2_{y|x_1, \ldots, x_k}$.

One way to construct a test of the hypothesis that the squared multiple correlation coefficient is zero, is to use the probability distribution of R^2. Notice that testing whether $\rho^2 = 0$, is identical to testing whether $\beta_1 = \beta_2 = \ldots = \beta_k = 0$. If all the β's are zero then the best predictor of each individual's response is $\bar{y}$, the sample mean. If some of the β's are not zero, then we would predict the response of individual i as

$$\hat{y}_i = b_0 + b_1 x_{i1} + b_2 x_{i2} + \ldots + b_k x_{ik} .$$

Given that $\beta_1 = \ldots = \beta_k = 0$, then we would expect that b_1, b_2 ... b_k (their estimates) all would be near zero and thus $\hat{y}_i$ should not differ from $\bar{y}$ more than that amount allowable by sampling variation. The test statistic for testing the hypothesis that $\beta_1 = \ldots = \beta_k = 0$ is a function of the sum of squared differences of the prediction, $\Sigma(\hat{y}_i - \bar{y})^2$. This

160

quantity is distributed[*] as $\sigma^2\chi_k^2$ when $\beta_1 = \ldots = \beta_k = 0$ and is independent of the sum of squared residuals $(n-k-1)\hat{\sigma}^2 = \Sigma(y_i - \hat{y}_i)^2$ which is distributed as $\sigma^2\chi_{n-k-1}^2$. Hence the statistic

$$F = \frac{\Sigma(\hat{y}_i - \bar{y})^2/k}{\Sigma(y_i - y_i)^2/(n-k-1)}$$

$$= \frac{\text{variance explained}}{\text{variance unexplained}}$$

has an $F_{k,n-k-1}$ distribution under the null hypothesis $\beta_1 = \ldots = \beta_k = 0$. Large values of this quantity, in relation to the percentage points F-tables (Table 6), provide evidence that some $\beta_j \neq 0$. Now the total variation of y around $\bar{y}$ can be decomposed as:

$$\Sigma(y_i - \bar{y})^2 = k \text{ (explained variance)} + (n-k-1) \text{ (unexplained variance)}$$

$$= \Sigma(\hat{y}_i - \bar{y})^2 + \Sigma(y_i - \hat{y}_i)^2$$

since $\Sigma(y_i - \bar{y})^2 = \Sigma([y_i - \hat{y}_i] + [\hat{y}_i - \bar{y}])^2$ and $\Sigma(y_i - \hat{y}_i)(\hat{y}_i - \bar{y}) = 0$ (the residuals are uncorrelated with the explanatory variables which make up $\hat{y}_i$).

From Chapter 4,

$$R^2 = 1 - \frac{s_{y|x_1 \ldots x_k}^2}{s_y^2}$$

so that

---

[*] When $\beta \neq 0$, $\Sigma(\hat{y}_i - \bar{y})^2$ has expectation $k\sigma^2 + n\beta'C_{xx}\beta$, with β being the column vector $(\beta_1, \ldots, \beta_k)'$ and C_{xx} being the matrix of covariances of the independent variables.

161

$$\frac{R^2}{1 - R^2} = \frac{\Sigma(y_i - \bar{y})^2 - \Sigma(y_i - \hat{y}_i)^2}{\Sigma(y_i - \bar{y})^2}$$

$$= \frac{\Sigma(\hat{y}_i - \bar{y})^2}{\Sigma(y_i - \bar{y})^2} .$$

It follows that

$$F = \frac{R^2/k}{(1-R^2)/(n-k-1)} , \quad R^2 = \frac{kF}{n-k-1 + kF}$$

is another way to write the statistic just derived, so that this transformation of R^2 has a $F_{k,n-k-1}$ distribution. Note that both F and R^2 are large at the same time.

This confirms that: the distribution of R^2 is simply related to the F-distribution; and that the statistic F used to test $\beta_1 = \ldots = \beta_k = 0$ is a monotone function of the same R^2 used to measure the strength of the relation between $x_1, \ldots, x_k$ and y.

The population squared multiple correlation coefficient ρ^2 is the ratio of explained variance divided by total variance and can in the notation of Section 4.3 be written as, for $\beta' = (\beta_1, \ldots, \beta_k)$,

$$\rho^2 = \frac{\beta'C_{xx}\beta}{\sigma^2 + \beta'C_{xx}\beta} .$$

The distribution of R^2 when $\rho^2 > 0$ is not expressible in closed form in terms of elementary functions. In particular, R^2 is not an unbiased estimator of ρ^2. (Even when $\rho^2 = 0$, $ER^2 = k/(n-1) > 0$; and $ER^2 > \rho^2$ for all ρ^2.) Many computer programs and texts advocate using an "adjusted" or "corrected" R^2 defined as

$$R^2(\text{adj}) = 1 - (1-R^2)\,\frac{n-1}{n-k-1}$$

whose expectation usually is closer to ρ^2 than the expectation of R^2.[*]
If $\rho^2 = 0$ then $ER^2(\text{adj}) = 0$, so that ρ^2 is the expectation of $R^2_{(\text{adj})}$
when $\rho^2 = 0$.

Partial Correlation and Partial-Multiple Correlation

A more complicated procedure is required to test the hypothesis
that $\beta_{r+1} = \ldots = \beta_k = 0$ with $1 \le r \le k-1$ while permitting $\beta_0, \ldots, \beta_r$
to range freely. The test is constructed as follows using a computer.

1. Perform one regression to estimate $\beta_0, \ldots, \beta_r$ by predicting
y linearly from $x_1, \ldots, x_r$. Let R^2_1 be the multiple correlation
coefficient obtained.

2. Perform an additional regression to estimate
$\beta_0, \beta_1, \ldots, \beta_r, \ldots, \beta_k$ by predicting y linearly from $x_1, \ldots, x_k$ and
let R^2_2 be the multiple correlation coefficient obtained. Clearly
$R^2_2 \ge R^2_1$. (Why?)

Then

$$F_{k-r,\,n-k-1} \equiv \frac{(R^2_2 - R^2_1)/(k-r)}{(1 - R^2_2)/(n-k-1)} \tag{6.13}$$

has an F distribution with $k-r$ and $n-k-1$ degrees of freedom if
$\beta_{r+1} = \ldots = \beta_k = 0$. This statistic would be expected to be larger
if some $\beta_i \ne 0$ for $r + 1 \le i \le k$. Hence this statistic can be used
to test the hypothesis that $\beta_{r+1} = \ldots = \beta_k = 0$. Furthermore,

---

[*] See M. G. Kendall and A. Stuart, *The Advanced Theory of Statistics*,
Vol. 2, pp. 337-342 (Hafner, 1961), for a discussion of the distribution
of R^2 when ρ^2 is not necessarily zero.

$$R^2_{2|1} = \frac{R^2_2 - R^2_1}{1 - R^2_1}$$

has the interpretation of being the squared "partial-multiple" correlation coefficient, "partial" because x_1, ..., x_r are first fit and "multiple" because the correlation between residuals from the first regression are correlated with the best linear combination of the multiple variables x_{r+1}, ..., x_k. Note that $R^2_{2|1}$ reduces to the squared partial correlation if $r = k-1$ and in this case

$F_{k-r,n-k-1} = F_{1,n-k-1}$ is just the squared t statistic for testing

$\beta_k = 0$, since

$$t_{n-k-1} = \sqrt{F_{1,n-k-1}} .$$

At the other extreme, $r = 0$, $R^2_{2|1}$ is the multiple correlation coefficient and $F_{k,n-k-1}$ is the F-test for the multiple correlation coefficient for predicting y from x_1, ..., x_k. Note that $R^2_{2|1}$ is the additional fraction of variance explained by x_{r+1}, ..., x_k beyond that explained by x_1, ..., x_r. More generally we use the complete notation $R^2_{y;\{r+1, ..., k\}|\{1, ..., r\}}$ to denote the squared partial multiple correlation $R^2_{2|1}$. If A and B are sets of variables, we have, analogous to (4.18),

$$1 - R^2_{y;AB} = (1 - R^2_{y;A})(1 - R^2_{y;B|A}) .$$

Summary of Facts About Distribution for Inference About Regression Models

If the specified model is true, then the following facts are true in connection with making inference about regression coefficients. The results are stated for β_1 and β_2 but they apply with the obvious subscript changes to arbitrary β_j.

$E(b_1) = \beta_1$ (unbiased)

$$Var(b_1) = \frac{\sigma^2/n}{var(x_1|x_2, \ldots, x_k)} = \frac{\sigma^2_{y|x_1, \ldots, x_k}}{n\, s^2_{x_1|x_2, \ldots, x_k}} = \frac{\sigma^2_{y|x_1, \ldots, x_k}}{n\, s^2_{x_1}(1-R^2_{x_1,(x_2, \ldots, x_k)})}$$

$Corr(b_1, b_2) = -\, r_{x_1, x_2|x_3, \ldots, x_k}$ (the negative partial correlation!).

If the errors are from a normal distribution, then the $\{b_j\}$ are (multi-variate) normally distributed.

(a) We get independent inferences for β_1, β_2 (etc.) only if x_1, x_2 (etc.) have zero partial correlation.

(b) Var b_1 decreases if:

    (i) n increases,

    (ii) $\sigma^2 = \sigma^2_{y|x_1, \ldots, x_k}$ decreases (e.g., if more variables are available to explain y),

    (iii) x_1 is chosen to increase its residual variance (ideally uncorrelated with x_2, $\ldots$, x_k, since then $R^2_{x_1,(x_2, \ldots, x_k)} = 0$).

(c) If $s^2_{x_1|x_2, \ldots, x_k}$ is nearly zero (the problem of multicollinearity) then β_1 is estimated inaccurately. If $var(x_1|x_2 \ldots x_k) = 0$, then β_1 cannot be estimated at all. (For example, if

$$x_1 = \alpha_0 + \alpha_1 x_2 \, ,$$

α_0 and α_1 known, then

$$E(Y) = \beta_0 + \beta_1 x_1 + \beta_2 x_2$$

$$= \beta_0 + \beta_1 \alpha_0 + (\beta_1 \alpha_1 + \beta_2) x_2$$

so the coefficient of x_2 obtained from the regression determines $\beta_1\alpha_1 + \beta_2$, but does not determine the individual values of β_1, β_2.) The parameters in this model are said to be "underidentified."

Chapter 7

ANALYSIS OF VARIANCE

7.1. THE ANALYSIS OF VARIANCE AS A SPECIAL CASE OF THE GENERAL LINEAR MODEL

Scheffe (*The Analysis of Variance*, 1959) gives the following rough definition of the analysis of variance (ANOVA): "The analysis of variance is a statistical technique for analyzing measurements depending on several kinds of effects operating simultaneously to decide which kinds of effects are important and to estimate the effects." The analysis of variance is very similar to regression, except that the independent variables are integers that represent categories instead of continuous variables. Our interest is in the effect of the presence or absence of a variable. In econometrics this method is often referred to as "regression on dummy variables" with the "dummy variables" being specified in a particular way. When some of the independent variables indicate categories and some are continuous, the resulting mixture of ANOVA and regression is called the analysis of covariance (ANOCOVA).

All of these methods fall within the scope of the general linear model (GLM) discussed in Chapter 4,

$$y_i = \sum_{j=1}^{k} x_{ij} \beta_j + e_i, \quad i = 1, 2, \ldots, n \quad (7.1)$$

with the e_i being independent, or at least uncorrelated, with mean 0 and variance σ^2 (whose value is usually unknown). Often for developing distributions in this model, the $\{e_i\}$ are taken to be normal. The x_{ij} are known and the parameters β_j are unknown. The number k of unknown

parameters is less than n in all interesting cases. In this chapter
we adopt the convention of not explicitly writing the constant term in
contrast to earlier chapters, e.g., Eq. (6.12). When a constant term
is needed, think of x_{i1} as being identically 1 in Eq. (7.1). Also, to
avoid switching back and forth between Greek and Roman letters and
lower case and capital letters, we use (y_i, e_i) to denote both random
variables and their values and do not use (Y_i, ε_i) as in Chapter 6
[e.g., (6.12)]. This slight abuse of notation should help the
exposition.

Written out, the GLM is expressed as

$$y_1 = \beta_1 x_{11} + \beta_2 x_{12} + \ldots + \beta_k x_{1k} + e_1$$

$$y_2 = \beta_1 x_{21} + \beta_2 x_{22} + \ldots + \beta_k x_{2k} + e_2$$

$$\vdots \qquad \vdots \qquad\qquad \vdots \qquad \vdots$$

$$y_n = \beta_1 x_{n1} + \beta_2 x_{n2} + \ldots + \beta_k x_{nk} + e_n$$

In matrix notation, using the convention that the expectation of
a vector or matrix is that vector or matrix of expectations, the GLM
is written

$$Ey = X\beta \ ,$$

where

$$y = (y_1, \ \ldots, \ y_n)', \ \beta = (\beta_1, \ \ldots, \ \beta_k)'$$

and X is an n by k matrix. In shorthand: $y^{n \times 1}$, $X^{n \times k}$, $\beta^{k \times 1}$. The
covariance matrix of the y vector is

$$Cov(y) = \sigma^2 I \qquad (I = n \times n \text{ identity matrix})$$

meaning

$$Cov(y_i, y_j) = Cov(e_i, e_j) = \begin{cases} 0 & \text{if } i \neq j \\ \sigma^2 & \text{if } i = j \end{cases} .$$

167

For ANOVA, the matrix

$$X^{nxk} = \begin{pmatrix} x_{11} & \cdots & x_{1k} \\ \vdots & & \\ x_{n1} & \cdots & x_{nk} \end{pmatrix}$$

is called the *design matrix*. Then the mean of the vector y is computed as the vector

$$Ey = X\beta = \begin{pmatrix} x_{11} & \cdots & x_{1k} \\ \vdots & & \\ x_{n1} & \cdots & x_{nk} \end{pmatrix} \begin{pmatrix} \beta_1 \\ \vdots \\ \beta_k \end{pmatrix}$$

$$= \begin{pmatrix} \beta_1 x_{11} + \cdots + \beta_k x_{1k} \\ \vdots \\ \beta_1 x_{n1} + \cdots + \beta_k x_{nk} \end{pmatrix}$$

We will assume that $k \leq n$ and that X is of full rank. This means that no column of X can be expressed as a linear combination of other columns. In this case, the matrix X'X has an inverse and the ordinary least squares estimate of the vector β is expressed in matrix notation as

$$\hat{\beta} = (X'X)^{-1}X'y .$$

The analysis of variance (with indicator or dummy explanatory variables), regression theory (with continuous explanatory variables), and the analysis of covariance (mixing both dummy and continuous explanatory variables) are expressible as special cases of the GLM, so can be solved in this way. Most regression computer schemes will determine the GLM estimates, at least provided one column (usually the first) of X is the constant vector $(1, 1, 1, \ldots, 1, 1)'$. As pointed out above, this corresponds to estimating a constant term. In fact, had computers been developed before statistics, ANOVA and ANOCOVA probably would have developed quite differently, because the primary justification for their separate development is that hand calculations are much simpler with dummy variables, and methods were developed to facilitate these computations in special cases. However, in "higher way ANOVA" (defined below), much of the computational simplicity is lost if the observations are not appropriately balanced. There are many practical situations, particularly in the social sciences, where this balancing is not possible, so ANOVA provides no special advantage. Meanwhile, the computer can make all computations quickly and easily, so it is natural now to introduce ANOVA as a special case of the GLM and of multiple regression methodology.

The main interest in this chapter will be in seeing how to parameterize the design matrix X so that an ANOVA may be carried out with the computational methods of earlier chapters. Familiar explicit formulas for ANOVA will also be given for comparison in special situations where hand calculations can be made easily.

7.2. THE ONE-WAY LAYOUT: GRAND MEAN AND MAIN EFFECTS

The initial ideas for ANOVA in this section will be presented in the context of an example using data on total money income for U.S. persons over 14 in 1968. More data on this subject will be introduced as needed later.

Example. Average total money income by region (in thousands of dollars) and regional effects. We can describe income by region or by a general mean plus regional *effects*. For comparison, the latter is often more useful.

169

Thus (for a sample of I_j individuals in region j), we observe each response y_{ij} as the sum of the regional mean β_j and an error term e_{ij}:

$$y_{ij} = \beta_j + e_{ij}, \quad j = 1, 2, 3, 4 \quad i = 1, \ldots, I_j. \quad (7.2)$$

Alternatively,

$$y_{ij} = \mu + \alpha_j + e_{ij} \quad (\mu = \overline{\beta}, \ \alpha_j = \beta_j - \overline{\beta}) \quad (7.3)$$

is a different parameterization that describes the same income process. We set $\Sigma \alpha_j = 0$ so that the parameters may be uniquely identified. This identification problem must always be considered in ANOVA. The α_j are called the "main effects" for the regions, or more simply the "regional effects."

For our data set, the least squares estimates of the β_j's and α_j's are:

	β_j	α_j
Northeast (NE)	4.9	+0.2
North Central (NC)	4.6	−0.1
South (S)	4.1	−0.6
West (W)	5.2	+0.5
Grand mean	4.7	

The assumption that $\{e_{ij}\}$ be independent, have mean 0 and variance $= \sigma^2$ means that individuals in the sample were chosen at random and that the variance of income does not depend on region.

In the next several pages we will see how to code the design matrix X so that the regression coefficients estimated by least squares are the grand mean and the main effects.

The design matrix also has several descriptions, one for each parameterization. If no constant term is required, then we may write

$$
Ey = E
\begin{pmatrix}
y_{11} \\
y_{21} \\
\vdots \\
y_{I_1,1} \\
y_{12} \\
y_{22} \\
\vdots \\
y_{I_2,2} \\
y_{13} \\
y_{23} \\
\vdots \\
y_{I_3,3} \\
y_{14} \\
y_{24} \\
\vdots \\
y_{I_4,4}
\end{pmatrix}
=
\begin{matrix}
x_1 & x_2 & x_3 & x_4
\end{matrix}
\begin{bmatrix}
1 & 0 & 0 & 0 \\
\vdots & \vdots & \vdots & \vdots \\
1 & 0 & 0 & 0 \\
0 & 1 & 0 & 0 \\
\vdots & \vdots & \vdots & \vdots \\
0 & 1 & 0 & 0 \\
0 & 0 & 1 & 0 \\
\vdots & \vdots & \vdots & \vdots \\
0 & 0 & 1 & 0 \\
0 & 0 & 0 & 1 \\
0 & 0 & 0 & 1 \\
\vdots & \vdots & \vdots & \vdots \\
0 & 0 & 0 & 1
\end{bmatrix}
\begin{pmatrix}
\beta_1 \\
\beta_2 \\
\beta_3 \\
\beta_4
\end{pmatrix}
\qquad (7.4)
$$

If we use the above design matrix then

$$y_i = \beta_1 x_{i1} + \beta_2 x_{i2} + \beta_3 x_{i3} + \beta_4 x_{i4} + e_i.$$

If $x_{i1} = 1$, for example, $y_i = \beta_1 + e_1$ [see (7.4)]. But more generally, for regression requiring a constant term, we must have a term β_0:

$$y_i = \beta_0 + \beta_1 x_{i1} + \beta_2 x_{i2} + \beta_3 x_{i3} + \beta_4 x_{i4} + e_i.$$

This is equivalent to adding a column of 1's to X. If we add this column in (7.4), we have an extra parameter, so we must eliminate one column. Convention is to delete the last row or x_{i4}, yielding:

171

$$Ey = X\beta = \begin{bmatrix} 1 & 1 & 0 & 0 \\ \vdots & \vdots & \vdots & \vdots \\ 1 & 1 & 0 & 0 \\ 1 & 0 & 1 & 0 \\ \vdots & \vdots & \vdots & \vdots \\ 1 & 0 & 1 & 0 \\ 1 & 0 & 0 & 1 \\ \vdots & \vdots & \vdots & \vdots \\ 1 & 0 & 0 & 1 \\ 1 & 0 & 0 & 0 \\ \vdots & \vdots & \vdots & \vdots \\ 1 & 0 & 0 & 0 \end{bmatrix} \begin{bmatrix} \beta_0 \\ \beta_1 \\ \beta_2 \\ \beta_3 \end{bmatrix} \qquad (7.5)$$

This is the usual specification for dummy variables in regression. If the computer automatically fits the constant term, only the last three columns of X are needed. The first column is automatically included. Now

$$y = \beta_0 + \beta_1 x_1 + \beta_2 x_2 + \beta_3 x_3 + e .$$

Letting b_0, b_1, b_2, b_3 be the estimated regression coefficients obtained from a computer regression program with X specified as in (7.5), then $\hat{\beta}_j = b_j$, $j = 0,1,2,3$. Our parameters μ, α_1 in (7.3) are estimated by:

$$
\begin{aligned}
&\text{NE region:} && x_1 = 1, \text{ hence } b_0 + b_1 = \hat{\mu} + \hat{\alpha}_1 \\
&\text{NC region:} && x_2 = 1, \text{ hence } b_0 + b_2 = \hat{\mu} + \hat{\alpha}_2 \\
&\text{S region:} && x_3 = 1, \text{ hence } b_0 + b_3 = \hat{\mu} + \hat{\alpha}_3 \\
&\text{W region:} && x_1 = x_2 = x_3 = 0, \text{ hence } b_0 = \hat{\mu} + \hat{\alpha}_4 .
\end{aligned}
$$

We also require $\hat{\alpha}_4 = -\hat{\alpha}_1 - \hat{\alpha}_2 - \hat{\alpha}_3$, and therefore these five relations can be solved for $\hat{\mu}$, $\hat{\alpha}_j$ from b_0, b_1, b_2, b_3.

If we are more clever, we circumvent the need to solve for the $\hat{\alpha}_j$ and get t-test statistics directly from regression, as follows:

$$Ey = \begin{array}{cccc} x_0 & x_1 & x_2 & x_3 \end{array}$$

$$Ey = \begin{bmatrix} 1 & 1 & 0 & 0 \\ \vdots & \vdots & \vdots & \vdots \\ 1 & 0 & 1 & 0 \\ \vdots & \vdots & \vdots & \vdots \\ 1 & 0 & 0 & 1 \\ \vdots & \vdots & \vdots & \vdots \\ 1 & -1 & -1 & -1 \end{bmatrix} \begin{bmatrix} \mu \\ \alpha_1 \\ \alpha_2 \\ \alpha_3 \end{bmatrix} \qquad (7.6)$$

Then the estimated regression equation

$$\hat{y} = b_0 + b_1 x_1 + b_2 x_2 + b_3 x_3$$

gives $b_0 = \hat{\mu}$, $b_j = \hat{\alpha}_j$, $(\hat{\alpha}_4 = -b_1 - b_2 - b_3)$. This then is the proper way to code the X matrix for the one-way layout in ANOVA as in (7.3). One category (here, region 4) is omitted and no variable represents it. The other variables are coded -1 throughout the row when this category occurs. When any other category occurs, the variable indicating the region is coded 1 and other variables are coded 0.

To test whether there is any regional effect, one uses an F-test on the $\hat{\alpha}_j$ (Chapter 6), or equivalently, tests whether R^2 is greater than 0 (recall this tests whether the regression coefficients, except b_0, are simultaneously zero). With k categories, the degrees of freedom in the F-test are the number of parameters (excluding μ) k-1 and n-k, respectively, so k-1 is the number of α_j's, not counting the one linear restriction $\hat{\alpha}_k = \sum_1^{k-1} \hat{\alpha}_j$.

The regression formulation for ANOVA presented in (7.6) also may be used in connection with other variables. For example, the X matrix in (7.6) could have other columns, say x_4, x_5, appended. For example these additional columns might be the age and the number of years of education of each respondent. The estimated constant term b_0 won't be the grand mean as in (7.6) unless x_4 and x_5 have had their means subtracted; i.e., x_4 and x_5 would have to be replaced by $x_4 - \bar{x}_4$ and $x_5 - \bar{x}_5$. The resulting representation is called the "analysis of covariance" (ANOCOVA). The idea is that the regional comparisons are made only after accounting for the effects of the covariates (age and education). While ANOCOVA is traditionally taught as an additional topic in courses on the analysis of variance, there is no important reason to distinguish between ANOVA and ANOCOVA when both are viewed as part of the GLM. The analyst is free to code his categorical variables as described for ANOVA and to include other variables in the model as he sees fit. Then the computer will calculate the correct estimates. The analyst's job is to specify the matrix properly so that the estimates he gets are those he wants. If he fails, he will be forced to compute linear combinations of the estimates provided by the computer, and this might be complicated. He will also have a difficult time obtaining needed t-statistics for main effects from most computer packages, unless he makes the proper specification at the outset.

The statistical model described above is for the *one-way layout*. The following table usually is presented when data arranged in a one-way layout have been analyzed by classical ANOVA methods.

Table 7.1

ANALYSIS OF VARIANCE FOR ONE-WAY LAYOUT (k GROUPS)

Source	SS	df	MS	F
Between groups	SS_B	$k-1$	$SS_B/(k-1)$	$\dfrac{(n-k)SS_B}{(k-1)SS_E}$
Within Groups (error)	SS_E	$n-k$	$SS_E/(n-k)$	–
Mean	$n\bar{y}^2$	1	–	–
Total	$\Sigma \Sigma y_{ij}^2$	n		

$$E(MS) \text{ between groups} = \sigma^2 + \Sigma_j I_j \alpha_j^2/(k-1)$$

$$E(MS) \text{ within groups} = \sigma^2$$

The quantities in this table are defined and computed as

I_j = number of data points at level j (in the jth group)

$n = \Sigma I_j$ = number of data points (y_{ij})

$\bar{y}_{.j}$ = mean of jth level = $(\sum_{i=1}^{I_j} y_{ij})/I_j$

$\hat{\mu} = \bar{y}$ = grand mean = $(\Sigma_j \Sigma_i y_{ij})/n = \Sigma_j I_j \bar{y}_{.j}/n$

$\hat{\alpha}_j = \bar{y}_{.j} - \bar{y}$

$SS_B = \Sigma I_j \hat{\alpha}_j^2$ (sum of squares between groups)

$SS_E = \Sigma_j \Sigma_i (y_{ij} - \bar{y}_{.j})^2$ (error sum of squares)

$\quad\quad = \Sigma_j (I_j - 1)s_j^2$

where $s_j^2 = \Sigma_i (y_{ij} - \bar{y}_{.j})^2/(I_j-1)$ is the standard deviation for level j.

Note that if a computer has been used, then the displayed F-statistic for testing whether any of the variables is statistically significant is $F = [(n-k)SS_B]/[(k-1)SS_E]$, and is related to the

175

multiple correlation coefficient R^2 by

$$F = \frac{n-k}{k-1} \frac{R^2}{1-R^2} \cdot$$

The case $k = 2$ is especially interesting since the F statistic in the table is the square of the two-sample t-test [see (6.5)], the statistic most commonly used for comparing the means of two populations (cf., Section 6.2). With $k = 2$,

$$t_{n-2}^2 = F = \frac{SS_B(n-2)}{SS_E}$$

$$= \frac{SS_B(n-2)}{(I_1-1)s_1^2 + (I_2-1)s_2^2} \cdot$$

Since $SS_B = I_1[\bar{y}_{.1} - \bar{y}]^2 + I_2[\bar{y}_{.2} - \bar{y}]^2 = I_1 I_2(\bar{y}_{.1} - \bar{y}_{.2})^2/n$ it follows that

$$t_{n-2} = \frac{(\bar{y}_{.1} - \bar{y}_{.2})}{\hat{\sigma}} \sqrt{I_1 I_2/n}$$

with $n = I_1 + I_2$ and $\hat{\sigma}^2 = [(I_1-1)s_1^2 + (I_2-1)s_2^2]/(n-2)$ the unbiased estimate of σ^2. This statistic is appropriate for testing the hypothesis $\mu_1 = \mu_2$ if the I_1 observations y_{i1} from group 1 have mean μ_1, variance σ^2, the I_2 observations from group 2 have mean μ_2 and the same variance σ^2, and all the $\{y_{ij}\}$ are independent. If the $\{y_{ij}\}$ are normally distributed then t_{n-2} has Student's t-distribution with n-2 degrees of freedom when $\mu_1 = \mu_2$, so tables of the t-distribution may be used to test this hypothesis. We discussed this in Section 6.2, Equation (6.5).

7.3. THE TWO-WAY LAYOUT: INTERACTIONS AND ADDITIVITY

Expanding on the previous regional data, we add sex as a second dimension (M = male, F = female) and get the mean income by region and sex as:

Table 7.2

MEAN 1968 TOTAL INCOME BY SEX AND REGION ($000)

	NE	NC	S	W	
M	$\hat{\mu}_{11} = 6.9$	$\hat{\mu}_{12} = 6.8$	$\hat{\mu}_{13} = 5.7$	$\hat{\mu}_{14} = 7.4$	6.7
F	$\hat{\mu}_{01} = 2.9$	$\hat{\mu}_{02} = 2.4$	$\hat{\mu}_{03} = 2.5$	$\hat{\mu}_{04} = 3.0$	2.7
	4.9	4.6	4.1	5.2	4.7

Sex and Region are called *factors* and the number of *levels* of each factor is 2 and 4, respectively.

The μ_{ij} notation is not as useful for comparisons as the following notation (analogous to the one-way layout) is:

Grand mean (as before): $\hat{\mu} = 4.7$.

Main effects:

Sex: Male effect = $\hat{\alpha}_1$ = 6.7 - 4.7 = 2.0

Female effect: $\hat{\alpha}_0$ = 2.7 - 4.7 = -2.0

Region (as before):

NE: $\hat{\beta}_1$ = 4.9 - 4.7 = 0.2

NC: $\hat{\beta}_2$ = 4.6 - 4.7 = -0.1

S: $\hat{\beta}_3$ = 4.1 - 4.7 = -0.6

W: $\hat{\beta}_4$ = 5.2 - 4.7 = 0.5

(Note that the regional effects are called β_1, β_2, β_3, β_4 here, whereas these effects were called α_1, α_2, α_3, α_4 in Section 7.2. Here α_1 and α_0 are used for the male and female effects.) The main effects are computed for each of the factors ignoring the other as in the one-way layout. Note that the main effects add to zero: $\Sigma \hat{\alpha}_i = 0$, $\Sigma \hat{\beta}_j = 0$, as before. If we were only given the $\hat{\alpha}_i$, $\hat{\beta}_j$, $\hat{\mu}$, we represent $\hat{\mu}_{ij}$ by $\hat{\mu}^*_{ij} = \hat{\mu} + \hat{\alpha}_i + \hat{\beta}_j$. This gives the following table for $\hat{\mu}^*_{ij}$ based on *additivity*:

Table 7.3

ADDITIVE EFFECTS

	NE	NC	S	W
M	6.9	6.6	6.1	7.2
F	2.9	2.6	2.1	3.2

The difference between Table 7.2 and this table is the table of *interactions* gotten by differencing Table 7.2 and Table 7.3. The interactions are:

$$\hat{\gamma}_{ij} = \hat{\mu}_{ij} - \hat{\mu} - \hat{\alpha}_i - \hat{\beta}_j \tag{7.7}$$

Table 7.4

INTERACTIONS

	NE	NC	S	W
M	0	0.2	-0.4	0.2
F	0	-0.2	0.4	-0.2

Note that interactions add to zero in both directions. Hence in this case there are only three freely chosen interactions. If interactions are zero, then the male-effect, for example, is the same in every region. The table of interactions shows that this is not true for our data, but randomness must be considered ($\hat{\gamma}_{ij}$ are in Table 7.4, the true γ_{ij} are unknown). When all interactions are zero, the model is said to be *additive*, so that

$$y_{ijt} = \mu + \alpha_i + \beta_j + e_{ijt}$$

where $t = 1, 2, \ldots, n_{ij}$ indexes repeated observations in a cell. It is very useful to know if a model has this form, for if it does, it may be described more simply by many fewer parameters; and any dominant main effect is dominant independent of the level of the other factor.

178

Example: Sample of 40 persons with income:

Table 7.5

TOTAL INCOME BY SEX AND RACE ($000)

	Male: $X_2 = 1$		Female: $X_2 = -1$	
Non-white $X_1 = 1$	1.1	0.7	1.0	1.1
	12.0	2.4	1.5	4.6
	4.7	0.4	0.6	4.1
	3.4	0.1	0.9	3.6
	0.5	7.5	3.2	4.4
White $X_1 = -1$	8.8	4.0	0.1	5.0
	7.2	4.0	0.4	0.5
	5.9	11.1	4.5	2.0
	10.6	7.6	18.0	6.9
	9.0	9.0	1.1	1.6

Parameter structure:

$$
\begin{array}{cc|c}
\gamma & -\gamma & \alpha \\[2mm]
\hline
-\gamma & \gamma & -\alpha \\[2mm]
\beta & -\beta & \mu
\end{array}
$$

$$y = \mu + \alpha X_1 + \beta X_2 + \gamma(X_1 X_2) + e. \qquad (7.8)$$

The values X_1 and X_2 are like dummy variables. Their values ± 1 have been chosen carefully to produce the relation (7.8). The product $X_3 = X_1 X_2$ is needed to produce a variable for the interaction (note there is only one parameter, γ, for the interaction in the 2×2 case). The design matrix X has four types of rows corresponding to the parameter vector

$$\begin{pmatrix} \mu \\ \alpha \\ \beta \\ \gamma \end{pmatrix}.$$

179

The rows are:

$$NW - M: \quad (1, \ 1, \ 1, \ 1)$$

$$NW - F: \quad (1, \ 1, -1, -1)$$

$$W - M: \quad (1, -1, \ 1, -1)$$

$$W - F: \quad (1, -1, -1, \ 1)$$

Again the first column of 1's for μ is automatic if the computer always fits the constant term, so only the columns for X_1, X_2, X_3 (α, β, γ) need be entered. The regression equation will yield

$$\hat{y} = b_0 + b_1 x_1 + b_2 x_2 + b_3 x_3$$

giving: $\hat{\mu} = b_0$, $\hat{\alpha} = b_1$, $\hat{\beta} = b_2$, $\hat{\gamma} = b_3$. This completes the discussion of the example.

7.4. THE CLASSICAL TWO-WAY ANOVA: EQUAL OBSERVATIONS PER CELL

In most textbooks, a two-way layout is assumed to have equal observations in each cell. The regression version of ANOVA permits unequal observations in each cell. However, when the number of observations is the same in every cell then $n_{ij} = n$ and the variances σ^2 are the same for every cell, the two-way ANOVA may be computed as follows.

Let r be the number of row factor levels and c be the number of column factor levels. Observations in cell i,j are y_{ijk}, k = 1(1)n.

1. Compute the means $\bar{y}_{ij}$ and the variances s_{ij}^2 for each of the rc cells. It is assumed here that s_{ij}^2 is the unbiased estimate of σ^2, so the denominator of s_{ij}^2 is (n–1) in each case.

2. Compute the pooled sum of squares for error:

$$SS_E = (n - 1) \ \underset{i \ j}{\Sigma \ \Sigma} \ s_{ij}^2 \ .$$

180

3. Compute the row means $\bar{y}_{i.} = \frac{1}{c} \sum_j \bar{y}_{ij}$ for $i = 1, \ldots, r$, the column means $\bar{y}_{.j} = \frac{1}{r} \sum_i \bar{y}_{ij}$ for $j = 1, \ldots, c$ and the grand mean

$$\bar{y} = \frac{1}{r} \sum_i \bar{y}_{i.} \; (= \frac{1}{c} \sum_j \bar{y}_{.j} = \frac{1}{rc} \sum_i \sum_j \bar{y}_{ij}) \; .$$

4. Let $\hat{\mu} = \bar{y}$ and compute the main effects

$$\hat{\alpha}_i = \bar{y}_{i.} - \bar{y}$$

$$\hat{\beta}_j = \bar{y}_{.j} - \bar{y}$$

and the interactions

$$\hat{\gamma}_{ij} = \bar{y}_{ij} - \bar{y}_{i.} - \bar{y}_{.j} + \bar{y}$$

$$(= \bar{y}_{ij} - \hat{\mu} - \hat{\alpha}_i - \hat{\beta}_j) \; .$$

5. Compute the sum of squares between rows

$$SS_\alpha = cn \sum_i \hat{\alpha}_i^2 \; ,$$

between columns

$$SS_\beta = rn \sum_j \hat{\beta}_j^2 \; ,$$

and due to interactions

$$SS_\gamma = n \sum_i \sum_j \hat{\gamma}_{ij}^2 \; .$$

The following table is appropriate for the two-way layout, and is usually displayed when data have been analyzed by this method.

Table 7.6

ANALYSIS OF VARIANCE FOR A TWO-WAY LAYOUT

Source	SS	df	MS	F
Between Groups				
rows	SS_α	r-1	$SS_\alpha/(r-1)$	MS_α/MS_E
columns	SS_β	c-1	$SS_\beta/(c-1)$	MS_β/MS_E
Interaction	SS_γ	(r-1)(c-1)	$SS_\gamma/(r-1)(c-1)$	MS_γ/MS_E
Within Groups (Error)	SS_E	rc(n-1)	$SS_E/[(rc)(n-1)]$	
Grand Mean	$nrc\,\bar{y}^2$	1		
Total	$\Sigma\Sigma\Sigma y_{ijk}^2$	nrc		

The F statistic is the ratio of the mean square of the parameter being tested to the error term. If the interactions are known to be zero, then the error sum of squares in the F statistic may be taken to be $SS_\gamma + SS_E$ (rather than SS_E) and the degrees of freedom (df) to be $nrc - r - c + 1$, the pooled df. If n = 1, the error term cannot be estimated unless the interaction is assumed to be zero. The F statistic has degrees of freedom equal to the df of the numerator sum of squares and the pooled df is used for the denominator sum of squares.

For example, suppose for the sex and region data that the means are as given in Table 7.2. Then the grand mean, main effects and interaction have already been given. Suppose n = 10 persons per cell and SS_E = 57.6 has been calculated. We have r = 2, c = 4, and there were originally 80 observations. The ANOVA table then is as in Table 7.7.

Table 7.7

ANOVA: INCOME BY SEX AND REGION

Source	SS	df	MS	F
Between Groups				
Sex	320.0	1	320.0	400.0
Region	13.2	3	4.4	5.5
Interaction	4.8	3	1.6	2.0
Within Groups	57.6	72	0.8	
Grand Mean	1767.2	1		
Total	2162.8	80		

182

The interactions are not significant at the 0.1 level (using $F_{3,72}$ table). However, since the interactions are nearly significant at this level, we are reluctant to combine them with the error term, being concerned that a larger sample might produce significant interactions. Furthermore, pooling adds only three degrees of freedom to the 72 already available from the within groups sum of squares. The effect of sex is statistically significant at a very high level of significance, and region is significant somewhere between the 0.005 and 0.001 levels (the two $F_{3,72}$ values are about 4.7 and 6.1 for these significance levels, so $4.7 < F_{region} = 5.5 < 6.1$).

7.5. TWO-WAY ANOVA VIA REGRESSION METHODS: UNEQUAL OBSERVATIONS PER CELL

In case the sample sizes n_{ij} are not equal, the preceding analysis cannot be carried out. However, the following specification of the X matrix will provide an appropriate analysis. If $n_{ij} = n$ the following reduces to the ANOVA just given, and if $r = c = 2$, it reduces to the regression example considered at the end of Section 7.3.

We want

$$\mu_{ij} = \mu + \alpha_i + \beta_j + \gamma_{ij} \tag{7.9}$$

$$\sum_i \alpha_i = \sum_j \beta_j = 0, \ \sum_i \gamma_{ij} = \sum_j \gamma_{ij} = 0$$

$i = 1, \ldots, r$, $j = 1, \ldots, c$. Then $\alpha_r = -\sum_{i=1}^{r-1} \alpha_i$, $\beta_c = -\sum_{j=1}^{c-1} \beta_j$ can be eliminated, as can $r + c - 1$ of the interactions. To make matters more concrete, suppose $r = c = 3$. Then the parameter vector may be taken to be

$$\begin{bmatrix} \mu \\ \alpha_1 \\ \alpha_2 \\ \beta_1 \\ \beta_2 \\ \gamma_{11} \\ \gamma_{12} \\ \gamma_{21} \\ \gamma_{22} \end{bmatrix}$$

with α_3, β_3, γ_{13}, γ_{23}, γ_{31}, γ_{32} and γ_{33} eliminated. The design matrix X of explanatory variables is arranged to give $\mu_{ij} = \mu + \alpha_i + \beta_j + \gamma_{ij}$ as follows.

Factor Levels		Design Matrix								Parameters
row i	col j	Grand Mean	Row Effects		Column Effects		Interactions			
1	1	1	1	0	1	0	1	0	0	0	μ
1	2	1	1	0	0	1	0	1	0	0	α_1
1	3	1	1	0	-1	-1	-1	-1	0	0	α_2
2	1	1	0	1	1	0	0	0	1	0	β_1
2	2	1	0	1	0	1	0	0	0	1	β_2
2	3	1	0	1	-1	-1	0	0	-1	-1	γ_{11}
3	1	1	-1	-1	1	0	-1	0	-1	0	γ_{12}
3	2	1	-1	-1	0	1	0	-1	0	-1	γ_{21}
3	3	1	-1	-1	-1	-1	1	1	1	1	γ_{22}

Of course there is one row in X for every individual with a response, say N rows. Then the vector y of responses has dimension N with responses recorded in corresponding positions.

In general the X matrix is simple to compute by the following rule. Designate r-1 columns of X as x_1, ..., x_{r-1} for the row effects α_1, ..., α_{r-1}, and the next c-1 columns as u_1, ..., u_{c-1} for the column effects β_1, ..., β_{c-1}. If a row of X corresponds to factor levels i and j respectively, proceed as follows. If $i \leq r-1$, set $x_i = 1$ and the other $x_t = 0$ for $t \neq i$. If $i = r$, set $x_1 = x_2 = \ldots = x_{r-1} = -1$, since $\alpha_r = -\alpha_1 - \alpha_2 - \ldots - \alpha_{r-1}$. If $j \leq c-1$, similarly set $u_j = 1$ and $u_t = 0$ if $t \neq j$. If $j = c$, set $u_1 = \ldots = u_{c-1} = -1$. The interaction terms are computed simply, the coefficient of γ_{ij} being the product of the coefficients for α_i and β_j. Recall that computing interactions as products was introduced in Chapter 4. If the interactions are known to be zero, the $(r-1)(c-1)$ interaction columns are omitted from X. In all cases, it is understood that the first column of X is the vector $(1, 1, \ldots, 1, 1)'$. In general, it is convenient to have a computer routine to compute the elements of X from specifying the factor levels i and j.

The result is that regression estimates for an equation of the form

$$\hat{y} = b_0 + b_1 x_1 + \ldots + b_{r-1} x_{r-1}$$

$$+ b_r x_r + \ldots + b_{r+c-2} x_{r+c-2}$$

$$+ b_{r+c-1}(x_1 x_r) + \ldots + b_{rc-1}(x_{r-1} x_{r+c-2}) \tag{7.10}$$

are obtained. Then

$$\hat{\mu} = b_0, \ \hat{\alpha}_1 = b_1, \ \ldots, \ \hat{\alpha}_{r-1} = b_{r-1}, \ \hat{\alpha}_r = -\hat{\alpha}_1 - \ldots - \hat{\alpha}_{r-1}$$

$$\hat{\beta}_1 = b_r, \ \ldots, \ \hat{\beta}_{c-1} = b_{r+c-2}, \ \hat{\beta}_c = -\hat{\beta}_1 - \ldots - \hat{\beta}_{c-1}$$

$$\hat{\gamma}_{11} = b_{r+c-1}, \ \hat{\gamma}_{12} = b_{r+c}, \ \ldots, \ \hat{\gamma}_{r-1,c-1} = b_{rc-1}$$

and the remaining $\hat{\gamma}_{ij}$ are computed using the restriction $\sum_i \hat{\gamma}_{ij} = \sum_j \hat{\gamma}_{ij} = 0$.

7.6. INFERENCE IN ANOVA

Most regression routines will provide t-statistics for the individual parameters, but do not automatically provide F-statistics for testing that interactions as a group are zero, or that the main effects for only one of the two factors are zero, etc. But this can be accomplished by running several regressions and keeping track of each R^2. The method for doing this, as we saw in Chapter 6, is perfectly general in that it may be used in any regression problems for testing that the effect of a subset of variables is zero. The lone drawback of this method is that it may not be computationally efficient and therefore can be more expensive. In addition, using a regression package to do ANOVA can also be more expensive than an ANOVA package. But we will sacrifice computer efficiency in favor of pedagogical efficiency and expect the student to recognize and use ANOVA without distinguishing it from regression.

Suppose there are N individuals (rows in the X matrix) and suppose $k = k_1 + k_2$, where k_2 is the number of variables to be tested, but the other k_1 variables are to be included in any event (k_1 includes the constant term, if there is one, so that k is the number of regression parameters). Perform the regression on all k independent variables and let R_{12}^2 be the multiple correlation coefficient. Perform the regression again, using just the k_1 independent variables and let R_1^2 be the multiple correlation coefficient. Let

$$F = \frac{R_{12}^2 - R_1^2}{1 - R_{12}^2} \frac{N-k}{k_2} .$$

Then F has an F distribution with $(k_2, N-k)$ degrees of freedom, respectively, and large values of F indicate that the extra k_2 variables add

186

to the explained variance in a statistically significant way. Therefore, small values of F are consistent with the hypothesis that the k_2 parameters are all zero while large values imply at least one of the k_2 parameters differs from zero. Of course, if $k_2 = 1$, the F statistic just derived is the square of the t-statistic for the parameter in question.

This method can be used to determine whether interactions are zero by simply leaving them out in one regression. The main effects of one factor may also be simultaneously tested after the interactions are eliminated. The act of testing one factor in the presence of the other with interactions suppressed is equivalent to pooling the interaction sum of squares with the sum of squares for error in the denominator of the F statistic. The interaction columns of the X matrix must be retained in the regression if they are not to be pooled with the error sum of squares when testing the main effects. With equal numbers of observations per cell, tests compare with those in Sections 7.2 and 7.4.

7.7. SUMMARY

The sums of squares for the main effects for rows SS_α, for columns SS_β, for interactions SS_γ, and for error SS_E are all independent if $n_{ij} = n$ is constant. If the n_{ij} are not constant this is not so, so tests cannot be made independently. This is not terribly important, although it does increase the desirability of having somewhat equal n_{ij}.

Three- and higher-way analyses of variance can be accomplished analogously to the two-way ANOVA just presented. The X matrix need only be expanded to include extra columns corresponding to the main effects of the additional factors, and perhaps interactions. As before, first-order interactions are computed as products, and second-order and higher-order interactions now appear because of more combinations of variables. Double products are still used to determine entries for first-order interactions, and triple products are needed to determine entries for X with third-order interactions.

The details of this method perhaps have been dwelled upon here at too great a length. The important point is that one can think of modeling a problem in terms of the general linear model by choosing the design matrix properly. Then standard computer regression packages will compute appropriate estimates and tests.

7.8. APPENDIX: ANOVA WITH GENERAL REFERENCE POPULATIONS

The ANOVA models considered in this chapter effectively define the grand mean, main effects, and interactions relative to a particular "reference population," where this population has equal probabilities in each cell. By a "reference population" in the one-way layout, we mean a set of probabilities $(\pi_j, \pi_2, \ldots, \pi_k)$, $\Sigma \pi_i = 1$ such that π_i is the probability of an observation in category i. If the means for the categories are μ_i, then with respect to the reference population, $\hat{\mu} = \Sigma \pi_i \mu_i$ is the grand mean, and $\alpha_i = \mu_i - \bar{\mu}$ is a main effect. Thus, $\Sigma \alpha_i \pi_i = 0$, but $\Sigma \alpha_i \neq 0$ unless $\pi_i = 1/k$ always. The correct coding of the matrix in this case is

$$X = \begin{bmatrix} 1 & 1 & 0 & 0 \\ 1 & 0 & 1 & 0 \\ 1 & 0 & 0 & 1 \\ 1 & -\pi_1/\pi_4 & -\pi_2/\pi_4 & -\pi_3/\pi_4 \end{bmatrix}$$

since then $\mu = X\beta$ satisfies $\mu_1 = \beta_0 + \beta_1$, $\mu_2 = \beta_0 + \beta_2$, $\mu_3 = \beta_0 + \beta_3$, $\mu_4 = \beta_0 - (\beta_1 \pi_1 + \beta_2 \pi_2 + \beta_3 \pi_3)/\pi_4$. Thus, $\Sigma \pi_i \mu_i = \beta_0$ so β_0 is the

188

grand mean, and it follows that β_1, β_2, β_3 are the main effects α_1, α_2, α_3 while $\alpha_4 = -(\beta_1\pi_1 + \beta_2\pi_2 + \beta_3\pi_3)/\pi_4$. Hence $\Sigma \pi_i\alpha_i = 0$. It should be clear how this example with four categories generalizes to any number k.

Note that if $\pi_1 = \pi_2 = \pi_3 = \pi_4 = 1/4$ then X reduces to the ANOVA specification of earlier chapters.

This specification has two properties: the column average of main effect variables in the X matrix, if computed with weights $(\pi_1, \ldots, \pi_k)$, is zero; and categories (other than the omitted one) can be combined by adding the corresponding columns in X, without disrupting the ANOVA specification.

Reference populations also may be specified for the two-way layout. We treat the 3×3 case for concreteness. The reference population is a set of probabilities $\{\pi_{ij}\}$, $\Sigma \Sigma \pi_{ij} = 1$.

Factor 2

π_{11}	π_{12}	π_{13}	$\pi_{1.}$
π_{21}	π_{22}	π_{23}	$\pi_{2.}$
π_{31}	π_{32}	π_{33}	$\pi_{3.}$
$\pi_{.1}$	$\pi_{.2}$	$\pi_{.3}$	1

Factor 1

Write $\pi_{i.} = \Sigma_j \pi_{ij}$ and $\pi_{.j} = \Sigma_j \pi_{ij}$, so $\Sigma_i \pi_{i.} = \Sigma_j \pi_{.j} = 1$. Let F_i and L_j indicate factor i and level j; then the coding for X is simplest to write first as

189

Cell		F1L1	F1L2	F2L1	F2L2	Interaction			
$(1,1)$	1	$1/\pi_{1.}$	0	$1/\pi_{.1}$	0	$1/\pi_{11}$	0	0	0
$(1,2)$	1	$1/\pi_{1.}$	0	0	$1/\pi_{.2}$	0	$1/\pi_{12}$	0	0
$(1,3)$	1	$1/\pi_{1.}$	0	$-1/\pi_{.3}$	$-1/\pi_{.3}$	$-1/\pi_{13}$	$-1/\pi_{13}$	0	0
$(2,1)$	1	0	$1/\pi_{2.}$	$1/\pi_{.1}$	0	0	0	$1/\pi_{21}$	0
$(2,2)$	1	0	$1/\pi_{2.}$	0	$1/\pi_{.2}$	0	0	0	$1/\pi_{22}$
$(2,3)$	1	0	$1/\pi_{2.}$	$-1/\pi_{.3}$	$-1/\pi_{.3}$	0	0	$-1/\pi_{23}$	$-1/\pi_{23}$
$(3,1)$	1	$-1/\pi_{3.}$	$-1/\pi_{3.}$	$1/\pi_{.1}$	0	$-1/\pi_{31}$	0	$-1/\pi_{31}$	0
$(3,2)$	1	$-1/\pi_{3.}$	$-1/\pi_{3.}$	0	$1/\pi_{.2}$	0	$-1/\pi_{32}$	0	$-1/\pi_{32}$
$(3,3)$	1	$-1/\pi_{3.}$	$-1/\pi_{3.}$	$-1/\pi_{.3}$	$-1/\pi_{.3}$	$1/\pi_{33}$	$1/\pi_{33}$	$1/\pi_{33}$	$1/\pi_{33}$

$X^* =$ (the matrix above)

and then realize that

$$X\beta = X^* \begin{bmatrix} \bar{\mu} \\ \alpha_1 \pi_{1.} \\ \alpha_2 \pi_{2.} \\ \beta_1 \pi_{.1} \\ \beta_2 \pi_{.2} \\ \gamma_{11} \pi_{11} \\ \gamma_{12} \pi_{12} \\ \gamma_{21} \pi_{21} \\ \gamma_{22} \pi_{22} \end{bmatrix}$$

with β the vector $(\bar{\mu}, \alpha_1, \alpha_2, \beta_1, \beta_2, \gamma_{11}, \gamma_{12}, \gamma_{21}, \gamma_{22})'$ consisting of the grand mean, main effects for each factor, and interactions. Note that the interaction specification in X^* is *not* constructed as a product of main effects unless the two factors are independent in the reference population.

If factor i occurs at level j a total of n_{ij} times then X will have n_{ij} rows identically equal to the (i,j) row of X^* multiplied by the diagonal matrix with elements $(1, \pi_{1.}, \pi_{2.}, \pi_{.1}, \pi_{.2}, \pi_{11}, \pi_{12}, \pi_{21}, \pi_{22})$.

190

This follows from the identity in the preceding paragraph. If a weighted least-squares regression program is available then the number of rows of X may be taken to be the number of free parameters (nine in this example) by using each distinct row of X only once, and letting n_{ij} be the weight. In this case, the y value for each cell is the within cell mean. The pooled within cell variance, derived from $SS_E = \Sigma(n_{ij}-1)s_{ij}^2$, must be computed outside of the WLS program, and in fact there often would be zero degrees of freedom for estimating σ^2 from the program. This procedure is much cheaper computationally if $N = \Sigma\Sigma\, n_{ij}$ is large.

The classical ANOVA has $n_{ij} = n_{i'j'}$ and $\pi_{ij} = \pi_{i'j'}$ for all i, j, i', and j'. The first six sections of this chapter kept $\pi_{ij} = 1/rc$ but permitted the n_{ij} values to differ. If π_{ij} is proportional to n_{ij} then the grand mean estimate is independent of the interactions and of the main effects, and the interactions are independent of the main effects. The main effects for the two factors are not independent in this case though, unless the π_{ij} values satisfy $\pi_{ij} = \pi_{i.}\pi_{.j}$.

If μ_{ij} is the mean in cell (i, j), then $\alpha_i = E_\pi(y|F1 = i) - E_\pi(y)$, $\beta_j = E_\pi(y|F2 = j) - E_\pi(y)$ and $\bar{\mu} = E_\pi(y) = \Sigma\Sigma\,\pi_{ij}\mu_{ij}$, where $E_\pi(y|F1 = i)$ $\equiv \Sigma_j\,\pi_{ij}\mu_{ij}/\pi_{i.}$, etc. Then $\gamma_{ij} = E_\pi(y|F1 = i, F2 = j) - [\bar{\mu} + \alpha_i + \beta_j]$ $= \mu_{ij} - \bar{\mu} - \alpha_i - \beta_j$. Note that $0 = E_\pi\alpha = E_\pi\beta = E_\pi(\gamma|i) = E_\pi(\gamma|j)$. Thus the ANOVA parameterization is defined with respect to expectations relative to a reference population π. Since some inferences depend on the choice of π (for example, t-statistics in regressions using economic data), it is useful to make π an object of conscious choice and not select it by default to be the classical choice: $\pi_{ij} =$ the reciprocal of the number of cells.

Chapter 8

LOGIT REGRESSION FOR BINARY DEPENDENT VARIABLES

We have discussed regression with continuous dependent variables
when the independent variables take continuous values or indicate
the presence of categories. In these cases the same method of analysis
is appropriate and computer software is available. Now we consider
binary (two valued) dependent variables, and again seek to explain them
from some independent variables. The usual regression theory will be
unsatisfactory for reasons to be given, and "logit regression" methods
will be needed. At this point the reader should return to Table 1.2
of Chapter 1 to review the statements there showing how the type of
data dictates the analysis.

Regression problems with binary dependent variables most often
arise when a probability is to be estimated as dependent on some independent variables that influence it. In the simplest case (one independent variable) n individuals have been observed to have one of
two responses (A or not A), measured by $y = 1$ or $y = 0$, each individual
at some different level x. The problem is to use the data (values of
$y = 1$ when A occurs) to determine how the probability of A depends on
x. Applications are numerous, and examples include

1. $P(y = 1)$ = Airmen reenlistment probability, x = opportunity
   cost of reenlistment, or age, or qualifying test score.

2. $P(y = 1)$ = Aircraft system failure probability, x = previous
   maintenance failure measure.

3. $P(y = 1)$ = Probability that an individual favors integration,
   x = amount of education or other socioeconomic variable.

4. $P(y = 1)$ = Survival probability, x = dosage of a drug. This is called a "bioassay" problem.

5. $P(y = 1)$ = Lung cancer probability, x = years of smoking.

6. $P(y = 1)$ = Probability a student graduates from college, x = college board exam score.

7. $P(y = 1)$ = Probability of rain, x = forecaster's probability of rain.

8. $P(y = 1)$ = Probability that a football kicker makes a field goal, x = length of attempt.

We will use the last example to illustrate logit regression in this chapter.

8.1. WHY LOGIT REGRESSION IS NECESSARY

Ordinary least-squares regression is inadequate, primarily because when y has a Bernoulli distribution,

(1) $Var(y) = p(1-p)$ depends on the mean p, whereas the variance should be constant.

(2) $Ey = p = \beta_0 + \beta_1 x$ is not constrained to $0 \leq p \leq 1$.

The logit method assumes the $\{y_i\}$ are independent Bernoulli variables, and that

$$P(y_i = 1 | x_i) = p(x_i) = \frac{e^{\beta_0 + \beta_1 x_i}}{1 + e^{\beta_0 + \beta_1 x_i}} . \qquad (8.1)$$

Equivalently

$$L(x_i) = \log\left(\frac{p(x_i)}{1 - p(x_i)}\right) = \beta_0 + \beta_1 x_i$$

where log is the natural logarithm. Thus $p(x)$, the probability that $y = 1$ at x, is not linear in x, but rather the logistic transformation

193

$L(x) = \log[p(x)/(1-p(x))]$ is. We still have a linear model, but this modification must be accounted for.

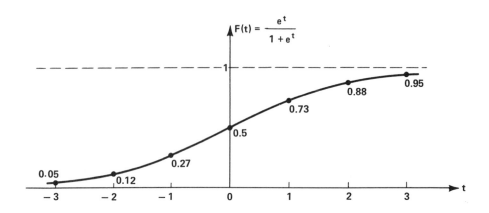

Figure 8.1 - Plot of the logistic distribution function
$$F(t) = e^t/(1+e^t) = 1/(1+e^{-t}).$$

The function $F(t) = e^t/(1+e^t)$ looks much like the normal distribution function with mean 0 and variance $\pi^2/3 = (1.8138)^2$, see Figure 8.1.

There is a "stimulus-response" curve interpretation for F: Let T be a variable with cumulative distribution function $F(t) = e^t/(1+e^t)$, and suppose Y = 1 if and only if $T \leq \beta_0 + \beta_1 x$. Then $P(Y=1) = P(T \leq \beta_0 + \beta_1 x)$ $= e^{\beta_0+\beta_1 x}/(1+e^{\beta_0+\beta_1 x})$. This argument is sometimes used to justify using the logit model in applications.

8.2. LOGIT REGRESSION FOR GROUPED DATA USING WEIGHTED LEAST SQUARES

A method for analyzing data with the logit model that works well, provided that there are repeated observations at most x_i's, is called "minimum logit chi-square." This method is simply to do (weighted) linear regression on the estimated logit $\log(p/(1-p))$, which depends

194

linearly on the unknown parameters, β_0, β_1. We use a modified estimate $\hat{p}_i = (y_i + 0.5)/(n_i + 1)$ of p_i (adding 0.5 and 1 protects against the cases $y_i = 0$, n_i), and set $L_i = $ logit at $i = \log(\hat{p}_i/(1-\hat{p}_i))$. The formula $\log(x) = 2.3026 \log_{10}(x)$ permits calculation of the natural log from log to the base 10.

Field goal kicking data for the entire National Football League (NFL) in the 1969 season are given below. Kickers attempted to kick field goals from various distances ranging from 10 to more than 50 yards from the goal. Longer kicks are more difficult--the reader unfamiliar with football should think of this as the problem of hitting a target from a given distance. Obviously the probability of a hit declines as the distance is increased.

The method discussed for estimating the regression coefficients in this section only works with grouped data. Thus the 461 attempts scattered between 10 and 55 yards are placed into five equally spaced intervals, with all attempts for each group assigned to a central value. The data appear in Table 8.1.

Note that we are only treating regression on one independent variable at first, but we shall soon consider more general cases. The method of this section is approximate, the approximation improving as the group sizes (n_i) increase, and reduces to the weighted least-squares method already studied. More general methods are treated later in the chapter.

As an example, we calculate for $i = 1$:

$$\hat{p}_1 = \frac{y_1 + 0.5}{n_1 + 1} = \frac{68.5}{78} = 0.878$$

$$\frac{\hat{p}_1}{1-\hat{p}_1} = 7.21$$

$$L_1 = \log\left(\frac{\hat{p}_1}{1-\hat{p}_1}\right) = 1.975.$$

195

Table 8.1

NFL FIELD GOAL KICKING PERFORMANCE, 1969

i	Yards	Center (x_i)	Successes (y_i)	Attempts (n_i)
1	(9.5, 19.5)	14.5	68	77
2	(19.5, 29.5)	24.5	74	95
3	(29.5, 39.5)	34.5	61	113
4	(39.5, 49.5)	44.5	38	138
5	(49.5, ∞)	52.0	2	38
Total or Average:		33.5	243 (.527)	461

The variance of L_i is estimated as

$$v_i = \frac{(n_i+2)(n_i+1)}{n_i(y_i+1)(n_i-y_i+1)} \qquad (8.2)$$

so for i = 1

$$v_1 = \frac{(79)(78)}{(77)(69)(10)} = 0.116.$$

The complete result is

Table 8.2

VALUES OF THE LOGITS AND THEIR VARIANCES

i	L_i	v_i
1	1.9755	0.1160
2	1.2427	0.0594
3	0.1582	0.0353
4	-0.9595	0.0358
5	-2.6810	0.3698

The weighted sum of squares to be minimized is

$$S = \sum_{i=1}^{k} (L_i - b_0 - b_1 x_i)^2 / v_i ,$$

the weights being the reciprocal values of v_i. Thus each L_i is an esti-mate of $\beta_0 + \beta_1 x_i$ and has variance v_i. By methods of calculus, or by com-pleting the square, it is not hard to show that the b_0, b_1 minimizing S satisfy the two equations:

196

$$\Sigma \frac{L_i}{v_i} = b_0 \Sigma \frac{1}{v_i} + b_1 \Sigma \frac{x_i}{v_i}$$

$$\Sigma \frac{L_i x_i}{v_i} = b_0 \Sigma \frac{x_i}{v_i} + b_1 \Sigma \frac{x_i^2}{v_i} .$$

The equations for this example are

$$8.3832 = 84.416 \, b_0 + 2898.2 \, b_1$$

$$-655.52 = 2898.2 \, b_0 + 108255 \, b_1$$

and are solved by $b_0 = 3.7993$ and $b_1 = -0.10777$. Therefore the estimate of p at yard line x is:

$$p(x) = \frac{\exp(3.8 - 0.108x)}{1 + \exp(3.8 - 0.108x)}$$

$$= 1/[1 + \exp(0.108x - 3.8)] .$$

We obtain the following table.

Table 8.3

PREDICTED PROBABILITY AT x FOR
$p(x) = 1/1 + \exp(-3.7993 + 0.10777x)$

Yard Line = x	Predicted Probability = p(x)
10	0.938
15	0.899
20	0.838
25	0.751
30	0.638
35	0.507
40	0.375
45	0.259
50	0.169
55	0.106

A useful fact for logit regression involving just one independent variable is that we always can reparameterize $p(x)$ to the convenient form

$$p(x) = \frac{1}{1+3^{[(m-x)/s]}}$$

with $m = -b_0/b_1$, $s = \log(3)/b_1 = 1.0986/b_1$. In this formulation the probability at $x = m$ is 0.50, while $x = m \pm s$ gives the quartiles (probabilities of 0.25 and 0.75) and $x = m \pm 2s$ the outer deciles (probabilities of 0.10 and 0.90). In the preceding example with $b_0 = 3.7993$, $b_1 = -0.10777$, then

$$p(x) = \frac{1}{1+3^{[(x-35.2)/10.2]}}$$

or approximately

$$p(x) = \frac{1}{1+3^{[(x-35)/10]}} \, .$$

Hence, the chance of making a field goal is estimated to be 0.5 from 35 yards; 0.75 and 0.25 from 25 and 45 yards; 0.90 and 0.10 from 15 and 55 yards respectively, c.v., Table 8.3. Those who know football know that these estimates are too high above 50 yards.

Values of $p(x)$ at the yard lines used for the data appear in Table 8.4 and may be compared with the binomial estimate y_i/n_i. The two kinds of estimates do not agree because estimates under the logit model use data from all intervals for estimating the probability at each. The only large discrepancy is at 52 yards when $p(x_i)$ may overestimate the correct value (we don't know the correct value).

Ordinarily one uses a weighted least-squares computer program to carry out regressions like the one just completed. This has the

198

Table 8.4

YARD LINES AND NFL SUCCESS PROBABILITIES

i	x_i	$p(x_i)$	y_i/n_i
1	14.5	0.904	0.883
2	24.5	0.761	0.779
3	34.5	0.520	0.540
4	44.5	0.270	0.275
5	52.0	0.141	0.053

advantage not only of being easier, but the variances of b_0, b_1 together

with t statistics are available. To carry this out, one must use the

L_i as the dependent variable and use weights $w_i = 1/v_i$. Standard weighted

least-squares computer packages then will provide the correct results.

The logits may be regressed on more than one explanatory variable.

For example, this method could be used to regress L_i on x_i (yards) and

x_i^2 (squared yards) to fit the curve

$$p(x, x^2) = \frac{1}{1+\exp(-b_0-b_1 x-b_2 x^2)} .$$

Or with American Football League (AFL) data for the same year from

Table 8.5, we have ten combinations of observations (five distances times

two leagues) and can fit the model

$$p(x, x^2, z) = \frac{1}{1+\exp(-b_0-b_1 x-b_2 x^2-b_3 z)}$$

with $z = 1$ for the AFL, $z = 0$ for the NFL. The result, obtained by

a computer, and dropping the x^2 term, is

$$p(x, z) = \frac{1}{1+\exp(-3.478 + 0.0984x - 0.0971z)}$$

with predicted values in Table 8.6. The result reduces the NFL per-

centage for the short kicks and increases the percentage for long

kicks (compare Table 8.4).

Table 8.5

AFL FIELD GOAL KICKING PERFORMANCE, 1969

i	x_i	Successes (y_i)	Attempts (n_i)
1	14.5	62	67
2	24.5	49	70
3	34.5	43	79
4	44.5	25	82
5	52.0	7	24
Total	32.0	186 (.578)	322

Table 8.6

YARD LINES, LEAGUE, AND SUCCESS PROBABILITIES FOR
AFL AND NFL, ESTIMATED FROM COMBINED DATA
(z = 1 for AFL, z = 0 for NFL)

i	x_i	z_i	$p(x_i, z_i)$	y_i/n_i
1	14.5	0	0.886	0.883
2	24.5	0	0.744	0.779
3	34.5	0	0.521	0.540
4	44.5	0	0.289	0.275
5	52.0	0	0.163	0.063
6	14.5	1	0.896	0.925
7	24.5	1	0.762	0.700
8	34.5	1	0.545	0.544
9	44.5	1	0.309	0.305
10	52.0	1	0.176	0.292

The method just used could be called the "Analysis of Covariance
for Logistic Regression," because differences between the distribution
of attempts are accounted for before the main effect between leagues
is estimated. The coefficient of the league indicator z is b_2 = 0.0971
(not -0.0971) which means that the AFL is estimated to be more successful.
However, this difference is not statistically significant because the
t-statistic is only 0.584. This value can be computed with tables of
the normal distribution, because sample sizes must be large for this
theory to apply, and hence the "degrees of freedom" for t are large

200

8.3. MAXIMUM LIKELIHOOD LOGIT

We have introduced the logit model, and a weighted least–squares method for estimating it that requires grouping of data. That method also is based upon certain approximations that require the sample sizes to be substantial in the groups. A more general estimating technique is needed, one that works whether data are grouped or not. The method considered here is "maximum likelihood logit." It is based on general maximum likelihood theory, which we have not discussed in this text, applied to the logit regression model, and hence its name.

The cost of getting better* and more general estimates is the additional computing required to solve for the best regression estimates. A sequence of familiar weighted regressions must be solved repeatedly until the desired solution is reached [formula (8.12) is used]. It often takes 3 to 6 iterations to do this, but in special cases it can be much more. With modern computational technology, this cost is not high and it decreases annually. Therefore logit regression, which because of the computational effort was almost never used for data analysis until the last decade, is taking its place alongside the classical linear model as one of the most frequently used methods in applied statistics.

We are considering the problem of explaining a dichotomous (binary, 0-1) dependent variable from a set of independent variables. The main logical problem with using ordinary least squares on the dependent variable is that its expectation (probability $Y = 1$) is constrained to be

---

*There have been differences of opinion among statisticians about the relative advantages of maximum likelihood logit and minimum logit chi-square methods of estimation for the past 30 years. See J. Berkson, "Minimum chi-square, not maximum likelihood!" (with discussion), *Annals of Statistics*, Vol. 8, 1980, pp. 457-487. The discussion and references are particularly illuminating.

between 0 and 1, but the predictions of it from a linear regression need not be, and often will not be. As discussed in Section 8.1, for this reason we introduced the logit curve which we write to emphasize the dependence on β as:

$$p(x_i,\beta) = \frac{1}{1 + \exp\left(-\Sigma_j \beta_j x_{ij}\right)} , \qquad i = 1, \ldots, N ; \quad j = 1, \ldots, k , \qquad (8.3)$$

which always lies in the interval [0,1]. Now we have specified a non-linear relation between Ey_i (the expected number of successes at $x_i = (x_{i1}, \ldots, x_{ik})'$) and $\beta = (\beta_1, \ldots, \beta_k)'$ as it enters into the expectation of y_i, a binomial $\left(n_i, p(x_i,\beta)\right)$ variable

$$Ey_i = n_i \left(p(x_i,\beta)\right) , \qquad i = 1, \ldots, N . \qquad (8.4)$$

There are a number of ways to estimate β. If the number of attempts n_i at each point x_i is fairly large, then an approximately unbiased estimate of $\Sigma_j \beta_j x_{ij}$ at each point is the "observed logit"

$$L_i = \log \frac{y_i}{n_i - y_i} , \qquad (8.5)$$

having approximate variance[*] (when $y_i \neq 0$ or n_i)

$$Var(L_i) \equiv v_i = \frac{n_i}{y_i (n_i - y_i)} . \qquad (8.6)$$

[The exact variance is given by (8.2).] This leads to the method of "minimum logit chi-square" just discussed in Section 8.2, but with the minor adjustments to L_i and v_i noted there to make these formulas more stable when y_i is quite near 0 or n_i. The minimum logit chi-square method reduces a nonlinear problem to a linear one by making the

---

[*]Formula (8.6) is justified by applying the "delta method" to (8.5).

202

nonlinear transformation (8.5) on the data. As in Section 8.2, β is chosen to minimize

$$\Sigma_i (L_i - \Sigma_j \beta_j x_{ij})^2 / v_i \qquad (8.7)$$

which may be carried out by regressing the $\{L_i\}$ on the matrix $X = (x_{ij})$ using weights $W_i = 1/v_i$. The STATLIB computing package has this capability.

But minimum logit chi-square provides satisfactory estimates only when there are sufficiently many repeated measurements n_i at each x_i, and when each y_i is sufficiently far from 0 and n_i so that it is approximately normally distributed. It is better not to linearize the statistics, but to solve certain nonlinear equations directly using iterative techniques that a computer can handle efficiently. The method described below is a special case of "maximum likelihood estimation" called "maximum likelihood logit estimation."

From inspection of (8.4) we might like to determine β so that each observed binomial variable y_i is equal to its expected value, that is

$$y_i = n_i p(x_i, \beta) \qquad (8.8)$$

for every $i = 1, \ldots, N$. In almost all cases this is impossible, however, because there are more equations (N) than unknown β values (k), and so we reduce the number of equations to k by asking only that the k linear combinations of formula (8.8)

$$\sum_i y_i x_{ij} = \sum_i n_i p(x_i, \beta) x_{ij}, \qquad j = 1, \ldots, k \qquad (8.9)$$

be equal to their expected values and instead solve (8.9). Except in special cases, a unique root $\hat{\beta}$ exists for (8.9), and it is called the "maximum likelihood estimator" for this problem. In vector and matrix notation, with $y = (y_1, \ldots, y_N)'$, $X = (x_{ij})$ an $N \times k$ matrix, and

203

$\mu(\beta) = \left(\mu_1(\beta), \ldots, \mu_k(\beta)\right)'$ where

$$\mu_i(\beta) = n_i p(x_i, \beta) ,$$

then (8.9) is written

$$X'y = E_{\hat{\beta}} X'y = X'\mu(\hat{\beta}) , \qquad (8.10)$$

and so $\hat{\beta}$ is the estimate necessary that $X'y$ be equal to its expected value at $\hat{\beta}$, exactly as in ordinary least squares.

Let $D(\beta)$ be the $N \times N$ diagonal matrix with i-th diagonal element $d_i(\beta) = Var(y_i) = n_i p(x_i, \beta)\left(1 - p(x_i, \beta)\right)$. When (8.10) is solved, the maximum likelihood estimator $\hat{\beta}$ has approximately mean β, covariance matrix $\left(X'D(\beta)X\right)^{-1}$ and is k-variate normal, i.e.,

$$\hat{\beta} \sim N_k\left(\beta, (X'D(\beta)X)^{-1}\right) , \qquad (8.11)$$

provided Σn_i is large (but not necessarily any one n_i). Note that $\hat{\beta}$ is used in $D(\beta)$ since β is unknown.

Proof of (8.11) requires advanced statistical theory, but the idea is that $X'y$ is approximately normal, from the central limit theorem, and therefore $\hat{\beta}$ is approximately normal as $N \to \infty$ because $\mu(\beta)$ is a smooth function of β. The "delta method" is used in this proof to determine an approximate mean and covariance matrix for $\hat{\beta}$.

A method of solving (8.10) is to apply the Newton-Raphson technique. This results in

$$\beta^{**} = \beta^* + \left(X'D(\beta^*)X\right)^{-1} X'\left(y - \mu(\beta^*)\right) \qquad (8.12)$$

if β^* is an initial guess and β^{**} is computed as the improved estimate. Observe that if $\beta^* = \hat{\beta}$ then $\beta^{**} = \hat{\beta}$ also. Except in special cases this method converges if iterated enough times. A good initial estimate of

β is the discriminant estimate given in the next section. Each time (8.12) is computed, the computer performs exactly the same operations required to solve a weighted least-squares problem.

This method is used in STATLIB to estimate $\hat{\beta}$ and the approximate distribution (8.11). Documentation is contained in the STATLIB manual. The "coefficients" recorded there are just the $\{\hat{\beta}_i\}$ from (8.10), the "estimated standard deviation" is the square root s_i of the i-th diagonal element of $\left(X'D(\beta)X\right)^{-1}$, and the t-statistic is the ratio $t_i = \hat{\beta}_i/s_i$. Strictly speaking, t_i does not have a t-distribution, but instead approximately has a normal distribution since the number of "degrees of freedom"

$$\delta = \sum_1^N n_i - k$$

must be large in order that (8.11) hold.

The "LOG LIKELIHOOD RATIO" (LLR) reported in the STATLIB printout may be doubled to get the equivalent of the "F-statistic." Strictly speaking,

$$2LLR/(k-1) \sim \chi^2_{k-1}/(k-1) \tag{8.13}$$

is distributed as $F_{k-1,\delta}$ with δ very large if $\beta_1 = \ldots = \beta_k = 0$. Unusually large values of 2LLR, in relation to tabled values of the chi-square distribution, indicate that at least one of the independent variables affects the probability of success.

A value R^2 also may be defined by analogy to more familiar situations (remember $R^2/(1-R^2) = (k-1)F/\delta$ from Chapter 6) as

$$R^2 = \frac{2LLR}{\delta + 2LLR} \quad . \tag{8.14}$$

In most logit situations R^2 will be very small because while one may "explain" a probability very well, unless the probability is near 0 or 1, the outcome y_i is not explained very well. A better notion of R^2 is needed [and is available, if each n_i is large, from the weighted least-squares output of the regression (8.7)].

8.4. DISCRIMINANT ANALYSIS

Oddly enough, although we don't regress binary variables $\{y_i\}$, y_i = 0 or 1 on the matrix X of independent variables to get meaningful estimates of the probabilities p_i = Ey_i, this procedure does lead to certain statistics needed for *discriminant analysis*, a type of logit regression that is appropriate in certain cases. Note that we insist y_i = 0 or 1, and do *not* let y_i range between 0 and n_i as it must for grouped data. The following method for estimating the regression coefficients in discriminant analysis only works for ungrouped data.

If we use an ordinary least-squares regression program to regress y on X linearly, then the prediction formula

$$\hat{y}_i = b_0 + b_1 x_{i1} + \ldots + b_k x_{ik} \tag{8.15}$$

is fitted and the estimated regression coefficients b = $(b_0, b_1, \ldots, b_k)'$ will be calculated and printed. For any new x the estimated formula for the "true" linear discriminant function is

$$\lambda(x) \equiv \log \frac{P(y=1|x)}{P(y=0|x)} = \log \frac{\pi_1}{\pi_0} + \sum_{j=1}^{k} b_j^* (x_j - \bar{x}_j^*) \tag{8.16}$$

with b_j^* = Nb_j/SSE, where SSE is the residual sum of squares from the regression. The meaning of π_0 and π_1 will be explained in the next paragraph. Let N_0 be the number of times that $y_i = 0$ and N_1 the number of times that $y_i = 1$. Then $N_0 = \Sigma(1-y_i)$, $N_1 = \Sigma y_i$, and $N = N_0 + N_1$ is

206

the total number of observations. Each $\bar{x}_j^*$ is a weighted mean of the values $x_{1j}, \ldots, x_{Nj}$ of the j-th independent variable. This average is

$$\bar{x}_j^* = \frac{1}{2} m_{j0} + \frac{1}{2} m_{j1} \tag{8.17}$$

with m_{j0} being the mean of the subset of N_0 values of x_{ij} for which $y_i = 0$ and m_{j1} being the mean of the subset of N_1 values of x_{ij} for which $y_i = 1$. Thus $m_{j0} = \sum_i (1 - y_i) x_{ij} / N_0$ and $m_{j1} = \sum_i y_i x_{ij} / N_1$. The logit function $\lambda(x)$ in (8.16) is the (estimated) linear discriminant function. This function is often used in medical and other applications to estimate the true probability [$\lambda^*(x)$ is defined in (8.19), the true logit],

$$p(x) \equiv P(y = 1 | x) = \frac{\exp(\lambda^*(x))}{1 + \exp(\lambda^*(x))}, \tag{8.18}$$

for a new value of x for which no y is observed, using historical data for which both y and X were observed. This is the analogue to the prediction problem using ordinary regression. Note here that even though the distribution of y depends on the X matrix, it is the logit function $\lambda(x)$, and not the probability $p(x)$, that is linear in x.

A medical application of this idea is as follows. In a sample of people the i-th person is coded as $y_i = 1$ if he has a certain kind of disease and $y_i = 0$ if not. The values $(x_{i1}, x_{i2}, \ldots, x_{ik})$ for that person would be certain laboratory measurements that might be used to diagnose the presence of the disease. A discriminant analysis (or maximum likelihood logit analysis) then estimates the probability that a person has the disease through formula (8.16). If a new patient asks for a diagnosis to determine whether he has the disease, the physician has the laboratory obtain the measurements x for that patient and then computes $p(x)$ from (8.18) and (8.16) as the probability that the patient has the disease. The values π_0 and π_1 that should be used are determined by the physician, where π_1 is defined as the probability that a patient chosen at random has the disease, and $\pi_0 = 1 - \pi_1$ is the probability he does not.

Discriminant analysis estimates for the logit curve are based on special assumptions about the distribution of the *independent* variables. We assume that $y_i = 1$ and 0 with probabilities π_1 and π_0, $\pi_0 + \pi_1 = 1$. Then the vector x_i (the i-th row of the X matrix, excluding the first column of units in X for a constant term) is assumed to be randomly distributed as multivariate normal with a covariance matrix Σ and mean vector μ_1 if $y_i = 1$ or μ_0 if $y_i = 0$. Although it is not apparent, the regression (8.15) computes what is needed to estimate μ_1, μ_0, and Σ from the data, b^* being the main quantity required. The regression does not determine the probability $\pi_1 = p(y = 1)$ in the absence of the data $(\pi_0 = 1 - \pi_1)$. If this value is not known, it sometimes is appropriate to choose $\pi_1 = N_1/N$, but only if the sample proportion of times $y_i = 1$ is thought to be an unbiased estimate of the true $P(y_i = 1)$.

The main point of the preceding paragraph is that the vector x is assumed to be stochastic (which it may be for logit regression also) and more importantly multivariate normal, or at least approximately so. (This can be quite restrictive.) Thus there is a risk in using discriminant analysis when the independent variables are not normal for reasons to be discussed soon.

An advantage of the linear discriminant function is that p(x) in (8.17) may be computed directly using OLS software, and for many examples the resulting estimates are quite close to the maximum likelihood logit estimates. The OLS t-statistics often are close to the logit t-statistics, and therefore it is cheaper to compute discriminant estimates for exploratory data analysis. And using b^* as the initial β^* in (8.11) leads to quickly converging estimates of the logit function. However, if $n_i > 1$ very often, the discriminant method can be more expensive computationally, since it cannot take advantage of this to reduce the number of rows of X. This difficulty can be corrected by estimating (8.15) using a weighted least-squares program.

Either (8.16) or maximum likelihood logit may be used to decide whether $y = 1$ or $y = 0$ is more likely for a given value of x, but (8.16) is better if the multivariate normal and other distributional assumptions for the x_i hold, while maximum likelihood logit is better if the x_i are nonnormal. In fact, for nonnormal x_i, (8.16) may be biased as $N \to \infty$; here the discriminant estimates are called "inconsistent," meaning that the estimation error does not become arbitrarily small as the sample becomes infinitely large. In this case, the variances of the regression coefficients b_j^* in (8.16) diminish to 0 as $N \to \infty$, but the b_j^* may not converge to the correct population values.

We will not prove the preceding assertions. However, the two-dimensional case with spherical symmetry $\Sigma = I$ (Figure 8.2) offers some insight, through geometry, that the discriminant procedure is a good one with normal populations. The "true" discriminant function for x having a multivariate normal distribution with covariance matrix Σ and mean μ_1 if $y = 1$, μ_0 if $y = 0$ is

$$\lambda^*(x) = \log(\pi_1/\pi_0) + (\mu_1 - \mu_0)'\Sigma^{-1}(x - \bar{\mu}), \quad \bar{\mu} \equiv (\mu_0 + \mu_1)/2 \qquad (8.19)$$

with

$$\pi_1 = 1 - \pi_0 = P(y = 1) .$$

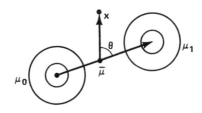

Figure 8.2 - Two (spherical) normal populations and
a new vector x to be classified as μ_0
or μ_1. $\Sigma = 1$.

Bivariate normal populations for x_1 and x_2 are plotted in Figure 8.2.
These explanatory variables are uncorrelated with the same variances.
It is clear in Figure 8.2 that x should be classified with the population that it is closest to (assuming that each normal population is
equally likely *a priori*, $\pi_0 = \pi_1 = \frac{1}{2}$). For x to be more likely to
have mean μ_1 than mean μ_0, the angle θ that the line $[\bar{\mu}, x]$ makes
with the line $[\bar{\mu}, \mu_1]$ must be between -90° and 90°, hence $\cos(\theta) > 0$.
Here $\bar{\mu} = (\mu_1 + \mu_0)/2$. $\cos(\theta) > 0$ means that the two-dimensional inner
product

$$(\mu_1 - \mu_0)'(x - \bar{\mu}) > 0 .$$

But this says $\lambda^*(x) > 0$ with λ^* given by (8.19) and $\pi_0 = \pi_1 = 0.5$.
Of course $\lambda^*(x) > 0$ if and only if $p(x) = P(y = 1|x) > \frac{1}{2}$ in (8.18).

8.5. A NUMERICAL EXAMPLE ILLUSTRATING FOUR KINDS OF LOGIT ESTIMATION

The data matrix in Table 8.7 was used to compute estimates of the
type A, B, C, D listed below.

A. Minimum logit chi-square estimates.

B. Maximum likelihood logit estimates (weighted data).

Here x_1 might be thought of as a scaled variable and x_2 as a dummy variable. Examples of data with this structure are (a) x_1 = yards away from goal, x_2 indicating AFL or NFL, and (b) x_1 = index of severity of chronic disease, x_2 indicating whether the respondent has a usual physician, Y = number of persons who refuse an offer to enroll in a Health Maintenance Organization, N = number asked. Note that N_0 = 17 and N_1 = 17 in this case.

Table 8.7

DATA FOLLOWING A PERFECT LOGIT RELATIONSHIP: $Y/(N-Y) = 2^{x_1+2x_2}$

i	Y_i (Successes)	N_i (Attempts)	x_{i1}	x_{i2}
1	1	9	-1	-1
2	2	3	-1	1
3	1	5	0	-1
4	4	5	0	1
5	1	3	1	-1
6	8	9	1	1

Since the total number of attempts is 34, we also may put this data set into a logically equivalent form with 34 rows with N = 1 in each case, by recording Y = 0 or 1 for each of the 34 persons, and retaining the x_1, x_2 for each. In this form the data are suitable for estimation by the

C. Maximum likelihood logit estimation method (unweighted data).

D. Discriminant analysis method.

Note that methods B and C theoretically give exactly the same results, but the others differ. The data were generated with

$$y_i/n_i = \frac{2^{x_{i1}+2x_{i2}}}{1 + 2^{x_{i1}+2x_{i2}}}, \quad i = 1, \ldots, 6$$

and therefore exactly follow a logit curve with $b_0 = 0$, $b_1 = \log(2) = 0.693$, $b_2 = 2 \log(2) = 1.386$. Methods B and C must give this answer, but method A won't because it adds half of a success and half of a failure to each of the six pairs (y_i, n_i), $i = 1, \ldots, 6$. Method D won't either because discriminant analysis is biased in this case, and the bias would not decrease even if the number of attempts were a million times larger for each group and the proportion of successes remained the same.

The results, obtained using the STATLIB regression packages, are summarized in Table 8.8.

Table 8.8

REGRESSION COEFFICIENTS, t-STATISTICS, R^2 and 2 LLR
(WHEN APPLICABLE) FOR DATA IN TABLE 8.7 OBTAINED
VIA FOUR ESTIMATION METHODS

Method	b_1	t_1	b_2	t_2	R^2	2 × LLR
A	0.614	75.68	1.114	167.68	0.999	–
B1 – iter 1	0.429	1.03	1.143	3.26	0.346	16.31
B3 – iter 3	0.693	1.33	1.386	3.26	0.359	16.93
C1 – iter 1	0.429	0.95	1.143	3.02	0.359	16.31
C3 – iter 3	0.693	1.23	1.386	3.02	0.359	16.93
D	0.773	1.22	2.061	3.88	0.445	–
True	0.693		1.386			

The results of both the first and final iterations for methods B and C are listed, as B1, B3, C1, C3 using STATLIB. If STATLIB started with discriminant analysis estimates then the initial values for method C would be those of D. "C-iter 1" or C1 (for iteration 1) shows that the first step isn't very close and is on the opposite side of the true values from the estimates using method D. The t-statistics for B3 and C3 should agree exactly (although they don't--perhaps for programming reasons). The comparison between D and C3 is adequate for exploratory purposes, at least in this case (compare the t-statistics). The t and R^2 statistics for A are much higher than the others, because the criterion there is to "explain" the ratios y_i/n_i (which is done nearly perfectly because of the way the data were generated) rather than the 34 individual 0-1 outcomes (which only can be done imperfectly). New statistics for method B could be defined and programmed to take advantage of this, but that hasn't been done. As it is, we have R^2 of about 0.36 when the fit is as perfect as possible because the data follow a perfect logit relationship. Note that although method A has a much higher R^2 than does B or D that it is not because of a better fit for A (the fit is worse) but because of different definition for R^2. The regression output for method D gave regression coefficients of 0.5 (constant term), 0.10714 (for x_1), and 0.28571 (for x_2). Since N = 34 observations and SSE was 4.7143, the values of b_1 and b_2 were computed as b_1 = 0.10714 × 34/4.7143 = 0.773 and b_2 = 0.28571 × 34/4.7143 = 2.061 from (8.16).

The cost comparisons are interesting for these data. Cost is roughly proportional to the number of observations times the number of iterations, this being the number of observations that the computer must process to obtain the estimates. This product is 6 for A, 18 for B, 102 for C, and 34 for D. Hence it is cheaper to work with grouped data when possible, and the requirement that discriminant analysis be run on 0-1 data when considerable grouping is possible causes it to be more costly than maximum likelihood logit on grouped data. Of course discriminant analysis is not intended for grouped data because that violates the assumption of normality for the independent variables.

Note that from (8.17) and the data in Table 8.7

$$m_{10} = -\frac{6}{17} \qquad m_{20} = -\frac{11}{17}$$

$$m_{11} = \frac{6}{17} \qquad m_{21} = \frac{11}{17} \qquad (8.20)$$

$$\overline{x}_1^* = \tfrac{1}{2} m_{10} + \tfrac{1}{2} m_{11} = 0, \ \overline{x}_2^* = \tfrac{1}{2} m_{20} + \tfrac{1}{2} m_{21} = 0$$

and since $N_1 = N_0 = 17$ we will estimate $\pi_1 = 0.5$. For methods B–C, at a new value of $x = (x_1, x_2)$, $P(y = 1 | x) = 0.5$ if (see Table 8.8)

$$0.693 \ x_1 + 1.386 \ x_2 = 0 \qquad (8.21)$$

and for method D, $P(y = 1 | x) = 0.5$ when $\lambda(x) = 0$ from (8.16) if

$$0.773 \ x_1 + 2.061 \ x_2 = 0 \ . \qquad (8.22)$$

The mean vectors (8.20) are plotted in Figure 8.3 as[*] $\mu_0 = (m_{10}, m_{20})$ and $\mu_1 = (m_{11}, m_{21})$ along with the lines that make classification equally likely, one for maximum likelihood (methods B–C, see (8.21)) and one for discriminant analysis (method D, see (8.22)). If a new value of $x = (x_1, x_2)$ falls above the line one decides that $y = 1$ because that is more likely, while below that $y = 0$ is more likely. The dashed line is perpendicular to the line connecting μ_0 and μ_1. Neither the line D nor B–C are perpendicular to $\mu_1 - \mu_0$ in this case. Note that the dashed line has intuitive appeal for the decision $y = 0$ or $y = 1$. In the shaded region, whether $y = 1$ or $y = 0$ is more likely depends on which estimation method was used. Lines where $P(y = 1 | x)$ is any number other than 0.5 are obtained by moving the lines D and B–C in

_____

[*]The mean vectors μ_0 and μ_1 are called the "centroids" for the $y = 0$ data and the $y = 1$ data, respectively.

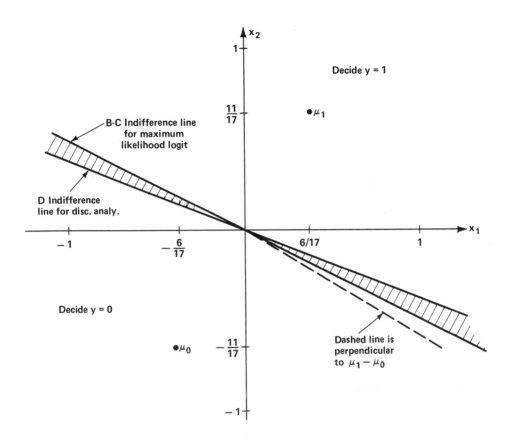

Figure 8.3 - Regions for classifying $x = (x_1, x_2)$ as corresponding to $y = 1$ or $y = 0$ using discriminant analysis (D) and maximum likelihood logit estimation (B–C).

Figure 8.3 up or down, but retaining the same slopes as shown here.

In concluding, logit analysis is the simplest example of a group of recently developed log-linear methods for analyzing binomial and multinomial distributions, and contingency tables. Like residual analysis, it is not yet a "classical" or widely used method, largely because it wasn't really feasible without high-speed computers. Most problems in applied statistics, at least in the social sciences and economics, involve relationships between many variables. When the dependent variable is categorical, these models and techniques now are available and ready for use. They are taking their place alongside the general linear model as the most important models for statistical analysis.

At this time the reader should refer to Figure 1.1 in Chapter 1 to determine the relationship of the various logit estimation techniques to the form and grouping assumptions made on the independent variables. That chart also provides an overview of the main models presented thus far.

Part III. ROBUSTNESS

Chapter 9

MODEL SPECIFICATION AND RESIDUAL PLOTTING

Thus far we have taken our model as given and made inferences about parameters in it. This chapter describes several tools that are useful to the data analyst who is using the data to decide on his statistical model rather than taking the model as given *a priori*. Section 9.1 covers probability plots; a method that is used to assess how consistent the data are with the assumed parent probability distribution. Section 9.2 presents methods, graphical and otherwise, for choosing how to specify the dependent variable and the independent variables in a regression model so that the required assumptions can be met. Section 9.3 covers weighted least squares fitting of regression models--a necessary step when the methods in Section 9.2 are not totally successful.

9.1. PROBABILITY PLOTTING

We turn now to examine the informal, but frequently informative method of probability plotting for fitting distributions to data. Probability plots give a visual measure of how consistent the data is with the assumed parent distribution. These plots may be used to check whether the data are normal and if not, how the true parent distribution differs from a normal distribution--heavier tailed, lighter tailed, bimodal, etc. The plots can also be used to estimate the parameters of the parent distribution.

The basic fact which motivates the use of probability plots is that as n increases the cumulative distribution function of the data converges to the actual cumulative distribution function (CDF) of the process from which the data are generated. For example, if Q is the p-th quantile (i.e., F(Q) = p where F is the true cumulative distribution function), then the proportion of sample items less than Q (call it $\hat{F}(Q)$) times n has a binomial distribution with parameters (n,p). Thus its variance is p(1-p)/n, and so it converges to p in probability as n becomes large. So a reasonable step in answering the question of whether F is the true cumulative distribution function from which the sample was drawn is to plot the empirical CDF and the actual CDF together: F(x) and F(x) against x, or F(x) against F(x). (The latter is called a P-P plot.) Each of these will convey important information.

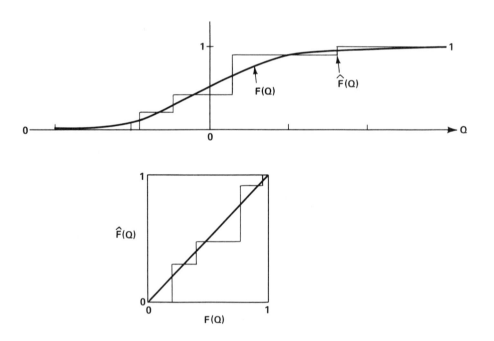

Figure 9.1 - Sample comparisons between theoretical and actual cumulative distribution functions.

These plots have two serious drawbacks which have precluded their heavy use as diagnostic tools. First, they test not only the shape of the empirical distribution, but also its location and scale as well. So, for example, if you want to decide if some data looks normal, you would first have to estimate μ and σ. The second disadvantage of these plots is that they do not adequately capture the variation in curves in the vicinity of $F(Q) = 0$ or $F(Q) = 1$. In practice, the major concerns frequently lie there. Outliers rarely contaminate more than 10% of the data. Thus, in investigating whether the empirical distribution is normal, the extent of its non-normality between the 0-5 and 95-100 percent points is of great interest. A plot which "unsquishes" the data in those regions would be more useful.

Probability plots do not have these drawbacks. They are plots, called Q-Q plots, of the empirical quantiles (y-axis) against the theoretical quantiles (x-axis): i.e., $\hat{F}^{-1}(p)$ against $F^{-1}(p)$ for $0 \leq p \leq 1$. Recall from Chapter 2 that some minor ambiguity remains due to the discreteness of $\hat{F}^{-1}$. Thus for simplicity, the sampled points $(\hat{F}^{-1}(p))$ are the order statistics $X_{(1)}$, $X_{(2)}$, ..., $X_{(n)}$ and the corresponding points of the true distribution $F^{-1}(p)$ are $F^{-1}(\frac{.5}{n})$, $F^{-1}(\frac{1.5}{n})$, ..., $F^{-1}(\frac{n-.5}{n})$.

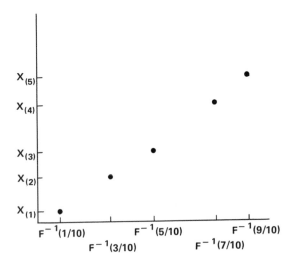

Figure 9.2 - Sample probability plot (Q-Q plot) for n = 5.

For those to whom the notation F^{-1} (reads the inverse function of F) is unfamiliar, think of the graph F^{-1} as the graph F except that the x and y axes are reversed. For example, F^{-1} for a normal distribution is given in Figure 9.3. The reasoning behind plotting the order statistics $X_{(1)} \leq X_{(2)} \leq \cdots \leq X_{(n)}$ against $F^{-1}(.5/n)$, $F^{-1}(1.5/n)$, ..., F((n-.5)/n) is as follows. If the data are a random sample from the distribution function F, the values of $X_{(i)}$ will (by definition), *tend* to line up exactly on the 45° line when $E_F(X_{(i)})$ is plotted on the x-axis. Probability plots use the approximate relationship

$$E_F(X_{(i)}) \cong F^{-1}(\frac{i-.5}{n}).$$

222

To see why this is true, take n = 15 and i = 8. The median of the sample is $X_{(8)}$, so we expect half the observations to be below $X_{(8)}$ which corresponds to $F^{-1}(\frac{1}{2})$ since $\frac{i-.5}{n} = \frac{7.5}{15} = \frac{1}{2}$.

All of the above assumes that there is a single distribution function F of interest. In most situations we use probability plots to determine whether the true parent distribution of the data is one of a parametric family--most commonly the normal family. Specifically, suppose we denote by Φ^{-1} the inverse cumulative distribution function of a standard normal distribution. Now if the data $X_{(1)} \le \cdots \le X_{(n)}$ are from a normal (μ, σ^2) distribution (F), then

$$E_F(X_{(i)}) = \mu + \sigma E_\Phi(Y_{(i)}) \doteq \mu + \sigma \Phi^{-1}(\frac{i-.5}{n})$$

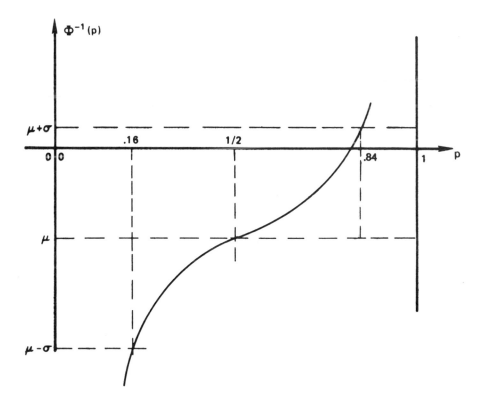

Figure 9.3 - The inverse normal distribution function.

223

where $Y_{(i)}$ are the order statistics from a standard normal distribution. Hence a plot of $X_{(i)}$ against $\Phi^{-1}(\frac{i-.5}{n})$ should tend to lie on a straight line with slope σ and intercept μ. Thus for simplicity we use the quantiles of a standardized form of the distribution on the x-axis. Judging whether the data lie on a straight line and hence are consistent with coming from the parameterized family is insensitive to the location and scale of the data. Clearly, the plots accomplish the objective of making the tails of distribution more visible.

Much has been written about normal probability plots. In particular, Shapiro and Wilk[*] have devised a formal test of normality with constant variance based on how far the "best" estimate of slope is from the usual estimate $\hat{\sigma} = \sqrt{\Sigma(X_{(i)}-\bar{X})^2/(n-1)}$. But in practice, it is extremely difficult to compute this "best" estimate of slope. Ordinary least squares is inappropriate because the order statistics are correlated and do not have constant variance (the largest observation is more variable than the median); computation of these variances and correlations is tedious, but necessary for this task. Also as indicated above, $F^{-1}(\frac{i-.5}{n})$ is not quite the right quantity upon which to regress $X_{(i)}$: the independent variable should be $E(X_{(i)})$ instead, which is much more difficult to compute. So, not surprisingly, most people do not opt for the formalism of testing in this situation.

---

[*] Shapiro, S. S., and M. B. Wilk, 1965, "An Analysis of Variance Test for Normality (complete samples)," *Biometrika*, Vol. 52, pp. 591-611.

A less formal, but highly informative, approach to interpreting normal probability plots is to ask whether the data (or residuals from a regression) lie roughly on a straight line. If they do, you declare the distribution to be normal, and proceed with fitting the linear model as prescribed by the assumption of errors being normally distributed with constant variance. If not, the question of what distribution the errors are most consistent with can be answered, as follows. First, if the quantiles curve toward the left and right-hand edges are as in Figure 9.4,

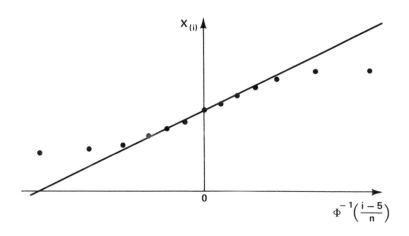

Figure 9.4 - Sample normal probability plot for a distribution with shorter-than-normal tails.

the distribution has shorter-than-normal tails, and thus, all observations can safely enter into the averaging; i.e., least-squares is perfectly acceptable. If the quantiles curve toward the top and bottom edges of the plot as in Figure 9.5,

225

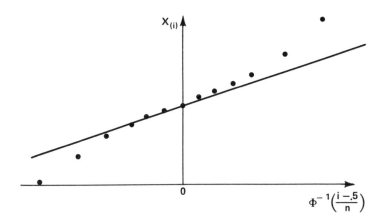

Figure 9.5 - Sample normal probability plot for a distribution with longer-than-normal tails.

the distribution has longer-than-normal tails, and it is dangerous

to average.   The procedures discussed in Chapter 10 can provide the

needed protection.

The above discussion has concentrated on deciding whether the

parent probability distribution is normal since normal models are

so widely used.   Probability plots are also available for a wide

variety of other distributions, one of which we consider below.

Example.   Interarrival time of fire alarms.[*]

Table 9.1 gives the time to the nearest minute between arrivals

of successive fire alarms at the dispatching center during one hour

of a particular period and region in New York City.

---

[*]See W. E. Walker et al. (eds.), *Fire Department Deployment
Analysis*, Chapters 7 and 8 (Elsevier-North Holland, 1979) for some
background on statistical models of fire alarm arrivals.

Figure 9.6 shows a normal probability plot of these points. It is difficult to make judgments about the shape of the distribution with only sixteen points that are rounded to the nearest integer. The points in Figure 9.6 do not seem to lie on a straight line, although it is difficult to tell. The theoretical reasoning given below leads to postulating an exponential distribution for the interarrival times. Figure 9.7 gives an exponential probability plot. Again it is hard to tell, but the points seem to lie closer to a straight line.

Table 9.1

SIXTEEN INTERARRIVAL TIMES OF FIRE ALARMS
(in minutes)

1	1	1	10
4	1	3	0
2	4	5	5
5	2	6	5

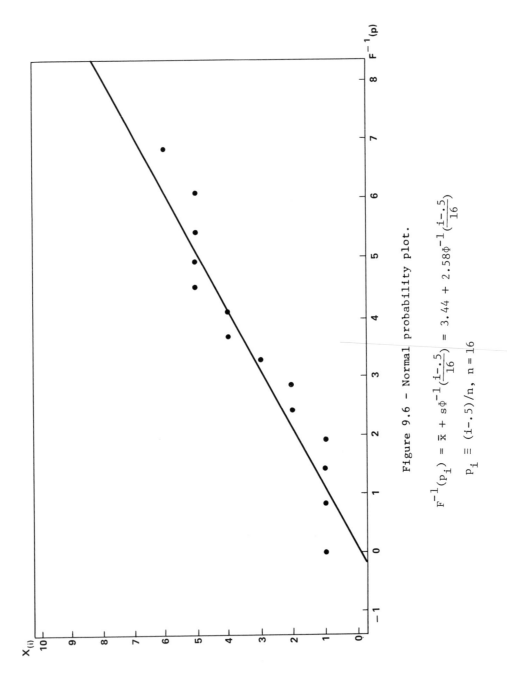

Figure 9.6 - Normal probability plot.

$$F^{-1}(p_i) = \bar{x} + s\Phi^{-1}(\frac{i-.5}{16}) = 3.44 + 2.58\Phi^{-1}(\frac{i-.5}{16})$$

$$p_i \equiv (i-.5)/n, \quad n = 16$$

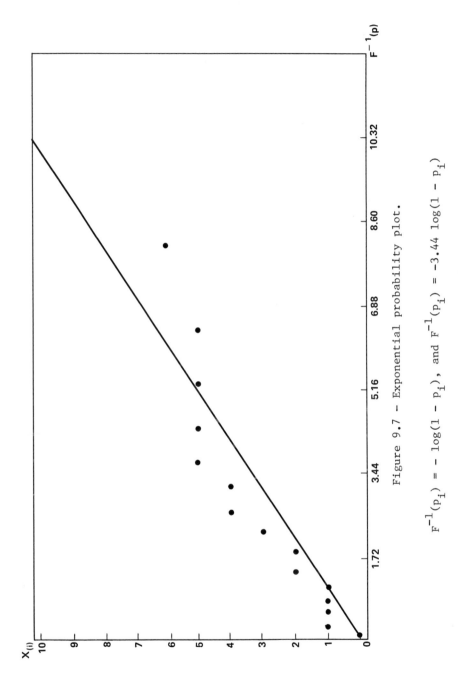

Figure 9.7 – Exponential probability plot.

$$F^{-1}(p_i) = -\log(1 - p_i), \text{ and } F^{-1}(p_i) = -3.44 \log(1 - p_i)$$

$$p_i \equiv (i - .5)/n, \quad n = 16$$

Some theoretical reasoning may lead us to something other than the normal distribution for interarrival times of fire alarms. Think of labeling the time of the last alarm as zero and dividing the time axis up into small intervals of length Δt. Assume that alarm arrivals in successive intervals are independent with the same distribution and that the probability of more than one alarm as an interval Δt is negligible. Thus, we think of flipping a coin with probability $\lambda \Delta t$ of heads independently for each little interval. If a head comes up, put an alarm in that interval. In order for the time until the next alarm to be greater than $k\Delta t$, we must have at least k tails in a row. This event has probability $(1-\lambda\Delta t)^k$. Thus, if X is the time until the arrival of the next alarm, the probability that X is greater than t is

$$P(X > t) = (1 - \lambda\Delta t)^{t/\Delta t}$$

A standard result, by working with the logarithm, is

$$\lim_{\Delta t \to 0} (1 - \lambda\Delta t)^{t/\Delta t} = e^{-\lambda t}.$$

Therefore,

$$P(X > t) = e^{-\lambda t}$$

and

$$P(X \leq t) = 1-e^{-\lambda t}.$$

We say a random variable has an exponential distribution with parameter λ if its distribution function is $1-e^{-\lambda t}$. The exponential density is obtained by differentiating so is given by

$$f_X(t) = \frac{\partial}{\partial t}(1-e^{-\lambda t}) = \lambda e^{-\lambda t} \; .$$

Therefore, if fire alarms are generated independently with constant probability, we would expect our data to be exponentially distributed rather than normally distributed. It is easy to show that the mean and variance of an exponential random variable X are given by

$$E(X) = 1/\lambda$$

$$Var(X) = 1/\lambda^2$$

A probability plot for the exponential distribution in standardized form will plot $X_{(i)}$ against $F^{-1}(\frac{i-.5}{n})$ where $F(x) = 1-e^{-x}$ so that $F^{-1}(\frac{i-.5}{n}) = -\log(1 - \frac{i-.5}{n})$. Figure 9.7 gives an exponential probability plot for the data. The slope of a line through the data estimates the scale parameter $1/\lambda = E(X)$ by $\overline{X} = 3.44$.

This section has given an alternative, probability plotting, to using formal significance tests in determining which particular parametric formula of distributions are consistent with data. When the data do not appear to be consistent with a member of the parametric family being the parent distribution, the plots give some indication of where to look for the correct parent distribution.

231

9.2. CHOOSING A REGRESSION MODEL

Thus far we have taken the form of our regression equation as given. That is, if we wish to predict a dependent variable Y from two independent variables x_1 and x_2, we fit a linear equation by least squares so that our prediction of Y is

$$\hat{Y} = b_0 + b_1 x_1 + b_2 x_2 .$$

In practice, the form of the fitted equation sometimes is determined by *a priori* considerations. For example, if we wish to predict adult weights from heights we might expect weight to be approximately proportional to the cube of height. This follows from thinking of weight as equivalent to volume and assuming that people are all proportioned about the same. Thus, we might let Y = weight and X = (height)^3 and fit the equation

$$Y = \beta X + \varepsilon$$

to our height-weight data. More frequently, however, we begin with data on several variables and we wish to model how one variable depends on the other variables. If we fit a least squares line and the data is a sample from a larger population, inferences based on the fitted line about the relationship between the variables in larger populations may be incorrect unless the usual assumptions for the multiple regression models are valid. Those assumptions (from Chapter 6) are: (i) the relationship is linear; (ii) the errors are independent with normal distributions having mean zero and constant variance. One goal of this section will be the selection of the appropriate independent variables and dependent variables so that these assumptions are sufficiently well satisfied that we can confidently use a linear regression model. Some relevant discussion is given in Section 4.5.

The model building process has many facets, most of which we will not attempt to cover in this section. The process may begin with constructing a statement of the problem which motivates the building of the model. The problem statement would include the specification of a possible response variable, and a list of potential independent variables. Frequently the potential independent variables are not directly measureable so substitutes or proxies must be found. If one is designing an experiment, the selection of levels at which to measure the independent variables must be made. The costs and benefits of measuring different possible variables must be weighed. After the data have been gathered, one must decide not only which variables to use in the model, but what functional form to use.

The analyst's goal should be to choose a model which appears to explain a body of data. He runs two kinds of risk in this effort, the first being that he chooses so few variables that the model fails to fit the data adequately. Alternatively, he might construct a model with so many parameters that the analysis is unnecessarily complex. We should opt for parsimony in parameters. The parsimony principle[*] says that parameters should be introduced sparingly in such a way that the maximum amount of resolution is achieved for each parameter introduced. We will

_____

[*] See J. W. Tukey, "Discussion, Emphasizing the Connection Between Analysis of Variance and Spectrum Analysis," Vol. 3, No. 2, *Technometrics*, May 1961, for a discussion of this idea.

see that suitable transformations of the data frequently yield a parsimonious parameterization.

Consider a sample of twelve individuals with their annual wages (the dependent variable), years of experience, and IQ scores. These data, their predicted values and residuals, are given in Table 9.2. After examination of various plots and cross-tabulations of the data, suppose we begin by fitting a least-squares line to the data, and obtain

$$\hat{y} = 106.5 + 23.768x_1 - 80.852x_2 \qquad (9.1)$$

Table 9.2

REGRESSION OF WAGES ON EXPERIENCE AND IQ
(Formula 9.1)

i	Wages (thousands \$) y_i	Experience x_{i1}	IQ/100 x_{i2}	Predicted Value $\hat{y}_i$	Residuals $\hat{e}_i = y_i - \hat{y}_i$
1	9.375	0	1.0	25.980	-16.600
2	15.880	0	1.2	9.874	6.005
3	14.370	0	1.2	9.874	4.499
4	24.730	0	1.4	-6.231	30.960
5	13.280	1	1.0	49.750	-36.470
6	19.070	1	1.2	33.640	-14.570
7	20.820	1	1.2	33.640	-12.820
8	31.690	1	1.4	17.540	14.150
9	166.300	2	1.0	73.510	92.780
10	26.610	2	1.2	57.410	-30.800
11	25.690	2	1.2	57.410	-31.720
12	35.900	2	1.4	41.300	-5.403

The Examination of Residuals

When performing the regression analysis in (9.1), we made certain assumptions about the errors. If the model is correct, the residual errors e_i should be (approximately) independent, have zero mean, constant variance, and be normally distributed. Thus, we examine the data

234

and fitted model to determine

(i) whether the data appear to be consistent with the model,

(ii) or not, whether the assumptions are violated in a particular way.
In case (ii), we will want to change the model. The primary tool for
checking the model's assumptions and for respecifying the model will be
plots of the residuals.

Outliers. Our first step is to make an overall plot of the residuals
as in Figure 9.8. If the model is correct, then the plot should resemble
12 observations from a normal distribution with a zero mean. It is apparent
from this plot that at least one residual, 92.78, is much different from
the other 11. This residual is called an *outlier* since it is a data point
that is not typical of the rest of the data. It is clear from the plot
that this outlier could not have come from the same normal distribution
as the other 11 residuals. Outliers should be examined closely to see if
reasons for their peculiarity can be found.

Figure 9.8 – Residual Plot.

In this case, it turned out that the decimal point in the ninth data
point was misplaced so that recorded annual wage was ten times the actual
wage. The reason for an outlier is not always this easy to determine and
correct. If an outlier cannot be corrected, it should be removed and the
data reanalyzed without it. But care should be exercised in the removal
of outliers lest the analyst's own biases creep into the decision to reject

235

it.[*] Outliers can provide important information when follow-up shows that they arise from an unusual combination of circumstances which are of vital interest to the research.[**]

After the outlier has been corrected or removed, the regression should be recomputed (Table 9.3) and another plot of residuals made (Figure 9.9). The new equation is given by

$$\hat{y}_i = -36.92 + 5.06x_{i1} + 44.19x_{i2} \tag{9.2}$$

and is substantially different from (9.1).

Table 9.3

RECOMPUTED REGRESSION OF WAGES ON EXPERIENCE AND IQ
(Formula 9.2)

i	Wages (thousands $) y_i	Experi- ence x_{i1}	IQ/100 x_{i2}	Predicted Value $\hat{y}_i$	Residuals $\hat{e}_i = y_i - \hat{y}_i$
1	9.375	0	1.0	7.271	2.104
2	15.880	0	1.2	16.110	-0.231
3	14.370	0	1.2	16.110	-1.737
4	24.730	0	1.4	24.950	-0.223
5	13.280	1	1.0	12.330	0.945
6	19.070	1	1.2	21.170	-2.102
7	20.820	1	1.2	21.170	-0.349
8	31.690	1	1.4	30.010	1.677
9	16.630	2	1.0	17.390	-0.762
10	26.610	2	1.2	26.230	0.382
11	25.690	2	1.2	26.230	-0.539
12	35.900	2	1.4	35.070	0.832

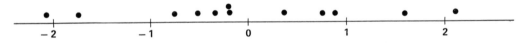

Figure 9.9 - Corrected Residual Plot.

---

[*] See Belsey, Kuh and Welch (1980) or Weisberg (1980) for a systematic account of how to analyze residuals from regression models.

[**] See R. Klitgaard and G. Hall (1973), *A Statistical Search for Unusually Effective Schools*, The Rand Corporation, R-1210-CC/RC, for an example of where the only points of interest were the outliers.

The new plot does not appear to be abnormal for a sample of twelve observations from a normal distribution. The reader can get more experience with how such a plot should look by drawing normal random deviates from a table (The Rand Corporation, *A Million Random Digits with 100,000 Normal Deviates*, The Free Press, Glencoe, Illinois, 1955) and plotting them. Formal tests also may be made, but usually this is unnecessary. When in doubt, normal probability plots (Sec. 9.1) of the residuals should be made.

Independent Variables

Plots of residuals against independent variables can be useful in improving the specification of the independent variables in the regression equation. Figure 9.10 shows three possible patterns the residuals might fall in when plotted against the values of a single independent variable. The plots indicate the following:

(i) Plot looks as it should.

(ii) The variance is not constant but increases with the value of x. As discussed later, a weighted least-squares analysis or a transformation of the dependent variable should be used.

(iii) A quadratic term in x is needed.

Figure 9.11 shows a plot of residuals from Table 9.3 against x_1, experience. We also can plot the residuals against the value of $x_2(IQ/100)$ as shown in Figure 9.12. The inclusion of x_2^2 *might* be suggested from this plot. Including x_2^2 and refitting yields a new equation

$$\hat{y}_i = 17.21 + 5.06x_{i1} - 47.30x_{i2} + 38.12x_{i2}^2 . \tag{9.3}$$

(Note that the coefficient of x_{i1} is unchanged from (9.2), because x_2 and x_2^2 are uncorrelated with x_1.) Replotting (not shown) of residuals against x_2 yields a satisfactory pattern.

237

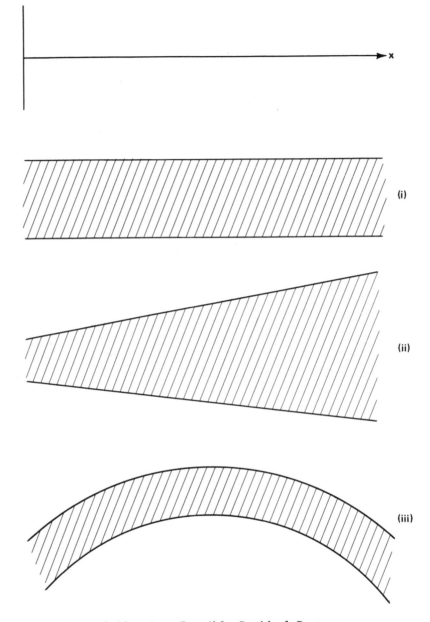

Figure 9.10 - Some Possible Residual Patterns.

238

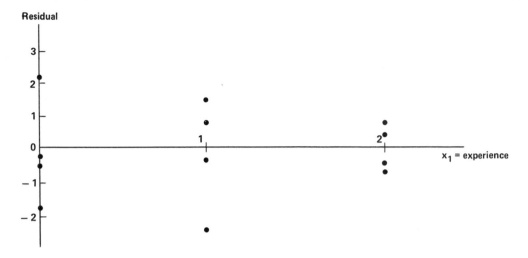

Figure 9.11 - Residual plot (using Equation (9.2),
Table 9.3) vs. experience.

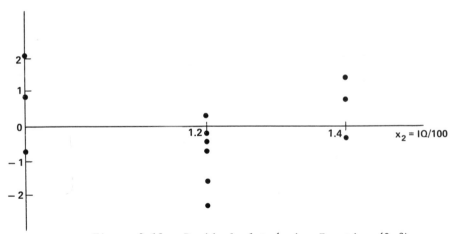

Figure 9.12 - Residual plot (using Equation (9.2),
Table 9.3) vs. x_2 = IQ/100.

Besides plotting residuals against the values of the independent variables, we can plot the residuals against the predicted values. This plot is shown in Figure 9.13 for Equation (9.3). The residuals seem mildly U-shaped, so as in Chapter 4, we might try including an interaction term. The resulting fitted equation is

$$\hat{y}_i = 23.09 - 0.82x_{i1} - 52.20x_{i2} + 38.12x_{i2}^2 + 4.90x_{i1}x_{i2} \ . \quad (9.4)$$

Table 9.4 summarizes this regression equation. The residuals from (9.4) are replotted in Figure 9.14. Note that all the residuals now are much smaller. The equation now is the best we can easily obtain without attempting to transform the dependent variable.

The Dependent Variable

The preceding model uses five parameters (b_0, ..., b_5) to predict y_i. Sometimes it is possible to use fewer parameters and simultaneously to get a better fitting equation that satisfies the assumptions by transforming the dependent variable (and possibly the independent variables, too). This was done in Chapter 4 when the log transformation was used successfully. Here we might hope to find a nonlinear transformation g so that

$$g(Y) = \beta_0 + \beta_1x_1 + \beta_2x_2 + \varepsilon$$

where ε has a normal distribution with mean zero and constant variance.

The most popular nonlinear transformations g of the dependent variable y if y > 0 are

$$g(y) = y^2, \ y \ (\text{no transformation}), \ \sqrt{y}, \ \log(y), \ \text{and} \ 1/y \ .$$

One may add a positive constant to y before making the transformation if this is necessary to make the transformation be properly defined

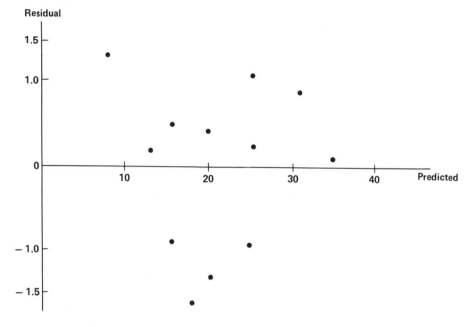

Figure 9.13 – Residual plot vs. predictions (Eq. 9.3).

Table 9.4

QUADRATIC REGRESSION WITH INTERACTION OF EXPERIENCE AND IQ
(Equation 9.4)

i	Wages (thousands $) y_i	Experi- ence x_{1i}	IQ/100 x_{2i}	Predicted Value $\hat{y}_i$	Residuals $\hat{e}_i = y_i - \hat{y}_i$
1	9.375	0	1.0	9.014	0.361
2	15.880	0	1.2	15.350	0.532
3	14.370	0	1.2	15.350	-0.974
4	24.730	0	1.4	24.730	-0.005
5	13.280	1	1.0	13.090	0.183
6	19.070	1	1.2	20.410	-1.339
7	20.820	1	1.2	20.410	0.414
8	31.690	1	1.4	30.770	0.915
9	16.630	2	1.0	17.170	-0.544
10	26.610	2	1.2	25.470	1.145
11	25.690	2	1.2	25.470	0.224
12	35.900	2	1.4	36.810	-0.910

and increasing throughout the range of y. Each of these possible choices for the dependent variable would be regressed on the independent variable, and the choice with the best residual plot against the predicted variable would be preferred to the others provided, although other factors may enter into this choice too.

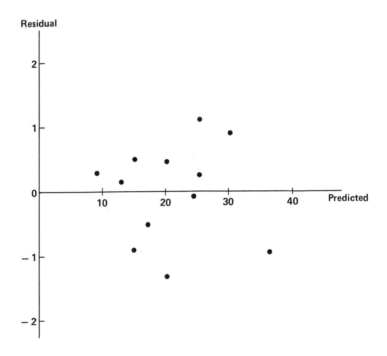

Figure 9.14 - Residual plot vs. predictions (Eq. 9.4).

We briefly digress to give a theoretical basis for choosing one or another of these transformations.[*] Suppose that a random variable Y has a distribution with mean m and variance $\sigma^2(m)$. Then if g is a transformation of Y, the random variable g(Y) has approximately a

_____

[*]The procedure described here is an informal version of the Box-Cox power transformation method. See G.E.P. Box and D. R. Cox, "An Analysis of Transformation," *Journal of the Royal Statistical Society*, Series B, 1964, for a complete description.

distribution with mean $g(m)$ and variance $(g'(m)\sigma(m))^2$. See C. R. Rao, *Linear Statistical Inference and Its Applications*, 2d Ed. (Wiley, 1965) for precise statement and proof. We are searching for a transformation $g(Y)$ that makes the variance independent of the mean (variance stabilizing). We want $\text{Var}(g(Y)) = c$, a constant. That is, $g'(m)\sigma(m) = \sqrt{c}$. If we restrict ourselves to standard deviations which are powers of the mean, $\sigma(m) = \text{constant} \times m^p$, we can solve for g to get:

$$g'(m) = \frac{\text{constant}}{m^p} \quad \text{or} \quad g(m) = \text{constant} \times \frac{m^{1-p}}{1-p} \quad \text{if } p \neq 1.$$

If $p = 1$, $g(m) = \text{constant} \times \log m$. Thus, the random variable $g(Y) = Y^{1-p}$ will have approximately constant variance, and our rule is to transform Y by $g(Y) = Y^{1-p}$ if $p \neq 1$ and $g(Y) = \log Y$ if $p = 1$. An easy way to remember this is to take $g(y) = \frac{Y}{\sigma(Y)} \propto \frac{Y}{Y^p} = Y^{1-p}$.

A nonlinear transformation of the dependent variable necessarily affects how the independent variables in the regression should be specified. Thus, the process of searching for a regression model specification is an iterative one. For each dependent variable determined by a transformation, the process of analyzing residuals and trying different specifications of the independent variables should be carried out. In deciding which dependent variable to use, the analyst should, all other things being equal, opt for the most parsimonious parameterization--the equation with the fewest independent variables.

We illustrate the interaction between choice of dependent variable and choice of independent variables by our example of 12 points. Suppose we start with Equation (9.12) and first look for alternative dependent variables while keeping x_{i1} and x_{i2} as the independent variables.

243

Table 9.5 gives the results of fitting each transformation to a regression on x_{i1} and x_{i2}. Plots of the residuals against the predictions are given for y, $\sqrt{y}$, and log y in Figure 9.15. Other plots shown can be made of residuals against x_{1i} and x_{2i} for each transformation. The independent variable plots for $\sqrt{y_i}$ looked acceptable (unlike those for y_i in Figures 9.11 and 9.12) or for the other transformations (not shown here). Also the plot of residuals against predicted values for $\sqrt{y_i}$ in Figure 9.6 looks quite consistent with the constant variance assumption--there is no apparent variation in the absolute size of the residuals and as the predicted values vary. Note that the corresponding plots for y_i and log y_i do not have as consistent a pattern.

Thus for our data, the square root transformation fits best:

$$\widehat{\sqrt{y_i}} = -1.85 + 0.56x_{i1} + 4.84x_{i2}.$$

This happens to correspond to the highest R^2 for the five different models. It is not unusual for the best R^2 to determine the best transformation of the dependent variable, but examination of the residual patterns and correctness of the model assumptions provide the overriding criterion.

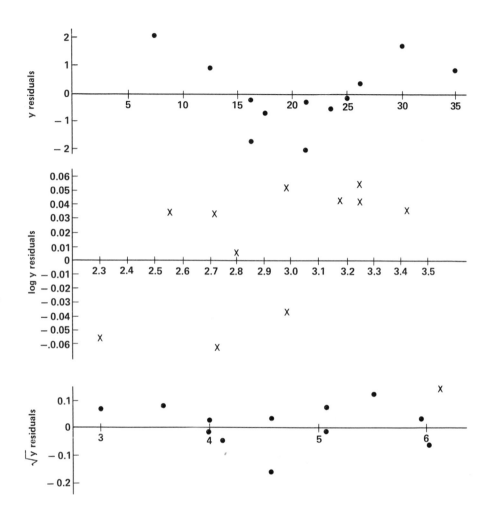

Fig. 9.15--Residuals vs. predictions for three dependent variables.

Table 9.5

REGRESSIONS AND RESIDUALS FOR FIVE DEPENDENT VARIABLES

y^2		y		$\sqrt{y}$		$\log(y)$		$1/y$	
Residual	Prediction	Residual	Prediction	Residual	Prediction	Residual	Prediction	Residual	Prediction
198	-110	2.10	7.27	0.0651	2.997	-0.0569	2.2949	0.0143	0.0924
-32	284	-0.23	16.11	0.0199	3.965	0.0353	2.7298	-0.0057	0.0687
-77	284	-1.74	16.11	-0.1743	3.965	-0.0646	2.7298	0.0009	0.0687
-66	678	-0.22	24.95	0.0395	4.933	0.0434	3.1646	-0.0045	0.0450
65	112	0.95	12.33	0.0866	3.557	0.0360	2.5503	-0.0027	0.0780
-142	506	-2.10	21.17	-0.1590	4.526	-0.0370	2.9852	-0.0019	0.0543
-72	506	-0.35	21.17	0.0370	4.526	0.0508	2.9852	-0.0063	0.0543
105	900	1.68	30.01	0.1352	5.494	0.0360	3.4200	0.0010	0.0306
-57	334	-0.76	17.39	-0.0403	4.118	0.0055	2.8057	-0.0035	0.0636
-20	728	0.38	26.23	0.0718	5.087	0.0407	3.2405	-0.0023	0.0399
-68	728	-0.54	23.23	-0.0181	5.087	0.0056	3.2405	-0.0010	0.0399
167	1122	0.83	35.70	-0.0633	6.055	-0.0947	3.6754	0.0117	0.0162

Prediction =
-2080 + 221.9x₁
+ 1970 x₂

SSE = 118.92

R² = 0.912378

Prediction =
-36.92 + 5.06x_{i1}
+ 44.19x_{i2}

SSE = 1.395

R² = 0.97469

Prediction =
-1.85 + 0.56x_{i1}
+ 4.8 x_{i2}

SSE = 0.10547

R² = 0.987852

Prediction =
0.12053 + 0.25583x₁
+ 2.17437x₂

SSE = 0.055535

R² = 0.983519

Prediction =
211 - 0.0144x₁
- 0.1185x₂

SSE = 0.00717

R² = 0.9157

Summary

   We have concluded that fitting the square root of wages with
three parameters is superior to fitting wages with five parameters.
The principle of parsimony of parameters dictates that the fit of
the square root with fewer parameters is preferable. In fact, these
data were generated using the model

$$\sqrt{y_i} = -2 + 0.5x_{i1} + 5x_{i2} + e_i \qquad\qquad (9.5)$$

where we drew the e_i from a table of random normal deviates with mean
zero and standard deviation = 0.1. Our final fitted equation

$$\sqrt{y_i} = -1.85 + 0.56x_{i1} + 4.84\ x_{i2} + \hat{e}_i$$

clearly reflects the method generating the data. Equation (9.5) shows
that a model could be fitted directly to the y_i's. The appropriate
representation of the y_i themselves follows from squaring (9.5) to get

$$y_i = (-2 + 0.5x_{i1} + 5x_{i2})^2 + 2(-2 + 0.5x_{i1} + 5x_{i2})e_i + e_i^2$$

$$= 4 - 2x_{i1} - 20x_{i2} + 0.25x_{i1}^2 + 25x_{i2}^2 + 5x_{i1}x_{i2}$$

$$+ (-4 + x_{i1} + 10x_{i2})e_i + e_i^2 . \qquad\qquad (9.6)$$

This model not only contains six parameters but the last two quantities,
which involve the errors, no longer satisfy the assumption of the errors
having mean zero and constant variance. The errors will have mean
zero if we use $e_i^2 - (0.1)^2$ rather than e_i^2 in the last term of (9.6) and
add the $(0.1)^2$ to the constant term. The reader should compare this
equation with the fitted Equation (9.4) to see how good an estimate
of (9.6) the data gave. The remaining barrier to fitting a six parameter least-squares line to the y_i's is the fact that the errors:

$$e_i' = (-4 + x_{i1} + 10x_{i2})e_i + e_i^2 - 0.01$$

are not normal and do not have constant variance since $Ee_i' = 0$,
$Var(e_i') = (-4 + x_{i1} + 10x_{i2})^2 \, 0.01 + 0.0002$. Another remedy to this
difficulty is to use weighted least squares in fitting the least squares
line. This is covered in Section 9.3. In this procedure, since some
of the observations are "less reliable" than the others, we fit the
line which minimizes a weighted sum of squared residuals. The weights
are proportional to the reciprocal variances and of course the formulas
for the regression coefficients, their standard errors, etc., are
changed.

In summary, we have covered a number of different methods for
fitting models to data. Plots of residuals are our main tool. Simple
plots help us detect outliers and check for normality. Plots of
residuals against the independent variables or the predicted values
may lead to transformations of independent variables or the inclusion
of additional functions of the independent variables such as higher
order terms or interaction terms in the equation. Transformations of
the dependent variable may also be suggested from the residuals and
predicted variables. The appropriate dependent variable transformations change the model and stabilize the residual variance. When the
model fits but the residual variance is not constant, weighted least
squares, as described below, can be used to fit the model to the data.
No all-purpose prescription dictates when and in what order to use
these methods, but considering many different plots of the data helps.
Common sense, past experience, and theory also are invaluable.

9.3. HETEROGENEITY OF VARIANCE AND WEIGHTED LEAST SQUARES

In the previous section, we used transformations of the dependent variable to produce regression equations with error terms having constant variance. It sometimes happens that we have the correct model for E(Y) but the errors have unequal variances. Transforming Y may tend to equalize the error variance but lead to an incorrect model for E(Y). We use *weighted least squares* to fit regression models when some of the observations are "less reliable" than others.

Example 1. Frequently data are reported by averaging all observations y corresponding to a particular value of the independent variable. A very simple model for the individual y's might be

$$Y_{ij} = \beta x_i + \epsilon_{ij}; \quad i = 1, \ldots, m; \quad j = 1, \ldots, n_i \tag{9.7}$$

where $\epsilon_{ij} \overset{ind}{\sim} N(0, \sigma^2)$ where i indexes the group determined by the independent variable x_i and j indexes the observations within the group. Letting $\bar{y}_i = \frac{1}{n_i} \sum_{j=1}^{n_i} y_{ij}$, the model for the averaged y's is

$$\bar{y}_i = \beta x_i + \epsilon_i \; ; \quad i = 1, \ldots, m \tag{9.8}$$

where $\epsilon_i \sim N(0, \sigma^2/n_i)$ since $\epsilon_i = \frac{1}{n_i} \sum_{j=1}^{n_i} \epsilon_{ij}$.

Use of ordinary least squares in (9.8) leads to estimating β by

$$\frac{\sum_i x_i \bar{y}_i}{\sum_i x_i^2} \; . \tag{9.9}$$

249

On the other hand, ordinary least squares applied to (9.7) leads to the alternative estimator, with $w_i = n_i$,

$$\hat{\beta} = \frac{\sum\limits_{i=1}^{m}\sum\limits_{j=1}^{n_i} x_i y_{ij}}{\sum\limits_{i=1}^{m}\sum\limits_{j=1}^{n_i} x_i^2} = \frac{\sum\limits_{i=1}^{m} n_i x_i \bar{y}_i}{\sum\limits_{i=1}^{m} n_i x_i^2} = \frac{\sum w_i x_i \bar{y}_i}{\sum w_i x_i^2} . \qquad (9.10)$$

The above example motivates the general definition of "weighted least squares" (WLS) where the objective is to minimize $\sum_{i=1}^{m} w_i(y_i - \hat{y}_i)^2$. The w_i's are called the weights. For the above example, choosing $w_i = cn_i$ for an arbitrary constant c leads to the estimator of β given by (9.10). In general, if $\mathrm{Var}(y_i) = \sigma^2/w_i$ then the *minimum variance unbiased estimator* of β is given by (9.10). In this example, using weighted least squares with m observations gives the same estimator of β as using ordinary least squares (OLS) with $N = \sum_{i=1}^{m} n_i$ observations. Even if the values of y_{ij} are available, using $\bar{y}_i$ can reduce the size of the problem considerably. All of the above relations are true for the general multiple regression model

$$Y_i = \beta_0 + \sum_{\ell=1}^{k} \beta_\ell x_{i\ell} + \epsilon_i$$

where $\epsilon_i \overset{ind}{\sim} N(0, \sigma^2/w_i)$.

The above model can be used to illustrate some general properties of weighted least squares. An important point to remember is, using ordinary least squares with heterogeneous variances, $\mathrm{Var}(\epsilon_i) = \sigma^2/w_i$, yields inefficient but still unbiased estimators of the regression coefficients. An idea of the relative efficiency of ordinary least squares (OLS)

and weighted least squares (WLS) is given below. Suppose $E(Y) = \beta x$
and $\sigma_i^2 = \text{Var}(Y_i) = kx_i^2$ so that $w_i \propto 1/x_i^2$. The OLS estimator of β is

$$\hat{\beta} = \frac{\Sigma x_i y_i}{\Sigma x_i^2}, \qquad i = 1, \ldots, m.$$

The appropriate weighted least squares can be derived for this model
very simply by transforming Y_i as

$$\frac{Y_i}{x_i} = \beta + \frac{\varepsilon_i}{x_i}$$

that is,

$$Y_i^* = \beta + \varepsilon_i^* \qquad\qquad (9.11)$$

so that $\text{Var}(\varepsilon_i^*) = (1/x_i^2)\text{Var}(\varepsilon_i) \propto (1/x_i^2)x_i^2\sigma^2 = \sigma^2$. Thus, we can use OLS
on (9.11) to estimate β by

$$\overline{y}^* = \frac{1}{m} \sum_{i=1}^{m} y_i/x_i = \frac{\sum_{i=1}^{m} w_i x_i y_i}{\Sigma w_i x_i^2}$$

where $w_i = 1/x_i^2$. Let $\beta^* = \overline{y}^*$ be the weighted least squares estimate of
β. The efficiency of $\hat{\beta}$ with respect to β^* is defined as $\text{Var}(\beta^*)/\text{Var}(\hat{\beta})$.
Here $\text{Var}(\beta^*) = \frac{1}{m^2} \Sigma (1/x_i^2)\sigma^2 x_i^2 = \sigma^2/m$ and

$$\text{Var}(\hat{\beta}) = \frac{\Sigma x_i^2 \sigma^2 x_i^2}{(\Sigma x_i^2)^2} = \sigma^2 \frac{\Sigma x_i^4}{(\Sigma x_i^2)^2} .$$

Thus, the efficiency is

$$\frac{(\sum x_i^2)^2}{m \sum x_i^4} \leq 1,$$

implying that OLS is worse than WLS. For evenly spaced x_i's taking

values 1, 2, ..., m, the efficiency is 5/9 for large m.

The above procedure for weighted least squares estimators is

general. That is, if $\text{Var}(\varepsilon_i) = \sigma^2/w_i$ the WLS estimator is gotten by:

(1) Defining $Y_i^* = Y_i/\sqrt{w_i}$

(2) and $x_{ij}^* = x_{ij}/\sqrt{w_i}$, including dividing $x_{i0} = 1$ for the
     constant term by $\sqrt{w_i}$ to get $x_{i0}^* = 1/\sqrt{w_i}$,

(3) and compute OLS estimates of the β's with the transformed
     variables.

Example 2. Besides the normal distribution, another distribution

that arises in modeling real world phenomena is the *Poisson distribution*.

As we shall see later, it is a distribution whose variance is a function

of the mean and hence in some situations it can lead to the use of

weighted least squares. We digress here to derive the Poisson distri-

bution and its properties as an example outside the normal distribution

context where statistical modeling can be carried out.

The derivation is similar in spirit to the exponential distribution

in Section 9.1 and is the limit of a familiar elementary distribution--

the binomial distribution. Suppose we are recording fire alarms[*] as

they occur as in Section 9.1 during a particular fixed time period--say

one hour. The fixed period is divided into n equal intervals--if n = 60,

each interval is one minute. We assume as earlier that for large n,

---

[*]See W. E. Walker et al. (eds.), *Fire Department Deployment
Analysis*, Chapter 7 (Elsevier-North Holland, 1979) for a more detailed
theoretical derivation than that given here.

(i) the probability of more than one alarm in an interval is negligible, (ii) that all intervals have the same probability of at least one alarm, and (iii) that the occurrence of an alarm in one interval is independent of occurrences in other intervals. Let p be the probability of at least one alarm in a single interval, then the number of intervals with at least one alarm in the one-hour period has a binomial distribution with parameters n and p. Put $\lambda = np$ and suppose that λ is not too large. Now the probability of zero alarms in the time period is

$$P(0) = (1 - p)^n = (1 - \lambda/n)^n . \qquad (9.12)$$

Now recall that if we keep λ fixed,

$$\lim_{n \to \infty}(1 - \lambda/n)^n = e^{-\lambda} . \qquad (9.13)$$

This can be made precise in that so long as λ^2 is much smaller than n, the approximation $P(0) = e^{-\lambda}$ is a good one. Now from the binomial distribution density function,

$$P(k) = \binom{n}{k} p^k (1-p)^{n-k}$$

where

$$\binom{n}{k} = \frac{n!}{k!(n-k)!} .$$

Thus, we have

$$\frac{P(k+1)}{P(k)} = \frac{\binom{n}{k+1} p^{k+1} (1-p)^{n-k-1}}{\binom{n}{k} p^k (1-p)^{n-k}}$$

$$= \frac{k!(n-k)!}{(k+1)!(n-k-1)!} \frac{p}{1-p}$$

and

$$\frac{P(k+1)}{P(k)} = \frac{1}{k+1} \frac{(n-k)p}{1-p}$$

$$= \frac{\lambda}{k+1} [(1 - \frac{k}{n}) \frac{1}{1-p}] .$$

For k small relative to n and p very close to zero, the term in square brackets is almost 1 so that

$$\frac{P(k+1)}{P(k)} \doteq \frac{\lambda}{k+1} . \tag{9.14}$$

From (9.13), (9.14), and (9.15) we have

$$P(1) \doteq \lambda P(0) \doteq \lambda e^{-\lambda}$$

$$P(2) \doteq \frac{\lambda}{2} P(1) \doteq \frac{\lambda^2}{2} e^{-\lambda}$$

$$P(3) \doteq \frac{\lambda}{3} P(2) \doteq \frac{\lambda^3}{3 \cdot 2} e^{-\lambda}, \text{ etc.}$$

The above has been derived assuming that n is very large. In the fire alarm example, this corresponds to breaking up the one-hour period into smaller and smaller intervals so that in the limit we are counting the number of alarms occurring in the period rather than the number of inter-vals with at least one alarm. Thus, given assumptions (i), (ii), and (iii), the number of alarms X, is said to have a Poisson distribution with parameter λ. The density function of X is

$$P(k) = P(X = k) = \frac{\lambda^k}{k!} e^{-\lambda} \tag{9.15}$$

$$\text{for } k = 0, 1, 2, \ldots$$

Since the Poisson distribution is the limit of the binomial distribution as $n \to \infty$ and $p \to 0$ in such a way that $np = \lambda$ is constant, one would expect the mean and variance of the Poisson distribution to be $np = \lambda$ and $np(1-p) \to \lambda$, respectively. Formally, this can be verified as follows. If X is Poisson,

$$
\begin{aligned}
E(X) &= \sum_{x=0}^{\infty} x P(X = x) \\
&= \sum_{x=0}^{\infty} x \frac{\lambda^x}{x!} e^{-\lambda} \\
&= \sum_{x=1}^{\infty} \frac{\lambda^x}{(x-1)!} e^{-\lambda} \\
&= \lambda \sum_{x=1}^{\infty} \frac{\lambda^{x-1}}{(x-1)!} e^{-\lambda} \\
&= \lambda \sum_{x=0}^{\infty} \frac{\lambda^x}{x!} e^{-\lambda} \\
&= \lambda \cdot 1 = \lambda .
\end{aligned}
$$

The last inequality holds because a density must sum to 1. The variance can be easily calculated by first observing that

$$
\begin{aligned}
E[X(X-1)] &= \sum_{x=0}^{\infty} x(x-1) P(X = x) \\
&= \sum_{x=2}^{\infty} x(x-1) \frac{\lambda^x}{x!} e^{-\lambda} \\
&= \lambda^2 \sum_{x=2}^{\infty} \frac{\lambda^{x-2}}{(x-2)!} e^{-\lambda} \\
&= \lambda^2 \sum_{x=0}^{\infty} \frac{\lambda^x}{x!} e^{-\lambda} = \lambda^2 .
\end{aligned}
$$

Now

$$\text{Var}(X) = E(X^2) - E^2(X)$$

$$= \left\{ E(X^2) - E(X) \right\} + \left\{ E(X) - E^2(X) \right\}.$$

The first term is λ^2 as calculated above so that

$$\text{Var}(X) = \lambda^2 + \left\{ \lambda - \lambda^2 \right\} = \lambda.$$

Thus, for the Poisson distribution, the variance is equal to the mean.

Note that for the binomial approximation above, $\lambda = np$ so that the binomial variance of npq is $\lambda(1 - \lambda/n)$, that is, approximately λ for large n.

When the dependent variable in a regression model has a Poisson distribution, the variance being equal to the mean, the methods described in this chapter are appropriate. In particular, either using the square root transformation on the dependent variable or weighted least squares is indicated.[*]

W. E. Walker et al. (eds.)[**] describe a regression analysis where the number of fire alarms per hour is the variable that is of interest to predict. The regression model presented there predicts the square root of fire alarms per hour as a linear function of several time variables. These time variables capture the effects of time of day, day of week, season and yearly trend. Using the square root of hourly alarms as the dependent variable insures that the residuals have constant variance, provided that (i) hourly alarms have a Poisson distribution whose mean λ varies only with time and (ii) the specification of the independent

---

[*]See D. R. Cox and P. A. W. Lewis, *The Statistical Analysis of Series of Events*, Chapter 3 (Methuen, London, 1966).

[**]See W. E. Walker et al. (eds.), *Fire Department Deployment Analysis*, Chapter 8.

variables capturing the time effect is correct. It turns out that the logarithm of Poisson distributed data as the dependent variable tends to have properties that are similar to the square root. Using the logarithm of $y/(n-y)$ where y is binomially distributed as the dependent variable was discussed in Chapter 8. There this dependent variable was fit by weighted least square to get the "minimum chi-square logit" method.

We next work through the details of a real-world example where weighted least squares is appropriate.

Example 3. Frequently, as in the earlier income prediction example and the Poisson example, the variance of the errors is a function of the x's or more simply of the expected value of Y. We work through a slightly more complicated case than the Poisson one; suppose

$$Y_i = \beta_0 + \beta_1 x_1 + \varepsilon_1 , \qquad (9.16)$$

where $\varepsilon_i \sim N(0, \sigma^2/w_i)$ and $w_i = (\beta_0 + \beta_1 x_i)^2$. In words, the variance of Y is proportional to the square of its mean. Figure 9.16 gives a typical example of this phenomenon,[*] showing expenditure on tea by 2,200 British working-class households in 1937-38 by 54 groups defined by income and household size. The variance of the error term of a behavior equation which relates expenditure on tea to income and household size within each group was investigated. The variance within each group is approximately proportional to the square of the mean expenditure within each group as shown in Figure 9.16.

Since the weights w_i are functions of the unknown parameters β_0 and β_1, weighted least squares estimates cannot be directly computed in one step. Here is an iterative procedure.

---

[*] S. J. Prais and J. Aitchison, "The Grouping of Observations in Regression Analysis," *Revue de L'Institut International de Statistique,* Vol. 22 (1954), pp. 1-22.

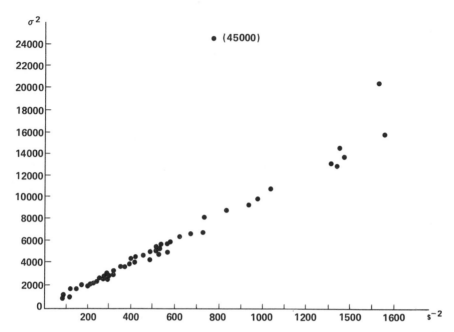

Figure 9.16 - Variance of expenditure on tea and square of
mean expenditure British working-class households, 1938 (units: pence per week).

(i)   First estimate β_0 and β_1 by OLS. Call these estimates $\hat{\beta}_0^{(0)}$
       and $\hat{\beta}_1^{(0)}$.

(ii)   Estimate $w_i = E(Y_i)^2$ by $\hat{w}_i^{(0)} = \left(\hat{\beta}_0^{(0)} + \hat{\beta}_1^{(0)} x_i\right)^2$.

(iii) Compute $\hat{\beta}_0^{(1)}$ and $\hat{\beta}_1^{(1)}$ by WLS using weights $\hat{w}_i^0$.

(iv)   Estimate w_i by $\hat{w}_i^{(1)} = \left(\hat{\beta}_0^{(1)} + \hat{\beta}_1^{(1)} x_i\right)^2$, etc., until conver-
       gence is achieved.

Prais and Aitchison give the following example to show how much these

estimates can improve in successive iteratives. The random variable

$Y = \beta_0 + \beta_1 x + \varepsilon$ was constructed by taking 20 equally spaced x's from

0 to 19 with $\beta_0 = \beta_1 = 7$ and ε being distributed as $N\left(0, \ (\beta_0 + \beta_1 x_i)^2/49\right)$.

Since the actual parameters are known, the correct weighted least

squares estimators and their variances are known. The following table

gives the comparisons.

	Correct WLS	OLS	First Iteration $\hat{\beta}^{(1)}$	Second Iteration $\hat{\beta}^{(2)}$
$\hat{\beta}_0$	$7.06 \pm .83$	11.24 ± 3.04	7.41 ± 1.16	$7.08 \pm .84$
$\hat{\beta}_1$	$7.32 \pm .27$	$6.73 \pm .48$	$7.23 \pm .29$	$7.32 \pm .28$

One can see that the second iteration produces results very close to the correct WLS estimators. The ratio of the variances between OLS and correct WLS is high in this example because of the relatively high correlation. Note that even with only one iteration, the situation is markedly improved over OLS.

In our example thus far, we have dealt only with the situation of the vector of errors $\underset{\sim}{\varepsilon}' = (\varepsilon_1, \ldots, \varepsilon_n)$ having a diagonal covariance matrix with diagonal elements $\sigma^2/w_1, \ldots, \sigma^2/w_n$. Hence the name "weighted least squares" since the estimation procedure depends only on the weights $w_1, \ldots, w_n$. More generally, we sometimes have models in which the errors are correlated. That is, the covariance matrix is not diagonal. The minimum variance unbiased estimates of the regression coefficients are gotten by using generalized least squares (GLS) in fitting the Y_i's. The idea is the same as in weighted least squares although the notation is somewhat more complicated. Theil[*] gives an excellent account of GLS in fitting regression models. We will not cover this generalization here and refer the interested reader to Theil. Many standard regression programs are capable of carrying out GLS fitting. Weighted least squares and more generally GLS are useful methods for data analysts.

---

[*]H. Theil, *Principles of Econometrics*, Wiley, 1971.

Chapter 10

ROBUST ESTIMATION

In this chapter, we will investigate the consequences of assumptions not being met in our statistical models. We will also present modified methods that perform "robustly" when the assumptions break down. For example, in earlier chapters, we have considered linear regression only in the case where errors were normally distributed. We have cautioned against fitting equations when one or more observations are extreme outliers, e.g., if the residual plot shows the existence of "wild shots." Our prescription has been to delete the outliers from the model. But not many bodies of data are really "vocal" about their errors, and the situation often arises in which the designation of observations as outliers is highly questionable. This chapter gives methods applicable to this situation.

Consider the simple problem of estimating the center of a distribution. Common assumptions are that we have a random sample $X_1, \ldots, X_n$ from a normal distribution with mean μ and variance σ^2. We then use the average $\overline{X} = (1/n) \sum_{i=1}^{n} X_i$ as our estimate of μ. Now suppose there are outliers or "gross errors" in the data. Since $\overline{X}$ can change drastically as a function of a single wild observation, an estimator that is less sensitive to one or two outliers may be preferable. One such alternative estimator is the *median* of $X_1, \ldots, X_n$. The median is the same whether the largest (or smallest) observation is near the main body of data or many standard deviations away. Although using the median provides protection from outliers, it is only 64 percent as efficient as $\overline{X}$ if $X_1, \ldots, X_n$ really constitute a random sample from a normal (μ, σ^2)

distribution. That is, the standard deviation of the median is the same as the standard deviation of $\overline{X}$, if $\overline{X}$ is calculated from only 64 percent as many observations. Thus, if we think of using the median as "taking out insurance" against the possible presence of gross errors, there is a 36 percent premium.

But for certain other distributions, this strategy can pay off handsomely. If, for example, the X_i's are a random sample from a very "heavy-tailed distribution," say a Student t-distribution with 1 degree of freedom (e.g., $X_i = \mu + \sigma Z_{i1}/Z_{i2}$, where Z_{ij} are independent standard normal random variables, $\overline{X}$ has infinite variance, while the median has variance $\pi^2 \sigma^2 / 4n$. Even in less extreme situations, the benefits can often be substantial: Table 10.1 shows the variances of the mean and median when observations are generated from Student t-distributions of various types. The fewer the degrees of freedom in a Student t-distribution, the more prone the distribution is to outliers. The table shows that the mean does as well or better than the median when the degrees of freedom exceed 4; otherwise, the median is preferable, and often by far.

Table 10.1

VARIANCES OF THE MEAN AND MEDIAN FOR
VARIOUS STUDENT PARENT DISTRIBUTIONS
AND LARGE n

	1	2	3	4	5	8	10	15	24	∞
n(var[mean])	∞	∞	3.00	2.00	1.67	1.33	1.25	1.15	1.09	1
n(var[median])	2.47	2	1.85	1.78	1.73	1.67	1.65	1.62	1.60	1.571

Thus, averaging is a good thing to do in the normal or near-normal cases. Otherwise it can be good practice to eliminate the extreme points at either end of the data. This idea leads to a set of estimators that are compromises between the mean and median: the trimmed mean and the Winsorized mean. To define these estimators we need some notation. Let $X_{(1)} \leq \ldots \leq X_{(n)}$ be the ordered sample. If we fix the amount of trimming or Winsorizing at g, the *g-trimmed mean* is defined as

$$\overset{\lfloor g \rfloor}{X_T} = \frac{X_{(g+1)} + \ldots + X_{(n-g)}}{n - 2g} , \tag{10.1}$$

while the *g-Winsorized mean* is defined as

$$\overset{\lfloor g \rfloor}{X_W} = \frac{(g+1)X_{(g+1)} + X_{(g+2)} + \ldots + X_{(n-g-1)} + (g+1)X_{(n-g)}}{n} . \tag{10.2}$$

Thus if $g = 0$, both estimates are $\bar{X}$, while if $g = \frac{n-1}{2}$ (assume n odd), both estimates are the median. We usually choose g to be small in order to protect against only a few gross errors, and thus pay a small premium. The appropriate value of g depends on the problem.

Table 10.2 gives the variances of the trimmed mean and of the Winsorized mean for various values of g and various Student t "parent" distributions of the data. For the optimal choice of g, variances are small. Also some choices of g do well over a wide range of distributions. Such choices are called "robust." Thus, trimmed and Winsorized means show promise of being robust.

Table 10.2

VARIANCES OF TRIMMED AND WINSORIZED MEANS FOR VARIOUS
STUDENT PARENT DISTRIBUTIONS AND n = 20

	Trimmed Means					Winsorized Means			
	g						g		
	1	2	3	5		1	2	3	5
$\nu = 1$	21.59	9.11	6.53	3.13	$\nu = 1$	39.55	15.48	9.27	5.90
2	2.74	2.21	1.97	1.82	2	3.47	2.77	2.36	1.94
3	2.06	1.79	1.64	1.61	3	2.37	2.14	1.68	1.66
4	1.63	1.49	1.45	1.44	4	1.81	1.64	1.57	1.48
8	1.23	1.23	1.23	1.31	8	1.28	1.25	1.25	1.28
∞	1.02	1.05	1.08	1.17	∞	1.02	1.04	1.06	1.14

Probability plots using the parameters in Table 10.2 that show
the effectiveness of various degrees of trimming/Winsorizing as functions of the parent distribution are useful in deciding what percentage
of the observations should be modified. In fact, a rule can easily be
constructed which translates curvature in the probability plot of the
data (or residuals) to an estimate of the degree-of-freedom parameter
of the Student distribution, which in turns leads to the selection of
the g that is most efficient for that distribution. Figure 10.1 provides plots of actual Student quantiles versus normal quantiles for
various ν, and is thus a basis for such a rule.

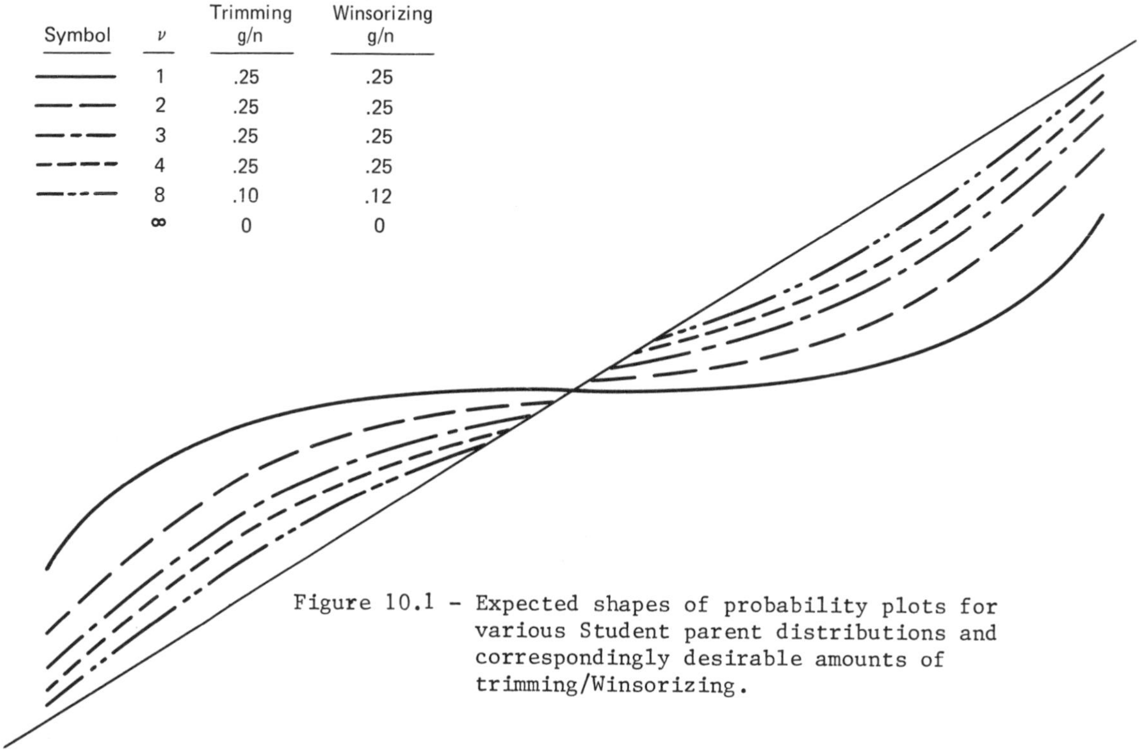

Symbol	ν	Trimming g/n	Winsorizing g/n
———	1	.25	.25
— —	2	.25	.25
—·—	3	.25	.25
— — —	4	.25	.25
—··—	8	.10	.12
	∞	0	0

Figure 10.1 – Expected shapes of probability plots for various Student parent distributions and correspondingly desirable amounts of trimming/Winsorizing.

The usual t-statistic based on $\overline{X}$ for getting confidence intervals for μ or for making hypothesis tests about particular values of μ is:

$$t = \frac{\sqrt{n}\ (\overline{X} - \mu)}{\sqrt{\dfrac{\Sigma\,(X_i - \overline{X})^2}{n-1}}} \tag{10.3}$$

where $X_1, \ldots, X_n$ are assumed to be a random sample from a normal distribution with mean μ and unknown variance σ^2. We now turn to characterizing how distribution of (10.3) changes when assumptions are relaxed.

If our problem is testing whether $\mu = 0$, we substitute the 0 for μ in (10.3) and look up the signficance probability in a t-table with n-1 degrees-of-freedom. For confidence intervals we look up the appropriate (e.g., 95%) cutoff points and convert the inequality on t to an inequality on μ. Besides the simple inference on a mean from a univariate sample, we use statistics with a t-distribution for testing whether regression coefficients are zero and for testing whether two means are the same. (See Chapter 6.) Thus our conclusions about the univariate case will apply with the obvious modifications to these other problems.

We observed in Chapter 6 that if our assumptions are met, the t-statistic has the same distribution as the ratio of two independent random variables; the numerator being a standard normal variable and the denominator being the square root of a chi-square variable with n-1 degrees-of-freedom divided by n-1. Besides the questions of efficiency of the various methods described earlier, we want to know how accurate the inferences we make are when assumptions are relaxed; that is, does the t-statistic defined in (10.3) still have approximately a Student's t-distribution with n-1 degrees-of-freedom? If the X_i's are not normal, the $\overline{X}$ in the numerator is still approximately normal by the Central Limit Theorem. With nonnormal X_i's the denominator is still a good estimate of σ, but it will not have the square root of a chi-square distribution, and the numerator and denominator will not be independent. If the nonnormality is in the direction of a "heavy-tailed" distribution, the actual probabilities will be larger than the values in the t-table. Thus, tests of hypotheses that assume a normal parent distribution will not reject often enough and the confidence intervals will be too large.

Table 10.3 displays the critical values taken from Yuen and Murthy.[*]
They use t-distributions with varying degrees-of-freedom as their heavy-tailed parent distributions. The row marked "∞" corresponds to the parent distribution being normal--the nominal assumption. Note that the departure of the test stability in (10.3) from the t-distribution is greatest for the heavier-tailed parent distributions.

If $\bar{X}$ is not a good estimate of μ it also makes little sense to use it as the basis for hypothesis tests or confidence intervals for μ. Our alternative to $\bar{X}$ as an estimator of location are the trimmed or Winsorized means. Confidence intervals based on these estimators should be centered around them, and the intervals themselves possibly determined in ways insensitive to outliers. In the case of g-Winsorizing, the appropriate modification is

$$
t_W^{\lfloor g \rfloor} = \frac{\sqrt{n}\ (\bar{X}_W^{\lfloor g \rfloor} - \mu)}{\sqrt{\Sigma (X'_{(i)} - \bar{X}_W^{\lfloor g \rfloor})^2 / (n-1)}}
\tag{10.4}
$$

where

$$
X'_{(i)} = \begin{cases} X_{(i)} & \text{if } g+1 \le i \le n-g \\ X_{(g+1)} & \text{if } i \le g \\ X_{(n-g)} & \text{if } i \ge n-g+1 \ . \end{cases}
$$

Thus, $t_W^{\lfloor g \rfloor}$ is the usual t-statistic calculated on the g-Winsorized data. If g is chosen to be large enough that the gross errors are "Winsorized away," and thus that the remaining center of the distribution is

---

[*] Karen K. Yuen and V. K. Murthy, "Percentage Points of the Distribution of the t-Statistic When the Parent is Student's t," *Technometrics*, 1974.

Table 10.3

TWO-SIDED PERCENTAGE POINTS OF THE T-STATISTIC WHEN
THE PARENT DISTRIBUTION ITSELF IS STUDENT'S T

n	degrees-of-freedom	$\alpha = 0.01$	$\alpha = 0.05$	$\alpha = 0.10$
5	3	4.046	2.540	1.999
	4	4.189	2.590	2.039
	5	4.260	2.631	2.055
	7	4.355	2.673	2.078
	9	4.400	2.690	2.086
	∞	4.604	2.776	2.132
10	3	2.990	2.148	1.773
	4	3.050	2.175	1.788
	5	3.088	2.197	1.803
	7	3.141	2.220	1.815
	9	3.164	2.228	1.818
	∞	3.250	2.262	1.833
15	3	2.773	2.070	1.720
	4	2.833	2.084	1.734
	5	2.863	2.100	1.743
	7	2.904	2.118	1.751
	9	2.922	2.127	1.756
	∞	2.977	2.145	1.761
20	3	2.701	2.030	1.700
	4	2.737	2.047	1.709
	5	2.777	2.062	1.719
	7	2.806	2.076	1.725
	9	2.822	2.082	1.727
	∞	2.861	2.093	1.729

approximately normal, then

$$\frac{n - 2g - 1}{n - 1} \frac{|\mathcal{E}|}{t_W}$$

has approximately a t-distribution with $n - 2g - 1$ degrees-of-freedom for $g \leq n/4$.[*] This result gives us a method for hypothesis testing and computing confidence intervals in the presence of outliers.

As a practical matter, it is probably desirable to select the amount of Winsorization (g) after looking at the data. The above approximation makes no allowance for such selection and will lead to false optimism about the accuracy of the result. For the interested reader, Dixon and Tukey[**] and Tukey and McLaughlin[***] give approaches for allowing for such selection. Stigler[****] discusses trimming for various data sets of historical importance.

We have made suggestions here for estimating the center of a distribution. The improvements claimed apply to symmetric cases, such as samples from t-distributions. In nonsymmetric cases, the population mean and median differ, and then it makes little sense to compare efficiencies of the sample mean and median, even for long-tailed distributions. An exponential distribution, for example, has mean 1.44 times its median, and the analyst must decide in such a case which parameter

---

[*] A description of this approximation and more precise tables are given in W. J. Dixon and J. W. Tukey, "Approximate Behavior of the Distribution of the Winsorized t (Trimming/Winsorization 2)," *Technometrics*, Vol. 10, No. 1, 1968, pp. 83-98.

[**] Ibid.

[***] J. W. Tukey and D. H. McLaughlin, "Less Vulnerable Confidence and Significance Procedures for Location Based on a Single Sample (Trimming/Winsorization)," *Sankhya*, Series A, Vol. 25, No. 3, 1963, pp. 331-352.

[****] S. M. Stigler, "Do Robust Estimators Work with *Real* Data?" *Annals of Statistics*, Vol. 5, No. 6, 1977, pp. 1055-1078.

he wishes to estimate, for he estimates a different function of the population distribution for each amount trimmed or Winsorized. The improvements to $\overline{X}$ suggested in this chapter are limited to *symmetric cases only*, while improvements to $\overline{X}$ may be impossible in general cases.

This chapter has introduced methods for protecting the analyst against the consequences of heavy-tailed symmetric parent distributions and the outliers they generate. The interested reader should consult Andrews et al.[*] for more complete coverage of these methods.

---

[*] D. F. Andrews, P. J. Bickel, F. R. Hample, P. J. Huber, W. H. Rogers, and J. W. Tukey (1972), *Robust Estimates of Location: Survey and Advances*, Princeton University Press.

Chapter 11

NONPARAMETRIC AND GOODNESS-OF-FIT INFERENCE

Chapter 10 introduced statistical methods that are "robust" against departures from the assumptions usually made in *parametric* statistical inference. Thus far we have always had a probabilistic description of how our data were generated with everything known except for the values of several parameters, e.g., regression coefficients. As an alternative to the modified parametric methods described in Chapter 10, we now introduce *nonparametric* methods of inference. Nonparametric procedures do not require a full probabilistic description of the process generating the sample. Their advantage over modified parametric methods is their applicability across a much wider range of models that may have generated the sample.

Our first set of procedures are "rank tests." These are test statistics which use only the relative order that observations come in rather than at their actual values. There are several reasons for using rank tests or, more generally, nonparametric procedures rather than parametric methods. Nonparametric procedures do not depend on an assumed form of the distribution of the observations. Thus if our data is given in an arbitrary scale or just as an ordering, nonparametric methods can be applied directly. There is no need to scale the data so that the assumptions for the parametric methods are met. In some situations, there may be so little data that using diagnostic methods as in Chapter 9 yields inconclusive results. Alternatively, we may know the appropriate parametric model perfectly but it is so complex that the construction of optimal parametric methods is an

enormous if not impossible task. Or since some nonparametric procedures are very easy to apply, we might try them first and if the
results are sufficiently accurate, dispense with the more powerful
parametric procedures.

We introduce rank tests in Section 11.1 by describing two nonparametric, one-sample rank tests of hypotheses about the center of
the parent distribution. Section 11.2 gives rank tests for the two
or more sample location problem--the nonparametric analogue of the
two-sample t-test and one-way analysis of variance. Section 11.3
describes rank correlation methods--the rank analogue to simple linear
correlation and regression. Section 11.4 digresses to describe multidimensional scaling--an alternative to pure nonparametric rank methods
for ordinal data. Finally, Sections 11.5 and 11.6 cover chi-square
goodness-of-fit tests for categorical data. The reader interested in
learning more about nonparametric inference, particularly rank tests,
is referred to Lehmann's (1975) excellent book on the subject.

11.1 ONE-SAMPLE RANK TESTS

The trimming and Winsorization procedures described in Chapter 10
are designed specifically as a guard against a particular departure
from normality--the presence of gross errors in an otherwise normally
distributed sample. We now look at nonparametric tests of the center
of a distribution. In general, we will use the median rather than the
mean of the distribution as our definition of center. For the normal
distribution or any symmetric distribution this makes no difference
since the mean is equal to the median.

271

From a random sample X_1, ..., X_n suppose we wish to test the hypothesis that the median of the distribution of X_i is equal to zero against the alternative hypothesis that it is greater than zero (0 is arbitrary). If we wish to make no assumptions other than that the X_i's are a random sample, the *Sign Test* is the appropriate procedure. Let S be the number of positive values of the X_i's. Then from Chapter 5, S has a binomial distribution with parameters n and $p = \text{Prob}(X_i > 0)$. It is easy to see that the value $p = \frac{1}{2}$ corresponds to our null hypothesis that the median is 0. The Sign Test rejects the null hypothesis for large values of S. That is, large values indicate that $p > \frac{1}{2}$ and therefore that the median of the distribution of X_i is above zero.

The Sign Test requires almost no assumptions, but in turn it "wastes information." Is there a "less wasteful" compromise between it and usual parametric test for this problem the t-test? The *Wilcoxon One-Sample Test* is such a compromise. The premium we pay for its added strength is making another assumption, namely: the distribution being sampled from is approximately symmetric about its center. To test whether the median is zero against the median of the distribution of X_i is avove zero, The Wilcoxon test ranks the X_i's in order of their distance from zero (zero is arbitrary). The test statistic W is the sum of the ranks which correspond to positive observations. For example, if $X_1 = 3$, $X_2 = -1.8$, $X_3 = 7$, $X_4 = -7.6$, $X_5 = 10.1$ and $X_6 = 3.6$, the observations and rank in order of distance from zero are:

Observation:	−1.8	3	3.6	7	−7.6	10.1
Rank:	1	2	3	4	5	6

The value of W is $W = 2 + 3 + 4 + 6 = 15$. To compute significance levels we need the distribution of W under the null hypothesis.

If the distribution of the X_i's is centered at zero, the symmetry assumption implies that any given rank is equally likely to come from a positive value as a negative value. If

$$I_r = \begin{cases} 1 \text{ if rank r is from positive observation} \\ 0 \text{ if rank r is from negative observation} \end{cases}$$

we can write W as

$$W = \sum_{r=1}^{n} r I_r \; .$$

Now $P(I_r = 1) = \frac{1}{2} = P(I_r = 0)$ under the null hypothesis. So

$$E(W) = \sum_{r=1}^{n} r \, E(I_r)$$

$$= \sum_{r=1}^{n} r \, P(I_r = 1)$$

$$= \frac{n(n+1)}{4} \; .$$

Likewise, since the values of I_1, I_2, ..., are independent (under the null hypothesis), the variance of W is

$$\sigma^2(W) = \sum_{r=1}^{n} r^2 \sigma^2(I_r)$$

$$= \sum_{r=1}^{n} r^2 P(I_r = 1) P(I_r = 0)$$

$$= \frac{n(n+1)(2n+1)}{6} \times \frac{1}{4} = \frac{n(n+1)(2n+1)}{24} \; .$$

There is a nice recursion formula to get the exact distribution of W (see D. Owen, *Handbook of Statistical Tables*, Addison-Wesley, 1962). We will use the normal approximation in this book. Thus to test the hypothesis that the distribution is centered at 0, we calculate:

$$\frac{W - E(W)}{\sigma(W)} = \frac{W - \dfrac{n(n+1)}{4}}{\sqrt{\dfrac{n(n+1)(2n+1)}{24}}}$$

and look up its value in a normal table. In the above example, n = 6, so that $E(W) = n(n+1)/4 = 10.5$ and $\sigma^2(W) = n(n+1)(2n+1)/24 = 22.75$. Using the normal approximation to the distribution of W, and making the "continuity correction" for the integer-valued random variable W by subtracting 1/2 from its observed value of W = 15, we obtain the significance probability for the "one-sided test" against the alternative hypothesis that the median is above zero as

$$P(W \geq 15) = P(W \geq 14.5) = P\left(\frac{W-E(W)}{\sigma(W)} \geq \frac{14.5-10.5}{\sqrt{22.5}}\right) \doteq P(Z \geq 0.83) = 0.80$$

from the table of the normal distribution. Hence these data are consistent with the hypothesis that the median of the distribution is zero.

If we regard nonparametric tests as insurance policies, we can ask: suppose all assumptions are satisfied and the t-test would have been best; how much have we lost? The other side of the coin is: how badly can the t-test do? More precisely, to get the same error probabilities with two tests we can calculate the ratio of necessary sample sizes between two tests. This is called the "relative efficiency" between the two tests. The following table shows that, relative to the t-test, which is designed for the normal distribution, for large samples the sign test has considerable power while the Wilcoxon test is nearly as powerful. In the worst cases for the t-test, for very long-tailed distributions, the sign test and Wilcoxon test are infinitely better.

Parent Distribution	Relative Efficiency of One-Sample Tests With Respect to the t-Test	
	Sign Test	Wilcoxon Test
Distribution giving minimum value	1/3	0.864
Normal distribution	$2/\pi = 0.636$	$3/\pi = 0.955$
Distribution giving maximum value	∞	∞

Example: Ballot Position

Returning to our election example of Chapter 6, we consider the problem of testing whether first position on the ballot affects the candidate's chance of winning. Suppose n two candidate elections are analyzed, and let p be the probability that the candidate listed first wins (we assume that elections with incumbents are excluded, for incumbents are usually listed first). In Chapter 6, we described the Sign Test (but did not use that name) for testing the hypothesis that ballot position is unimportant; that is $p = \frac{1}{2}$. We saw that a sample size of n = 10,000 was needed to differentiate between p = 0.50 and p = 0.52 using the Sign Test.

In order to get a more powerful test which uses more of the information contained in the data, we can construct a parametric test for this problem. We think of the following model of how the vote totals are generated in our sample of n two candidate elections. For the i-th election, let p_i be the proportion of votes the candidate in the first position receives. Even if there were no positional bias, we would not expect p_i to be exactly 1/2 because (i) of the sampling variance in the actual proportion of votes in election i even if each voter votes independently for the first position candidate with probability 1/2 each, and (ii) of other factors which we are not controlling for, so that p_i should be regarded as itself being sampled from a distribution. Since elections we are considering typically have more than 10,000 votes, the standard deviation of the observed population is less than $1/200 = 0.5$ percent. The observed variation in p_i itself is typically at least an order of magnitude above this, so we will ignore the sampling variance in this example. Summarizing, our model is that p_i is sampled from a distribution.

The null hypothesis is that this distribution is centered (has its median) at $1/2$ while the alternative hypothesis is that the median is greater than $1/2$. If we assume that the p_i's are normally distributed the appropriate statistic for testing this hypothesis is

$$t = \frac{\sqrt{n}\,(\bar{p} - \frac{1}{2})}{\sqrt{\Sigma_{i=1}^{n}(p_i - \bar{p})^2/(n-1)}}$$

which has a t-distribution with $n-1$ degrees of freedom under the null hypothesis. If the data are nonnormal we prefer the Wilcoxon or Sign test. Suppose the data in Table 11.1 came from $n = 25$ two-candidate elections in which the ballot position is determined by alphabetical order of last name or by lot. The student should try a Winsorized t-test (from Chapter 10), a Wilcoxon one-sample test (subtract .500 from each observation before applying the preceding theory), the Sign test, and the t-test to decide whether positional bias exists for the elections in Table 11.1. Use size of $\alpha = .01$ and a one-sided test. Also construct a histogram to see whether nonnormality of the p_i's seems likely.

Table 11.1

25 ELECTION PROPORTIONS FOR CANDIDATE IN
FIRST BALLOT POSITION

.254	.480	.551	.593	.648
.384	.496	.556	.594	.676
.452	.505	.557	.603	.680
.456	.528	.570	.612	.697
.468	.536	.582	.626	.775

$$\bar{p} = .555, \quad \sum_{1}^{25} (p_i - \bar{p})^2 = .283$$

11.2. TWO-SAMPLE RANK TESTS AND GENERALIZATIONS

Both the Sign test and the Wilcoxon one-sample test are rank tests. Both were based on the ranks of the positive (or negative) observations when we lined up all observations in order of their distance from the central value to be tested (zero in our case). The Sign test assigned a value of one to each observation, then summed the values coming from positive observations. The Wilcoxon one-sample test assigned each observation the value of its rank (in distance from zero), then summed the values coming from the positive observations.

One motivation in using functions of the ranks rather than actual values is to give tests which are insensitive to gross error or extreme values. More generally, we want tests whose calculated significance probabilities are valid for observations from any (or at least any plausible) distribution. We will describe a number of different tests based on ranks.

For testing whether two populations have the same center, we commonly have two random samples drawn from the two different populations. We wish to know whether a value from one population typically tends to be larger (or smaller) than a value from the other. We look at a specific data set to develop the methods.

Example. Listed below are two random samples (with replacement) of size ten each of 1969 crime rates per 1,000 population in U.S. metropolitan area. Sample A is taken from U.S. metropolitan areas of population 250,000 or more (119 total) while Sample B is from areas with populations of less than 250,000 (127 total). The question of interest can be put as: Are crime rates in smaller areas typically lower than those of the larger areas?

A (over 250,000)	B (under 250,000)
13.88	41.17
17.64	25.90
14.34	19.19
21.17	34.22
37.01	11.83
24.15	13.59
31.64	19.95
32.83	23.96
34.67	13.68
29.03	14.69

The parametric method for this question that we covered in Chapter 6 and Chapter 7, which was a special case of analysis of variance, is the two-sample t-test. It can be thought of as one-way analysis of variance with only two levels. The square of the t-statistic given below thus will be the usual F-test of no effect. If samples of size n and m from populations A and B are denoted by $X_1, \ldots, X_n$ and $Y_1, \ldots, Y_m$, respectively, the appropriate t-statistics with m+n-2 degrees of freedom is

$$\frac{(\overline{X} - \overline{Y})/(\frac{1}{n} + \frac{1}{m})^{.5}}{\sqrt{\frac{\Sigma(X_i-\overline{X})^2 + \Sigma(Y_i-\overline{Y})^2}{n + m - 2}}} = .933$$

(where $\overline{X} = 25.636$, $\overline{Y} = 21.818$, $\Sigma(X_i-\overline{X})^2 = 662.19$, $\Sigma(Y_i-\overline{Y})^2 = 844.45$) and is significant at the 0.18 level for the above data, i.e., we cannot reject the null hypothesis of equality.

Recall from regression (or ANOVA) that our observations must have normal distributions with equal variances. If these assumptions are violated (particularly the equal variance one) our significance probabilities are in error.

The two-sample analogue to the Sign test of Section 11.1 is the *Median test*. It is carried out by first ranking all the observations from lowest to highest. Then count the number of observations from population A in the upper half (above the median) of the combined sample. Note this is the same as assigning value 0 to the observations below the median (ranks less than $(m+n-1)/2$) and the value 1 to observations above the median and summing these values for those observations in sample A. The "arrowed" values below are from sample A.

Rank	Observations	Rank	Observations
1	11.83	11	23.96
2	13.59	12	24.15 ←
3	13.68	13	25.90
4	13.88 ←	14	29.03 ←
5	14.34 ←	15	31.64 ←
6	14.69	16	32.83 ←
7	17.64 ←	17	34.22
8	19.19	18	34.67 ←
9	19.95	19	37.01 ←
10	21.17 ←	20	41.17

The computed value is $M = 6$.

We use the median as our measure of typical value. The test is of equality of distributions against one distribution "stochastically" larger than the other, i.e., for all points t, $P(X > t) > P(Y > t)$. Note there is no assumption of normality or of equal variances. The probability distribution of M is the "hypergeometric distribution,"

279

$$P(M=k) \;=\; \frac{\dbinom{n}{k}\dbinom{m}{\frac{N}{2}-k}}{\dbinom{N}{\frac{N}{2}}}\;, \qquad N = m+n\,,$$

if the null hypothesis that the distributions are the same holds. Thus if $M = c$, the significance probability is

$$P(M \geq c) \;=\; \sum_{k=c}^{N/2} \frac{\dbinom{n}{k}\dbinom{m}{\frac{N}{2}-k}}{\dbinom{N}{\frac{N}{2}}} \;=\; \sum_{k=6}^{10} \frac{\dbinom{10}{k}\dbinom{10}{10-k}}{\dbinom{20}{10}} \;=\; \frac{60{,}626}{184{,}756} \;=\; .328\,,$$

since for our data, $m = n = 10$, $N = 20$, $c = 6$. Or we may use the tables in Owen to find this value. This is significant at the 0.33 level, and hence the null hypothesis cannot be rejected. For large m and n, $M \sim N(\frac{m+n}{2}, \frac{m+n}{4})$.

Alternatively, we can use the *Wilcoxon two-sample test*. For this test, again rank all the observations. Sum the ranks of all observations from sample A and reject if this sum is large. This test is the same as the median test except that the possible values to be summed are 1, 2, ..., 20 rather than 0, ..., 0, 1,..., 1. The value for these data is

$$W = 4 + 5 + 7 + 10 + 12 + 14 + 15 + 16 + 18 + 19 = 120\,.$$

The assumptions are the same as for the median test. To calculate significance probabilities one can use a clever recurrence relation to get the exact probability distribution and it is tabled (see Owen). We will use the normal approximation to the distribution of W. One can easily show that the mean of W is $n(N+1)/2$ and the variance is $mn(N+1)/12$, where $N = m+n$. For our data:

$$\text{Mean} = 10(21)/2 - 105$$

$$\text{Variance} = 10(10)(21)/12 = 175$$

$$\text{Standard Deviation} = 13.22$$

$$\frac{W-E(W)}{\sigma(W)} = \frac{120-105}{13.22} = 1.134.$$

Therefore, the data are significant at the 0.13 level. We do not reject.

As in the one-sample problem, we can regard two-sample nonparametric tests as insurance policies against a nonnormal parent distribution with some premium that must be paid. The relative efficiencies of the Median Test and Wilcoxon Two-Sample Test with respect to the t-test are given below.

Relative Efficiencies of Two-Sample Tests to t-Test

	Median	Wilcoxon Two-Sample
Minimum value	1/3	0.864
Normal Distribution	$2/\pi = 0.64$	$3/\pi = 0.995$
Maximum value	∞	∞

One can think of the Wilcoxon Two-Sample and One-Sample Tests or the Median and Sign Tests as replacing the observations by their ranks or 0 and 1's, respectively. This may be verified algebraically.

A Nonparametric Alternative to One-Way Analysis of Variance

We saw in Chapter 7 that testing whether all effects are zero in a one-way analysis of variance can be regarded as a generalization of the two-sample t-test. The one-way analysis of variance model is

$$y_{ij} = \mu + \beta_j + e_{ij}; \qquad j = 1, \ldots, k,$$
$$i = 1, \ldots, n_j,$$

where the errors e_{ij} are values drawn from a $N(0,\sigma^2)$ distribution. The two-sample problem is $k = 2$, and the two-sample t-test tests whether $\beta_1 = \beta_2 = 0$.

The *Kruskal-Wallis H-test* is a generalization of the two-sample Wilcoxon test. The model is the same as above except that the error distribution need not be normal; it is arbitrary but symmetric. The problem is to test whether $\beta_1 = \beta_2 = \ldots = \beta_k = 0$ against the alternative hypothesis that they are unequal. That is, do all k-populations have the same distribution? The two-sample Wilcoxon test is equivalent to ranking the combined samples and substituting the ranks in the two-sample t-statistic. For the Kruskal-Wallis H-test we combine all observations and rank them. Then we substitute those ranks into the usual F-statistic formula. With some simplification this formula results in

$$H = \frac{12}{N(N+1)} \sum_{j=1}^{k} \frac{R_j^2}{n_j} - 3(N+1)$$

where k = number of samples

n_j = number of observations in j-th sample

$N = \sum_{j=1}^{k} n_j$

R_j = sum of ranks in j-th sample .

The test rejects the null hypothesis of equality if H is large. Under the hypothesis of equality, H has approximately a chi-square distribution with k-1 degrees of freedom. This approximation is reasonably good when $n_j > 5$. Thus we can use chi-square tables to look up significance probabilities for the value of H. The reader should verify $H = 9/7 = (1.134)^2$ is the square of the Wilcoxon statistic for the crime data, for which $k = 2$, $n_1 = n_2 = 10$, $N = 20$.

282

For all the rank tests thus far described, when two or more observations have the same values, the usual practice is to assign the mean of the ranks to the tied observations to each observation. There are minor corrections which can be made for the effect of ties in the Kruskal-Wallis H-test as well as the other rank tests presented in this chapter. We will not cover the corrections here; the interested reader should see Lehmann (1975).

11.3. RANK CORRELATION TESTS

The usual correlation coefficient ρ measures how well Y can be predicted linearly from X. We turn now to rank correlation measures-- measures which do not vary with changing the scale of measurement. We will see that (X,Y), (log X, Y), (X^3, log Y), etc., all have the same rank correlation. The best known rank correlation coefficients are *Spearman's* r_s and *Kendall's* τ.

We present only Spearman's r_s here. Suppose we have an X measurement and Y measurement taken on each of n units (no tied X's or tied Y's) and we desire a measure of the correlation between X and Y. Arrange the n pairs of numbers in increasing order of the X variable. If X and Y are independent and hence uncorrelated the corresponding Y sequence a priori is equally likely to be any one of the n! (n factorial) possible permutations of the n values of Y. If the X and Y are "monotonically" related, the Y values should tend to form an increasing (or decreasing) sequence so that any statistic that measures the degree of this increase (or decrease) can be used to test for nonzero correlation. As we did for the Wilcoxon two-sample test and the Kruskal-Wallis H-test, we replace the observations by their ranks and calculate the usual parametric measure. Doing this for the ordinary correlation coefficient results in Spearman's r_s. Let Y_i be the rank in the Y sample of the Y observation corresponding to the i-th smallest value of X. The rank correlation coefficient ("Spearman's r_s") is the ordinary correlation of these numbers,

$$r_s = \frac{\frac{1}{n}\left[\sum_1^n i\, Y_i - \left(\frac{n+1}{2}\right)^2\right]}{\frac{n^2-1}{12}} \qquad (11.1)$$

Note that the mean of the first positive integers is $\frac{1}{2}(n+1)$ and the variance of a randomly chosen one is $(n^2-1)/12$, so (11.1) is *exactly the linear correlation coefficient with observations replaced by ranks.* A more convenient formula for calculation is

$$r_s = 1 - \frac{6}{n(n^2-1)} \sum_{i=1}^n (Y_i-i)^2 . \qquad (11.2)$$

To test for independence of X's and Y's, or more accurately for non-monotonic correlation, we use the same methods as linear correlation. Define

$$t = r\sqrt{\frac{n-2}{1-r^2}} \qquad (11.3)$$

The distribution of t in (11.3) is approximately a t-distribution with n-2 degrees of freedom for large n. The exact distribution of r_s for small samples can be found in Owen.

Thus far we have presented some of the simplest nonparametric rank tests. For more complicated situations, there is a growing literature in nonparametric multivariate analysis. Typically, the data points are replaced by their ranks and then the usual multivariate models

are applied to the ranks. See Puri and Sen for a fairly complete
description of the methods available and their underlying theory.

11.4. MULTIDIMENSIONAL SCALING

Thus far in this chapter, we have concentrated on taking numeri-
cally valued data, reducing it to ranks and developing the resulting
nonparametric rank tests. We now briefly discuss the "reverse" problem.
Suppose the original data are poorly scaled or unscaled and we wish to
transform it into multivariate scaled data so that we can then apply
the usual multivariate techniques such as regression analysis. The
subject of multidimensional scaling has developed empirically and
there are a number of different formulations and methods available.
We describe here a special case of a method due to Kruskal.[*]

Suppose there are n individuals to be scaled--people, products,
cars, etc. The data come in the form of a symmetric $n \times n$ input matrix
Δ where

$$\Delta = \begin{bmatrix} \delta_{11} & \cdots & \delta_{1n} \\ \vdots & & \\ \delta_{n1} & \cdots & \delta_{nn} \end{bmatrix}.$$

The δ_{ij}, $i \neq j$, represent the observed dissimilarity (or similarity)
between individuals i and j. There are many possibilities of how Δ
might arise, but we will treat the case where δ_{ij} = rank correlation
(Spearman's r_s) of the n objects where there are N observations on
each individual. This means that we have N poorly scaled or unscaled

---

[*]See J. B. Kruskal, "Multidimensional Scaling by Optimizing Goodness of Fit to a Non-Metric Hypothesis," *Psychometrika*, Vol. 29 (March
1964), pp. 1-27. See also S. J. Press, *Applied Multivariate Analysis*,
Holt, Rinehart and Winston, 1972, Chapter 15, for a good summary.

sets of values for the n objects. These might be the values or ratings assigned them by N different raters. Then δ_{ij} is defined as the value of r_s calculated from the N pairs of values for the i-th and j-th objects. We will never use δ_{ii}, the diagonal elements. Given Δ, we now order the $M = n(n-1)/2$ distinct values of δ_{ij} (assuming no ties)-- the nondiagonal entries of Δ--as

$$\delta_{i_1 j_1} < \delta_{i_2 j_2} < \ldots < \delta_{i_M j_M} \tag{11.4}$$

where $(i_1, j_1), \ldots, (i_M, j_M)$ are all pairs of unequal subscripts of δ_{ij}, e.g., (2, 3), (2, 5), (3, 6), etc.

The objective is to create a scale for the n objects. For concreteness we describe the case when we want a scale in two-dimensional space (x_1, x_2). The idea is to assign a location (x_{1j}, x_{2j}) in the plane to the j-th object (j = 1, 2, ..., n) so that the distances between the objects are as close as possible in ordering to the ordering of their rank correlations. Suppose we have constructed a scale. The distance between the i-th and j-th objects using this scale is the $d_{ij} = \sqrt{(x_{1i} - x_{1j})^2 + (x_{2i} - x_{2j})^2}$. A plot of the M points (d_{ij}, δ_{ij}) might look like Figure 11.1. This scale as shown is poor because monotonicity is not preserved. That is, as the distance between objects increases, their dissimilarities do not always increase. We need a "goodness of fit" measure of how well a given scale preserves monotonicity. With such a measure we then choose the scale that minimizes this distance. Kruskal's goodness of fit measure, called the *stress*, of a scale $(x_{11}, x_{21}), \ldots, (x_{1n}, x_{2n})$ is defined as

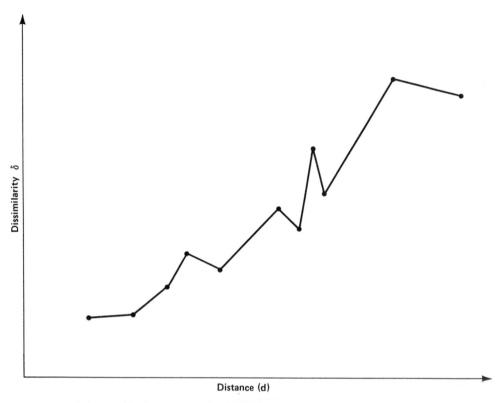

Figure 11.1 - Plot of Distances and Dissimilarities.

$$S(x) = \sqrt{\frac{\sum\limits_{\ell=1}^{M} (d_{i_\ell j_\ell} - \hat{d}_{i_\ell j_\ell})^2}{\sum\limits_{\ell=1}^{M} d^2_{i_\ell j_\ell}}}$$ (11.5)

where $\hat{d}_{i_1 j_1} \leq \hat{d}_{i_2 j_2} \leq \cdots \leq \hat{d}_{i_M j_M}$ is chosen to minimize the quantity in the numerator. Recall by (11.4) the ordering in the $\hat{d}$ agrees perfectly with the given matrix Δ. Thus the stress $S((x_{11}, x_{21}), \ldots, (x_{1n}, x_{2n}))$ is a measure of how consistent the scale $(x_{11}, x_{21}), \ldots, (x_{1n}, x_{2n})$ is

287

to the input data Δ. The $\hat{d}_{ij}$'s give a set of numbers matching the input data which are close to the d_{ij}'s. The stress measures how much monotonicity is preserved by the ordering. The solution is then to choose a scale x which minimizes the stress. There is a computer program developed by Kruskal to do just this.

Once a scale has been selected then regression analysis or any other kind of analysis may be done on the n points $(x_{11}, x_{21}), \ldots, (x_{1n}, x_{2n})$. Hence the circle is complete; scaled data to ranked data and back to scaled data.

11.5. CATEGORICAL DATA: SIMPLE ONE-WAY CHI-SQUARE TESTS

We turn now to the earliest nonparametric test in statistics; the chi-square test dates back to a paper in 1900 by Karl Pearson. The test is used when the data are classified into k mutually exclusive categories. The simplest form of the chi-square goodness of fit test is for testing the null hypothesis $(p_1^0, p_2^0, \ldots, p_k^0)$ that n data points are sampled independently with probability p_i^0 each of falling into category i. The data may already come to the analyst in categories with no numerical scale (e.g., brown hair, black hair, blond hair, etc.) or the categories may be created from a numerical scale (e.g., ages 16-20, 21-25, etc.). The chi-square goodness of fit test is nonparametric since any null hypothesis $(p_i^0, \ldots, p_k^0)$, can be tested subject only to the constraint $p_i^0 \geq 0$ and $\Sigma_{i=1}^k p_i^0 = 1$.

We begin by giving the theory for the simple one-way chi-square test and applying it to some data. Think of the n observations as being a random sample from a population classified into k categories with each observation having probability p_i of falling into the i-th category. If N_1, N_2, ..., N_k ($\Sigma_{i=1}^{k} N_i = n$) are the number of observations in each of the categories, the vector $(N_1, N_2, \ldots, N_k)$ is said to have a *multinomial distribution* with parameters n and $(p_1, \ldots, p_k)$. The joint distribution of $(N_1, \ldots, N_k)$ is

$$\text{Prob}(N_1 = x_1, N_2 = x_2, \ldots, N_k = x_k)$$

$$= \binom{n}{x_1, \ x_2, \ \ldots, \ x_k} p_1^{x_1} p_2^{x_2} \cdots p_k^{x_k}$$

(11.6)

where

$$\binom{n}{x_1, \ \ldots, \ x_k} = \frac{n!}{x_1!, \ \ldots, \ x_k!}$$

is the number of ways to divide n objects into k categories with x_1 objects in one category, x_2 objects in another category, etc. For the case k = 2, Equation (11.6) reduces to

$$P(N_1 = x_1, N_2 = n - x_1) = \frac{n!}{x_1!(n - x_1)!} p_1^{x_1} (1 - p_1)^{n - x_1}$$

(11.7)

that is to say, N_1 has a *binomial distribution* with parameters n and p_1. The same combinatorial reasoning given in Chapter 5 to get (11.7) as the binomial formula can be used to get Equation (11.6) as the multinomial formula. That derivation will not be repeated here. Notice that for any given category, say the i-th, N_i has a binomial distribution with parameters n and p_i so that $E(N_i) = np_i$ for i = 1, ..., k. The null hypothesis being tested is that each p_i is a known number p_i^0. We later cover the case where p_i^0 is a function of unknown parameters to be estimated.

Thus under the null hypothesis the expected value of N_i is np_i^0 which we will designate by e_i. The chi-square test criterion proposed by Karl Pearson is to reject if $\chi^2 = \Sigma_{i=1}^{k} (N_i - e_i)^2 / e_i$ is large. This test statistic has approximately a chi-square distribution with k-1 degrees of freedom. It is the same for a variable to have a χ^2 distribution with one degree of freedom as for it to be distributed as the square of a standard normal random variable. A little algebra shows that

$$z^2 = \frac{(N_1 - np_1^0)^2}{np_1^0(1 - p_1^0)} = \frac{(N_1 - np_1^0)^2}{np_1^0} + \frac{\left(N_2 - n(1 - p_1^0)\right)^2}{n(1 - p_1^0)} .$$

Since the right-hand side is the χ^2 criterion, the normal approximation to the binomial implies that z^2 has a limiting χ_1^2 distribution. A generalization of this argument proves Pearson's assertion.

We now apply the chi-square test to some data and characterize how large the sample size should be in order to use the chi-square approximation to the distribution of χ^2. In 1971, a survey at The Rand Corporation revealed that the research staff had the age distribution given in Table 11.2. There was some interest in whether the Rand age distribution was in any sense "typical." One natural age distribution to compare it to is that of the U.S. labor force in the relevant age range. A chi-square test using the proportions from the U.S. labor force as the null hypothesis will tell us whether the ages of the Rand research staff are consistent with being a random sample from the U.S. labor force--at least down to the level of aggregation of our categories. From the chi-square table with k-1 = 4 degrees of freedom, our value of 95.42 has a significance probability of less than 0.0001. That is, it is significant at the 0.0001 level. The contributions to the chi-square statistic are dominated by that of category 1 (age 20-24). Perhaps the chi-square statistic is large because Rand may not hire many 20-22 year old researchers (since researchers normally have

Table 11.2

CHI-SQUARE TEST OF 1971 RAND AGES

Category	Rand Observed N_i	U.S. Labor Force Proportion	U.S. Labor Force Expected e_i	Chi-Square Contribution $\dfrac{(N_i - e_i)^2}{e_i}$
1 (20–24)	10	0.147	64.83	46.37
2 (25–34)	150	0.235	103.64	20.74
3 (35–44)	120	0.227	100.11	3.95
4 (45–54)	128	0.235	103.64	5.73
5 (55–64)	33	0.156	68.80	18.63
Total	441	1.000	441	95.42

at least a bachelor degree). To check this we might compare the age 25 and older Rand professionals to the age 25 and older labor force.

Since the chi-square distribution is only the limiting distribution of χ^2 as n goes to infinity, some guidelines are needed as to when the approximation is accurate for the values of n which occur in practice. The customary recommendation is that the expected number in each category (e_i) be at least 5. Yarnold[*] has shown that this standard can be relaxed considerably without the accuracy of the approximation deteriorating. Yarnold's rule is: If the number of categories k is 3 or more, and if r denotes the number of expectations (e_i) less than 5, then the minimum expectation can be as small as 5r/k.

If the analyst is able to construct his own categories, he is faced with two conflicting forces. One force is to create a few large categories so that he can use the chi-square approximation

---

[*] J. K. Yarnold, "The Minimum Expectation in χ^2 Goodness of Fit Tests and the Accuracy of Approximation for the Null Distribution," *Journal of the American Statistical Association*, Vol. 65, 1970, pp. 864–886.

with confidence. The other force is to create many small categories so that differences are not aggregated away by the large categories; he wants to have as many degrees of freedom as possible. This is particularly important if one suspects that the observations differ from the hypothesized distribution only in the upper or lower tail of the distribution where there is very little probability and hence a low expected value.

11.6. CHI-SQUARE TESTS WITH FITTED EXPECTED VALUES: GOODNESS-OF-FIT AND INDEPENDENCE

In the more common situation when the p_i^0 are functions of unknown parameters to be estimated from the data, the chi-square test must be modified. We give the modification and then apply it to two common problems: (i) chi-square tests of whether the data comes from a distribution of some particular parametric form and (ii) two-way chi-square tests of independence for contingency tables.

The null hypothesis is that the p_i's are known functions of s unknown parameters. Thus under the null hypothesis the sample has been drawn from a population having distribution determined by $p_i = p_i(\alpha_1, \ldots, \alpha_s)$ for some set of values of the parameters $\alpha_1, \ldots, \alpha_s$. If the true values of the parameters were known we would calculate

$$\chi^2 = \sum_{i=1}^{k} \frac{(N_i - np_i(\alpha_1, \ldots, \alpha_s))^2}{np_i(\alpha_1, \ldots, \alpha_s)} \tag{11.9}$$

and apply the usual chi-square test with k-1 degrees of freedom. If the unknown $\alpha_1, \ldots, \alpha_s$ in Equation (11.9) are replaced by estimated values, then the p_i's are no longer constants and the theorem giving the limiting χ^2 distribution will no longer apply. The distribution of χ^2 with the estimated parameters will in general depend on the method of estimation. The estimators of $\alpha_1, \ldots, \alpha_s$ obtained by minimizing (11.9), known as the minimum chi-square estimates, when substituted into (11.9) give (11.9) a limiting chi-square distribution with k-s-1 degrees of freedom. This theorem allows us to use the theory presented in the last section to test various composite hypotheses. The same considerations about the expected number of observations per category (Yarnold's rule) also apply in this case. Strictly speaking, other estimates of $\alpha_1, \ldots, \alpha_s$ make the distribution of (11.9) stochastically larger than a chi-square variable with k-s-1 degrees of freedom under H_0, meaning one is too likely to reject. However, any of the common methods of estimating $\alpha_1, \ldots, \alpha_s$ gives χ^2 a limiting chi-square distribution with k-s-1 degrees of freedom, so is valid.

Application 1. To use a chi-square test to test whether a given sample is drawn from *some* normal distribution, divide the data into k classes $(-\infty, c_1), [c_1, c_2), \ldots, [c_{k-1}, \infty)$. Under the null hypothesis that the parent distribution is normal,

$$P_i(\mu, \sigma^2) = \frac{1}{\sqrt{2\pi}\,\sigma} \int_{c_{i-1}}^{c_i} e^{-\frac{(x-\mu)^2}{2\sigma^2}}\, dx \qquad (11.10)$$

is the area between c_{i-1} and c_i (where $c_0 = -\infty$, $c_k = \infty$) under a normal curve with mean μ and variance σ^2 (see Figure 11.2) Written another way

$$P_i(\mu, \sigma^2) = \Phi\left(\frac{c_i - \mu}{\sigma}\right) - \Phi\left(\frac{c_{i-1} - \mu}{\sigma}\right)$$

where $\Phi(x)$ is the area to the left of x under a standard normal curve.

293

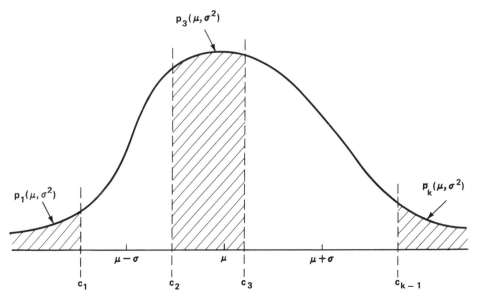

Figure 11.2 - The Normal Curve.

To illustrate the application, Table 11.3 from Cramer's book[*] gives mean June temperature in Stockholm for a 100-year period grouped into 10 intervals. There are two parameters, μ and σ^2, so $s = 2$ and we have $k-s-1 = 7$ degrees of freedom. The parameters are estimated directly from the data, in this case assuming that the observations occurred at the midpoints of the intervals. The value of χ^2 has a significance probability of 0.35 so that the data are consistent with assuming normality.

This method can be used in general for testing whether a sample is consistent with coming from any parametric distribution.

---

[*]H. Cramer, *Mathematical Methods in Statistics* (Princeton University Press, 1945).

Table 11.3

DISTRIBUTION OF MEAN TEMPERATURE FOR JUNE IN STOCKHOLM 1841-1940

Degrees Centigrade	Observed	Expected
-12.4	10	12.89
12.5-12.9	12	7.89
13.0-13.4	9	10.20
13.5-13.9	10	11.93
14.0-14.4	19	12.62
14.5-14.9	10	12.08
15.0-15.4	9	10.46
15.5-15.9	6	8.19
16.0-16.4	7	5.81
16.5-	8	7.93

$\bar{x} = 14.23$
$s = 1.574$
$\chi^2 = 7.86$ (7 degrees of freedom)
$p = 0.35$

NOTE: $\bar{x}$ and s are calculated from the ungrouped data.

Application 2. Two-Way Chi-Square Tests in Contingency Tables

Suppose now that n observations are classified into two-way categories to get a contingency table. We studied such tables in Chapter 2 where the observations were U.S. cities and the classifying variables were region and percent government employees.

Table 11.4

A CONTINGENCY TABLE

Arguments	1	J	Total
1	N_{11} N_{12} $\cdots$	N_{1J}	N_{1+}
2	N_{21} N_{22}		$\cdot$ $\cdot$ $\cdot$ $\cdot$
I	N_{I1} N_{I2} $\cdots$	N_{IJ}	N_{I+}
	N_{+1} $\cdots$	N_{+J}	n

We often wish to test the hypothesis that the two variables are independent. Let p_{ij} be the probability that a randomly chosen observation belongs to row i and column j. The independence hypothesis is that there exist $I + J$ nonnegative constants p_{i+}, p_{+j} the row and column probabilities, such that

$$p_{ij} = p_{i+}p_{+j}$$

$$\sum_i p_{i+} = \sum_j p_{+j} = 1 .$$

Under this hypothesis the distribution of the observations has $(I-1) + (J-1)$ unknown parameters since the last requirements of summing to 1 means that two of the constants say p_{I+}, p_{+J} can be expressed in terms of the remaining $I+J-2$.

For a chi-square test we calculate

$$\chi^2 = \sum_i \sum_j \frac{(N_{ij} - np_{i+}p_{+j})^2}{np_{i+}p_{+j}}$$

where p_{i+}, p_{+j} are replaced by estimates from the data. These estimates are $\hat{p}_{i+} = \frac{N_{i+}}{n}$, $\hat{p}_{+j} = \frac{N_{+j}}{n}$ so that under the null hypothesis of independence,

$$\chi^2 = \sum_{i=1}^{I} \sum_{j=1}^{J} \frac{(N_{ij} - \frac{N_{i+}N_{+j}}{n})^2}{\frac{N_{i+}N_{+j}}{n}}$$

(11.11)

$$= n \sum_{i=1}^{I} \sum_{j=1}^{J} \frac{N_{ij}^2}{N_{i+}N_{+j}} - n$$

has a chi-square distribution with $IJ-(I-1)-(J-1)-1 = (I-1)(J-1)$

degrees of freedom. Note that χ^2 is one of the measures of unrelated-ness for the city data in Chapter 2. Table 11.5 reproduces these data together with the chi-square value of 33.456 on $(I-1)(J-1) = (4-1)(6-1) = 15$ degrees of freedom. This is significant at the .005 level, but not at the .001 level.

As was pointed out in Chapter 2, testing for independence is the same as testing whether the two distributions are the same. Thus, saying that the percentage government employees is not independent of region is the same as saying the distribution of percentage government employees varies with region. More generally, this χ^2 test of inde-pendence is also a valid test of whether the J dimensional vectors $\underset{\sim}{p}_1, \ldots, \underset{\sim}{p}_I$ of I different, independent multinomial distributions (see formula (11.6) with $\underset{\sim}{p}_1 = (p_{11}, \ldots, p_{1J}), \ldots, \underset{\sim}{p}_I = (p_{I1}, \ldots, p_{IJ})$) are all the same.

Remarks: (i) If the number of degrees of freedom in a chi-square test is too large to be in the table, the normal approximation to the chi-square distribution can be used. Under the null hypothesis, $E(\chi^2) = k-s-1$ and $\sigma(\chi^2) = \sqrt{2(k-s-1)}$ so that for large values of $k-s-1$

$$Z = \frac{\chi^2 - (k-s-1)}{\sqrt{2(k-s-1)}}$$

has approximately a standard normal distribution. (ii) There are many possibilities for parameterizing the category probabilities besides the two applications given here. The logit model described in Chapter 8 is one such possibility. There our emphasis was on the method of estimation rather than on the χ^2 test.

Table 11.5

PERCENTAGE FEDERAL GOVERNMENT EMPLOYEES VS. REGION IN 125 U.S. CITIES

	Counts for Observed Percentage Categories of Government Employees (expected counts immediately below)						
Region	1 $(0 - \frac{1}{2})$	2 $(\frac{1}{2} - 1)$	3 $(1 - 1\frac{1}{2})$	4 $(1\frac{1}{2} - 2)$	5 $(2 - 2\frac{1}{2})$	6 $(2\frac{1}{2} -)$	Totals
NE 1	11 (6.480)	11 (7.920)	3 (6.480)	3 (2.880)	1 (2.160)	1 (4.080)	30
S 2	5 (8.856)	11 (10.824)	9 (8.856)	5 (3.936)	4 (2.952)	7 (5.576)	41
Cent 3	11 (6.480)	6 (7.920)	8 (6.480)	3 (2.880)	1 (2.160)	1 (4.080)	30
W 4	0 (5.184)	5 (6.336)	7 (5.184)	1 (2.304)	3 (1.728)	8 (3.264)	24
Totals	27	33	27	12	9	17	125

$$\chi^2 = \Sigma (0 - E)^2/E = 33.456$$

Part IV. ADDITIONAL TOPICS

Chapter 12

BAYES AND EMPIRICAL BAYES INFERENCE

12.1. BAYESIAN INFERENCE

We begin with an example that motivates the usefulness of Bayesian statistical inference. In the example, classical statistical methods give answers that clash with intuition.

Example 1. There are two observations from a uniform distribution centered at θ.

$$X_1, X_2: \quad U(\theta - \tfrac{1}{2}, \theta + \tfrac{1}{2}) \ .$$

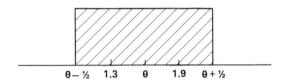

$$\theta - \tfrac{1}{2} \quad 1.3 \quad \theta \quad 1.9 \quad \theta + \tfrac{1}{2}$$

We desire an interval estimate for θ. An easy interval to calculate is the smallest and largest values of X_1 and X_2. If $L =$ maximum (X_1, X_2) and $S =$ minimum (X_1, X_2) then $\text{Prob}(S \le \theta \le L) = \tfrac{1}{2}$, so that (S, L) is a 50 percent confidence interval for θ. To see this, note that $P(S > \theta) = \tfrac{1}{2} \times \tfrac{1}{2} = \tfrac{1}{4}$, $P(L < \theta) = \tfrac{1}{2} \times \tfrac{1}{2} = \tfrac{1}{4}$, and $1 - \tfrac{1}{4} - \tfrac{1}{4} = \tfrac{1}{2}$. Now suppose the two observed values of X_1 and X_2 are 1.3 and 1.9. Since 1.3 and 1.9 are 0.6 units apart, both cannot be on one side of θ. Therefore $1.3 < \theta < 1.9$ with probability 1. Calling (1.3, 1.9) a 50 percent confidence interval seems misleading.

The distinction between Bayesian methods and non-Bayesian methods (covered thus far) is that non-Bayesian calculations are made before any data are taken. A (non-Bayesian) $1 - \alpha$ confidence level means that the probability of the random interval (S to L in the example) covering

the parameter is 1-α. It does not allow any attributes of the observed

values of S and L to modify the confidence statement. Bayesian calcu-

lations are made from the distribution of the parameters given the

actual observations. These ideas are elaborated below.

Example 2. Coin Tossing. Suppose a coin is about to be tossed.

It is known that the coin either has a head on both sides, a tail on

both sides, or a head on one side and a tail on the other. We observe

the outcome of one fair toss of the coin--either side is equally likely

to land face up. The problem is to decide what kind of coin was tossed.

If A = number of heads on the coin, the question is: does A = 0, 1, or

2? Letting X = the number of heads on the toss (0 or 1), then X = 1

implies A = 1 or 2, while X = 0 implies A = 0 or 1. To make a Bayes esti-

mate of the parameter A, we a priori assign probabilities to the possible

values of A. Suppose the assigned *prior* distribution consists of

probabilities 1/12, 1/2, 5/12 to the values 0, 1, and 2, respectively.

The *posterior* (a posteriori) probability distribution of A after ob-

serving X is defined as the conditional distribution of the parameter A

given the observation X. This distribution gives all the information

in the data about the parameter of interest. It is given below for

this example:

A	Prior Distribution P(A)	Conditional Distribution of X given A		Posterior Distribution	
		$P(X=1\|A)$	$P(X=0\|A)$	$P(A\|X=0)$	$P(A\|X=1)$
0	1/12	0	1	1/4	0
1	1/2	1/2	1/2	3/4	3/8
2	5/12	1	0	0	5/8

"Bayes theorem" allows us to calculate the last two columns. For example,

$$P(A=2|X=1) = \frac{P(A=2, X=1)}{P(X=1)}$$

$$= \frac{P(X=1|A=2)P(A=2)}{P(X=1|A=2)P(A=2)+P(X=1|A=1)P(A=1)+P(X=1|A=0)P(A=0)}$$

$$= \frac{1 \times 5/12}{1 \times 5/12 + \frac{1}{2} \times \frac{1}{2} + 0 \times 1/12} = \frac{5/12}{8/12} = \frac{5}{8} \ .$$

The above formula is called Bayes theorem. It gives a way of getting from $P(X|A)$ and $P(A)$ to $P(A|X)$. From the calculations, for our prior distribution we should infer that A is 1 if X = 0, but infer that A is 2 if X = 1.

   Example 1--Revisited. Suppose we assign a prior distribution on θ which is uniform between 0 and 10.

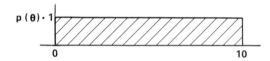

Thus the density of θ is given by

$$p(\theta) = \begin{cases} 0.1 \text{ if } 0 \leq \theta \leq 10 \\ 0 \text{ otherwise,} \end{cases}$$

and the density of θ given X_1 and X_2 (the posterior density) is

$$p(\theta|X_1 = 1.3, X_2 = 1.9) = \frac{p(\theta \text{ and } (X_1, X_2))}{p(X_1, X_2)}$$

$$= \frac{p(X_1, X_2|\theta)p(\theta)}{p(X_1, X_2)}$$

$$\propto 1 \times .1 \text{ if } 0 \leq \theta \leq 10$$

$$\text{and } \theta-\tfrac{1}{2} \leq X_i \leq \theta+\tfrac{1}{2}$$

$$= 0 \qquad \text{otherwise}$$

where "$\propto$" is read "is proportional to." That is,

$$P(\theta \,|\, X_1 = 1.3,\ X_2 = 1.9) \propto 0.1 \text{ if } \max(0,\ 1.4) \le \theta \le \min(10,\ 1.8)$$

$$= 0 \text{ elsewhere.}$$

The constant of proportionality is 2.5 in this case and is easily computed from the fact that a probability density must have unit area underneath it.

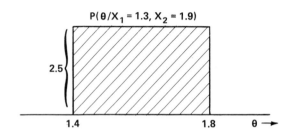

METHODS OF ASSIGNING PRIOR DISTRIBUTIONS

In some applications, extensive experience may make available a frequency distribution of values of θ in the past and it may be clear that θ itself is generated by a stable random process.

Sometimes no solid objective basis for choosing a prior distribution exists. One can subjectively choose a prior distribution by assigning betting odds to the possible parameter values. Although there are psychological difficulties for some people in giving odds, if one is willing to assign them consistently, a prior distribution can be assessed.

Inference About Proportions--The Binomial Distribution

In the coin tossing example the observed random variable X had a binomial distribution with n = 1 and unknown parameter p. The prior distribution put probabilities of 1/12, 1/2, 5/12 on the values for p of 0, 1/2, 1, respectively. We turn now to the more general situation

304

of making inferences about p when X has a binomial distribution. In general, any prior distribution of parameter p can be assessed. We will usually use a *prior density function* f, defined for $0 \leq p \leq 1$ to characterize the prior distribution on p.

The relation between a prior density of p and the posterior density g(p) given a binomial sample is given by

$$g(p) \propto f(p)p^{x}(1-p)^{n-x} \tag{12.1}$$

where n is the number of trials and x is the number of successes. We will frequently use the proportionality sign ($\propto$) rather than the equals sign when writing down densities to avoid having to recalculate normalization constants. The expression

$$\int_{0}^{1} f(p)p^{x}(1-p)^{n-x}$$

is omitted as a denominator in Equation (12.1).

For ease in calculations, it is convenient to use the beta family of distributions to generate prior densities on p. The beta density has two parameters (a,b) to be specified. The formula for a beta density with parameters a and b is

$$f(p) = \frac{p^{a-1}(1-p)^{b-1}}{\beta(a,\ b)} \tag{12.2}$$

where $\beta(a,\ b)$ is the beta function defined as

$$\beta(a,\ b) = \int_{0}^{1} p^{(a-1)}(1-p)^{(b-1)}dp\ .$$

Figure 12.1 shows the graphs of some beta densities with parameters a and b. The three densities in the top row have a = b beginning with the uniform (1, 1) density. The bottom three all have (b-1) = 2(a-1),

305

have a maximum at p = 1/3 and like the top densities get narrower from left to right. The posterior distribution of p given x successes and y = n-x failures in n Bernoulli trials with a prior (a, b) density is

$$g(p) \propto f(p)p^x(1-p)^y$$

$$\propto p^{a-1}(1-p)^{b-1}p^x(1-p)^y \tag{12.3}$$

$$= p^{a+x-1}(1-p)^{b+y-1} \ .$$

Thus, we see that a prior (a,b) density becomes a posterior (a+x, b+y) density.

In order to make inferences about p from the posterior distribution and to understand the implications of our choice of a prior distribution, formulas for the mean and variance of a beta distribution as a function of the parameters are useful. If p has an (a, b) density, its mean, variance and mode are given by:

$$\text{mean} = \frac{a}{a+b} \ ,$$

$$\text{variance} = \frac{ab}{(a+b)^2(a+b+1)} \ , \tag{12.4}$$

$$\text{and mode} = \frac{a-1}{a+b-2}$$

These formulas hold both for prior densities and for posterior densities. For moderate sizes of a and b the normal approximation to the beta distribution can be used to compute probabilities and the interpretation given below is valid. We write the mean and variance as

$$\mu = E(p) = \frac{a}{a+b}$$

$$\tag{12.5}$$

$$D = \sigma^2(p) = \frac{\mu(1-\mu)}{a+b+1} = \frac{\mu(1-\mu)}{N+1}$$

where $\mu = a/(a+b)$ and $N = a+b$. The motivation of the μ,N notation comes from the observation that if you start with a uniform prior opinion

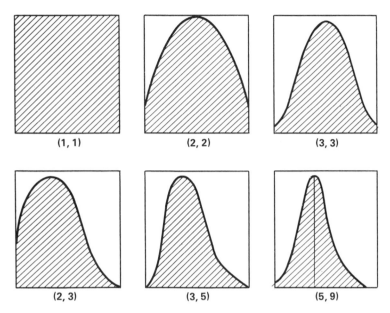

(1, 1)   (2, 2)   (3, 3)

(2, 3)   (3, 5)   (5, 9)

Figure 12.1 – Graphs of Beta Densities.

about p (a (1, 1) density), your posterior opinion after observing x
successes and y failures is a (x+1, y+1) density so that the posterior
distribution has mean approximately x/n (actually (x+1)/(n+2)) and
variance approximately [x/n(1-x/n)]/n (the sample estimate of the
variance of x/n). Thus our prior opinion is roughly equivalent in its
effect to getting "a" successes out of a+b = N trials. The "Bayes estimator" of p is the mean of the posterior distribution, (a+x)/(a+b+n).

The Simple Coin Tossing Example Revisited

Suppose we begin with a (1, 1) density--a uniform prior opinion
and observe 1 head. Our posterior opinion has a (2, 1) density or

307

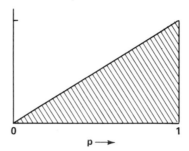

$$g(p) \propto p \text{ for } 0 \leq p \leq 1$$

so that the probability of p being 1/2 or less is 1/4. The posterior

mean is 2/3 and the posterior variance is $2/(3^2 \cdot 4) = 1/18$.

The posterior density formula (12.3), together with (12.4), yields

another intuitive fact: If the prior distribution is spread out and

the sample is fairly large, the posterior distribution will be about

the same for any choice of a prior distribution.

The mean and variance formulas show that a prior beta distribution

may be thought of as having $N = a{+}b$ previous observations with proportion

μ of successes ($a = x_0 = \mu N$ successes). The *Bayes estimator* is defined

as the mean of the posterior distribution or

$$E(p|x) = p^* = \frac{x_0 + x}{N+n} = \left(\frac{n}{N+n}\right)\frac{x}{n} + \left(\frac{N}{N+n}\right)\frac{x_0}{n_0} \tag{12.6}$$

which is a mixture of two estimators weighted by the amount of data

relevant to each. The prior distribution is treated like a sample

from a $Bin(N,\mu)$ distribution.

Example 3. Roberto Clemente's 1970 Batting Average. Early in the

1970 baseball season, Roberto Clemente was batting 0.400 after his first

45 times at bat, but despite his considerable ability, few thought he

was a "true 0.400 hitter." Taking Clemente's batting average for the

308

remainder of the 1970 season to be indicative of his "true ability" our problem is as follows. Given Clemente's batting average of 0.400 for the first 45 at bats (x = 18, y = 27), estimate his whole season average. Based on our knowledge of Clemente's hitting prowess during the 10 years prior to 1970, we will take our prior distribution on p, Clemente's true average, as having a mean μ = 0.322 and standard deviation of 0.014. This corresponds approximately (Eq. 12.4) to using a (358,755) prior density. Thus our posterior distribution has an (a+x, b+y) = (376,782) density. Our estimate of Clemente's average for the *remainder* of the season, based on his first 45 at bats, is 0.325 = 376/1158 = (a+x)/(a+b+n). Since he batted 412 times in all, our estimate of his complete season's batting average is

$$\frac{45}{412} (0.400) + (1 - \frac{45}{412}) 0.325 = 0.333 \ .$$

In fact, his 1970 average was 0.352, and he hit 0.346 for his last 367 at bats.

Credibility Intervals

Bayesian inferences are made conditional on the actual *data observed*. As we saw in the uniform distribution example at the beginning of the chapter, the usual confidence intervals are not conditional on the data. Posterior distributions allow one to make statements about an unknown parameter θ like

$$P\big(L(x) \leq \theta \leq U(x) | X = x\big) = 0.95 \ .$$

The interval (L,U) is called a *95 percent credibility interval*. A credibility interval of level 1-α is usually constructed from the posterior distribution by taking the $\alpha/2$ and 1-$\alpha/2$ quantiles as L and U, respectively. Credibility intervals for proportions can be constructed by using the normal approximation to the beta distribution for large "a+b" = N. For smaller values of (a, b) we can find credibility intervals from tables of the F-distribution using the following relationships. *If X has a beta distribution with parameters (a, b) then Y = bX/a(1-X) has an F-distribution with (2a, 2b) degrees of freedom, and Z = a(1-X)/bX has an F-distribution with (2b, 2a) degrees of freedom.* The above implies that X = Ya/(aY+b) or X = a/(bZ+a). Thus, if ℓ and u are the $\alpha/2$ and $1-\alpha/2$ quantiles of an F(2a, 2b) distribution where the posterior distribution of p is beta (a, b),

$$P(\ell \leq \frac{bp}{a(1-p)} \leq u) = 1 - \alpha$$

and therefore

$$P(\frac{a\ell}{a\ell+b}) \leq p \leq \frac{au}{au+b}) = 1 - \alpha \ .$$

Thus, (aℓ/(aℓ+b), au/(au+b)) is a $1-\alpha$ credibility interval for p.

Example. Credibility intervals for Roberto Clemente's true batting average. Let p be Clemente's "true average." Then using the normal approximation, *a posteriori*, a 90 percent credibility interval for p is $0.325 \pm 1.645 \times \sqrt{0.325(1 - 0.325/1159}$ or 0.302 to 0.348.

If on the other hand, you had been totally ignorant about the game of baseball, you might choose a (1, 1) prior. In this case, your posterior density is (18 + 1, 27 + 1) = (19, 28). Here the F-tables might be used for 38 and 56 degrees of freedom. Some interpolation is necessary to get $(\ell, u) \doteq (1/1.67, 1.63) = (.60, 1.63)$. Thus a 90 percent credibility interval is (19 × .60/(19 × .60 + 28), 19 × 1.63/(19 × 1.63 + 28)) or $P(0.289 \leq p \leq 0.525) = 0.90$.

Note that inferences about the "true batting average" and what actually occurred for the remaining 367 at bats can be different. For the latter, the additional binomial randomness must be accounted for.

12.2. EMPIRICAL BAYES ESTIMATORS

The Bayes estimate of Clemente's batting average was a better estimate of his average over the remainder of the season than his initial 0.400 average because "good" values of a and b were used. An analyst with no knowledge of baseball would use a much smaller value of a+b (relative to n) to represent his ignorance, however. He therefore could not expect to do better than 0.400 as an estimate were he to use a Bayes rule. If he chose a large value of a+b coupled with an inappropriate value of $\mu = a/(a+b)$, he would get a disastrously bad estimate.

If he is simultaneously estimating the batting averages of many players, then the analyst may be able to provide much better estimates than the unbiased estimates by using an *empirical Bayes* estimator. He can do this even with no knowledge of baseball as the following example will illustrate.

Given in Table 12.1 are the batting averages of eighteen major league players through their first 45 official at bats in the 1970 season.[*]

The problem is to predict each player's batting average p_i over the remainder of the season using the data of the 2nd column and no other data or knowledge of baseball or these players. By using X_1

---

[*]See B. Efron and C. Morris, "Stein's Paradox in Statistics," *Scientific American*, May 1977, pp. 119-127, and B. Efron and C. Morris, "Data Analysis Using Stein's Estimator," *Journal of the American Statistical Association*, June 1975, pp. 311-319.

as the estimate of p_i, the average squared error of prediction is

$$\frac{1}{18} \Sigma (X_i - p_i)^2 = 0.077 .$$

The unbiased estimate X_i of p_i has variance $= p_i(1-p_i)/n$, where $n = 45$ at bats.

Table 12.1

1970 BATTING AVERAGES FOR EIGHTEEN PLAYERS

i		p_i	X_i	$\hat{p}_i$
1	Clemente (Pitts, NL)	0.346	0.400	0.294
2	F. Robinson (Balt, AL)	0.298	0.378	0.289
3	F. Howard (Wash, AL)	0.276	0.356	0.284
4	Johnstone (Cal, AL)	0.221	0.333	0.279
5	Berry (Chi, AL)	0.273	0.311	0.275
6	Spencer (Cal, AL)	0.270	0.311	0.275
7	Kessinger (Chi, NL)	0.263	0.289	0.270
8	L. Alvarado (Bos, AL)	0.210	0.267	0.265
9	Santo (Chi, NL)	0.269	0.244	0.261
10	Swoboda (NY, NL)	0.230	0.244	0.261
11	Unser (Wash, AL)	0.264	0.222	0.256
12	Williams (Chi, AL)	0.256	0.222	0.256
13	Scott (Bos, AL)	0.304	0.222	0.256
14	Petrocelli (Bos, AL)	0.264	0.222	0.256
15	E. Rodriquez (KC, AL)	0.226	0.222	0.251
16	Campaneris (Oak, AL)	0.285	0.200	0.251
17	Munson (NY, AL)	0.319	0.178	0.246
18	Alvis (Mil, NL)	0.200	0.156	0.242

NOTE: Col 1: Batting average for remainder of season.
       Col 2: Batting average after first 45 at bats.
       Col 3: Empirical Bayes Estimates of batting average.

Assume again that the unknown values p_i are independently sampled from the same beta distribution with parameters a and b, with a,b unknown. We will estimate the common values of a and b from the data on all 18 players. Then our *empirical Bayes* estimates of p_i will be the mean of the posterior distribution with (a,b) replaced by estimates $(\hat{a},\hat{b})$. Thus, the empirical Bayes estimate of p is

$$\hat{p}_i = \frac{x_i + \hat{a}}{n + \hat{a} + \hat{b}}$$

where $n = 45$ and x_i is the number of hits the i-th player has in his first 45 at bats. The method of maximum likelihood can be used to estimate a and b. That is $(\hat{a}, \hat{b})$ are defined as those values of (a, b) for which the probability of observing x_i hits out of n at bats $(i = 1, \ldots, 18)$ is maximized, but the mathematics in calculating $(\hat{a}, \hat{b})$ are beyond the scope of the chapter and an alternative method works as well in this case, based on the method of moments.[*] We do not estimate a, b directly, but write the Bayes estimator as

$$\hat{p}_i = \frac{x_i + a}{n + a + b} = (1 - B)\frac{x_i}{n} + B\mu \tag{12.7}$$

where $\mu = a/N$ and $B = (a+b)/(a+b+n) = N/(N+n)$ as in (12.6). But B also may be written[**]

$$B = \frac{E[Var(x_i/n|p_i)]}{Var(x_i/n)} = \frac{Ep_i(1 - p_i)/n}{Var(x_i/n)} \tag{12.8}$$

and we approximate $Ep_i(1 - p_i)/n$, which does not depend on i, by $\mu(1-\mu)/n$ with $\hat{\mu} = \bar{p} = \sum_1^k (x_i/n)/k$, the average of the $k = 18$ estimates x_i/n of p_i. The variance of x_i/n is the marginal variance, with both p_i random and x_i random given p_i. This variance would be estimated by the unbiased estimate

$$\frac{S}{k-1} = \frac{1}{k-1}\sum_1^k \left(\frac{x_i}{n} - \bar{p}\right)^2 , \tag{12.9}$$

---

[*]See Carter and Rolph (1973), Appendix C, for the maximum likelihood approach.

[**]This approach and calculation is explained more fully in Efron and Morris (1977), (1975, and (1973).

but since it is the reciprocal variance that must be estimated, it is better to estimate

$$\frac{1}{\text{Var}(x_i/n)} \quad \text{by} \quad \frac{k-3}{S} \ .$$

Thus we estimate B by $\hat{B} = [(k-3)\bar{p}(1-\bar{p})/n]/S$. For these data, $\bar{p} = 0.265$, $S = 0.08251$, $n = 45$, $k = 15$, so

$$\hat{B} = \frac{(k-3)\bar{p}(1-\bar{p})/n}{S} = \frac{15(0.00433)}{0.08251} = 0.787$$

and thus from (12.6)

$$\hat{p}_i = 0.213 \frac{x_i}{n} + 0.787 \ (0.265) \ .$$

Table 12.1 gives these values for $\hat{p}_i$. The approximate corresponding values of $(\hat{a}, \hat{b})$ are (44, 122). Thus a single player's prediction is gotten by taking his performance with 45 at bats and adding an additional 166 at bats with 44 hits, a proportion of 0.265.

The empirical Bayes estimates are better than X_i for 16 of 18 batters, as a comparison of the estimates in Table 12.1 with the values p_i shows.

Clemente's empirical Bayes estimate 0.294 is not as close to his actual performance 0.346 as the Bayes estimate 0.325, but it is still slightly closer than $X_i = 0.400$. Effectively, for him, a = 44, b = 122 is not as good as a = 364, b = 756 for the Bayes case. But less was assumed known than for the Bayes case. This empirical Bayes method assumes that although the prior distribution is unknown, it can be estimated using the performance of all the players. If the players' performances are not comparable, then a+b will be estimated to be near zero ($\hat{a}+\hat{b} \ll n$), and the empirical Bayes estimate will differ little from the unbiased estimator X_i. By letting the data determine if the performances were comparable, we avoid the possible bad estimates that can result

314

from using bad a priori information. This comparison applies to all
18 players, of course. Our knowledge that Clemente was an usually fine
hitter is ignored when we estimate his performance using all players'
data.

EMPIRICAL BAYES ESTIMATES OF THE PROBABILITY THAT A FIRE ALARM IS A STRUCTURAL FIRE*

When a fire alarm is reported by someone pulling a street alarm
box, the dispatcher must decide how many (and which) units to send to
the alarm. Particularly in busy periods, he may vary the number of
units depending on his assessment of the probability that an alarm is
a serious fire. For our purposes we will take serious fire to mean
fire in a structure.

The obvious estimate of the conditional probability of a box-
reported alarm being a structural fire is the proportion of box-reported
alarms in the past several years at that location that were structural
fires. If "location" means alarm box, insufficient data is a problem
for the application below since at a given alarm box there are typically
5 to 40 box-reported alarms per year with about 15 percent of them
signaling structural fires. Using "location" to mean neighborhood
(4 to 30 boxes) solves the sample size problem but raises the issue:
Do alarm boxes in the same neighborhood have the same conditional probability of a box-reported alarm signaling a structural fire? As we
will see, the answer is frequently so.

Using data for 1967-1970 from the borough of the Bronx in New York
City, alarm boxes were grouped into neighborhoods using the following

---

*This application is taken from G. M. Carter and J. E. Rolph,
"Empirical Bayes Methods Applied to Estimating Fire Alarm Probabilities," *Journal of the American Statistical Association*, December 1974,
pp. 880-885.

rules:

(1) Neighborhoods should be geographically connected.

(2) Boxes with obvious geographical properties, e.g., those
around parks, or on highways, should be grouped together.

(3) Each neighborhood should have at least 100 box-reported
alarms in the period 1967-1969.

(4) Each neighborhood should have at least 4 alarm boxes.

For each of the 216 neighborhoods, empirical Bayes estimators were
used to get individual box estimates of the "probability structural."
Suppose there are k boxes in the neighborhood; let X_i be the proportion
of box-reported alarms that signal structural fires at box i; i = 1,
..., k and n_i is the total number of box-reported alarms at box i.
Then if p_i is the unknown "probability structural" we have

$$\text{Var}(X_i) = \sigma_1^2 = \frac{p_i(1-p_i)}{n_i}$$

which depends on i. Using a modification of the constant variance
empirical Bayes estimator given in the baseball example and using

$$\hat{\sigma}_i^2 = \frac{\hat{p}(1-\hat{p})}{n_i}, \quad \text{where } \hat{p} = \frac{\Sigma n_i X_i}{\Sigma n_i},$$

the values of p_i were estimated.

Data from 1967-1969 were used for estimating p_i and these estimates
were compared to 1970 data. Hypothesis tests showed that a statistically
significant improvement resulted from using the empirical Bayes estimators rather than estimating each p_i separately using only the data
from box i.

The General Approach

The baseball player and fire alarm examples illustrate how empirical Bayes estimators are related to their Bayes counterprarts. Our theory and examples have been for the binomial distribution, but the approach is more widely applicable, particularly to the normal distribution. See the Efron and Morris references for the normal distribution case. Aside from their appeal as Bayes estimators with data based prior distributions, empirical Bayes estimators have desirable properties in their own right. In practice, one commonly used estimation procedure is the preliminary-test estimator. In the baseball batting average example, this procedure consists of first testing the hypothesis that all the p_i's are equal and then estimating p_i by either X_i/n_i of $(\Sigma X_i)/(\Sigma n_i)$, depending on whether the test rejects or accepts the hypothesis of equality. This procedure can be written as the estimator

$$\hat{p}_i^* = (1-B) \frac{X_i}{n_i} + B \frac{\Sigma X_i}{\Sigma n_i}$$

where

$$B = 1 \text{ if } \chi^2 < c$$

$$= 0 \text{ if } \chi^2 \leq c$$

with χ^2 being the chi-square statistic for testing equality of the p's with $17 = 18-1$ degrees of freedom and c the cutoff point for the test. The empirical Bayes estimator is

$$\hat{p}_i = (1-\hat{B}) \frac{X_i}{n_i} + \hat{B} \frac{\hat{a}}{\hat{a}+\hat{b}}$$

where

$$\hat{B} = \frac{\hat{a}+\hat{b}}{n_i+\hat{a}+\hat{b}} \quad \text{and} \quad \frac{\hat{a}}{\hat{a}+\hat{b}} \doteq \frac{\Sigma X_i}{\Sigma n_i} .$$

317

Roughly speaking, $\hat{a}+\hat{b}$ and hence $\hat{B}$ is a smooth function of the chi-square statistic. Thus, the empirical Bayes estimator can be viewed as a smooth compromise between using X_i/n_i and $\Sigma X_i/\Sigma n_i$ rather than using either one or the other depending on whether χ^2 is above or below c as the preliminary test estimator does. It can be shown that this smooth compromise gives smaller average squared error of estimation than the preliminary test estimator does.[*] Thus, an empirical Bayes estimator is preferable to the common procedure of first using a hypothesis test and then estimating.

Empirical Bayes estimators can be used in many other situations, perhaps in different forms. A particularly important application to revenue sharing stabilizes income estimates based on local sampling of incomes from the decennial census. Fay and Herriot (1979) describe in that application how one shrinks sample estimates toward regression estimates using empirical Bayes methods. They show with census data that federal revenues are shared more equitably when income estimates are based on empirical Bayes methods.

---

[*] See S. Sclove, C. Morris, and R. Radhakrishnan, "Non-optimality of Preliminary-Test Estimators for the Mean of a Multivariate Normal Distribution," *Annals of Mathematical Statistics*, Vol. 43, No. 5, 1972, pp. 1481-1490.

Chapter 13

SAMPLING METHODS

In the previous chapters our observations have consisted of a
simple random sample from some probability distribution. In multiple
regression the errors were sampled randomly from a Normal (0, σ^2) distribution while in logit regression the y_i's were randomly sampled
Bernoulli variables. When we wish to estimate certain characteristics
of a population, we frequently can improve the accuracy of the estimate
for a given sample size by drawing something other than a simple random
sample from the population. In this chapter we describe several ways
of sampling from populations that either reduce standard errors of
estimates or reduce costs.

13.1. FINITE AND INFINITE POPULATIONS

We distinguish first between sampling from *finite* and *infinite*
populations. Examples of finite populations are Los Angeles residents
at a specific time, U.S. Air Force officers, and the students in a
given university. An example of an infinite population is the result
of all hypothetically possible coin tosses from the nickel in your
pocket, i.e., by sampling (removing) one toss, we do not diminish or
change the population in any way. We may regard the exam scores of
10 students as a sample from the hypothetical infinite population of
exam scores that students (present, past, future, etc.) might achieve
on this exam. In an infinite population, each successive randomly
selected item is governed by the same probability with the selections
being independent of one another. Note that successive randomly

sampled items from a finite population can have this property only if the sampled item is replaced after each selection so that it may be selected again. This is called sampling *with replacement*. Note that sampling with replacement from an infinite population is the same as sampling without replacement since there is no chance of drawing the same object twice. The reader may want to review additional material on sampling from Chapter 5.

Sampling from a finite population is best done without replacement. Sampling from an infinite population has the same characteristics as sampling with replacement from a finite population. Thus far we have considered sampling with replacement (or equivalently, sampling from an infinite population). These ideas have been abstracted in past chapters and presented as (i) observing Bernoulli trials and estimating the unknown p from the sample using the binomial distribution and (ii) sampling from a normal distribution and estimating some quantity like the mean or regression coefficients from the sample. Now let us consider finite population theory.

13.2. THE FINITE SAMPLING CORRECTION FACTOR FOR THE VARIANCE

Suppose there are N items in the population with values v_1, ..., v_N. The population mean is defined as

$$\mu = \frac{v_1 + \ldots + v_N}{N} = \bar{v} \, ,$$

while the population variance is

$$\sigma^2 = \frac{1}{N} \left[(v_1 - \bar{v})^2 + \ldots + (v_N - \bar{v})^2 \right] \, .$$

The problem is to estimate the unknown mean μ.

A sample of size n is drawn at random from the population, with each of the N items having the same chance of being selected on any given draw. Let X_1 be the v value of the first item drawn first, X_2 be the v value of the second item drawn, etc. Thus X_1, X_2, ..., X_n all have the same distribution. Their common mean, and variance is given by

$$E(X_1) = \ldots = E(X_N) = \overline{v} = \mu$$

$$Var(X_1) = \ldots = Var(X_N) = \sigma^2 .$$

The usual estimator of μ is $\overline{X} = \frac{1}{n} \Sigma_{i=1}^{n} X_i$.

It is always true that the mean of the sum of random variables is the sum of the means. Thus,

$$E(\overline{X}) = E(\frac{X_1 + \ldots + X_n}{n})$$

$$= \frac{1}{n} \sum_{i=1}^{n} E(X_i) = \frac{1}{n} n\mu = \mu .$$

That is, $\overline{X}$ is an unbiased estimator of the population mean.

If we sample *with replacement*, X_1, X_2, ..., X_n all have the same distribution and are independent of one another. A consequence of this independence is that the variance of their sum is the sum of their variances. Thus, if sampling is with replacement,

$$Var(\overline{X}) = Var(\frac{1}{n} \sum_{i=1}^{n} X_i)$$

$$= \frac{1}{n^2} Var(\sum_{i=1}^{n} X_i)$$

$$= \frac{1}{n^2} \sum_{i=1}^{n} Var(X_i)$$

$$= \frac{1}{n^2} n\sigma^2 = \frac{\sigma^2}{n} .$$

Recall that we used this fact in Chapters 5, 6 for inferences about binomial and normal means.

When sampling *without replacement* the values X_1, X_2, ..., X_n are dependent. Intuitively the variance of $\overline{X}$ should be smaller in the case of no replacement. For example, the extreme situation occurs when we sample the entire population, so n = N. Then

$$\text{Var}\left(\frac{1}{N} \sum_1^N X_i\right) = 0$$

since $\overline{X} = \mu$. It is shown below that the formula for the variance generally must be modified by the factor (N-n)/(N-1) to correct for the dependence. Thus, for sampling without replacement

$$\text{Var}(\overline{X}) = \frac{N-n}{N-1} \times \frac{\sigma^2}{n}$$

with (N-n)/(N-1) called the *finite sampling correction factor*. To prove this, let $c = \text{Cov}(X_i, X_j)$ be the covariance between X_i and X_j when $i \neq j$. Note that the symmetry of sampling without replacement permits us to assume that c is independent of i, j. If i = j then $\text{Cov}(X_i, X_j) = \text{Cov}(X_i, X_i) = \text{Var}(X_i) = \sigma^2$ which is also independent of i. We may now solve for c from σ^2 using the fact that $\sum_1^N X_i = N\mu$ so $\text{Var}(\sum_1^N X_i) = 0$. The formula $\text{Var}(\sum_1^N X_i) = \text{Cov}(\sum_1^N X_i, \sum_1^N X_j) = \sum_1^N \sum_1^N \text{Cov}(X_i, X_j)$ always applies and gives $0 = N\sigma^2 + N(N-1)c$ since there are N terms where i = j and N(N-1) terms where $i \neq j$. Thus $c = -\sigma^2/(N-1)$. Knowing c, we now repeat the argument on X_1, ..., X_n as follows to find $\text{Var}(\overline{X})$.

$$\text{Var}(\overline{X}) = \text{Var}(\frac{1}{n} \sum_1^n X_i) = \frac{1}{n^2} \text{Var}(\sum_1^n X_i)$$

$$= \frac{1}{n^2} \text{Cov}(\sum X_i, \sum X_j) = \frac{1}{n^2} \sum \sum \text{Cov}(X_i, X_j)$$

$$= \frac{1}{n^2} \{ n\sigma^2 + n(n-1)c \}$$

$$= \frac{1}{n^2} \{ n\sigma^2 - n(n-1)\sigma^2/(N-1) \}$$

$$= \frac{\sigma^2}{n} \{ \frac{N-n}{N-1} \} .$$

Note that this formula gives correct answers at $n = 1$ when $\overline{X} = X_1$, so $\text{Var}(\overline{X}) = \sigma^2$, and for a census at $n = N$ when $\overline{X} = \mu$ so $\text{Var}(\overline{X}) = 0$.

Example. Fifty Los Angeles residents are sampled without replacement and their heights are measured. The variance of

$$\overline{X} = \frac{X_1 + \cdots + X_{50}}{50} \text{ is } \frac{3,000,000-50}{3,000,000-1} \times \frac{\sigma^2}{50}$$

where σ^2 is the variance of the height of a randomly sampled resident of Los Angeles (population = 3,000,000). In this case the finite sampling correction factor makes little difference, and it only matters if n/N is not small, that is, if a significant portion of the population is sampled.

Beside the decision of whether to replace when sampling, there are other ways of increasing the accuracy of estimates, involving the choice of the sample. The two sampling procedures considered here are (a) to stratify the population into homogeneous groups, each of which has a smaller variance of the measured variable and; (b) to use an auxillary variable to decrease the variance of the parameter estimates via ratio or regression estimates. Finally, procedures such as cluster sampling can be used to decrease sampling costs, although these gains tend to be offset somewhat by attendant increases in variance. Our goal here will always be to estimate the unknown mean of the population. This choice of objective has a profound effect on the sample design. Other objectives are considered in the next chapter.

13.3. STRATIFIED SAMPLING: PROPORTIONAL AND OPTIMAL ALLOCATION

It often is possible to divide the population into a number of *strata* (subpopulations) where the values of v within each stratum are more homogeneous (more nearly constant) than in the population as a whole. For example, dividing Los Angeles residents into male and female makes height distributions more nearly constant within each stratum. If we then draw well-chosen small samples from each stratum separately, our estimator will be more accurate than taking a single large sample from the entire population.

Suppose there are L strata of size N_1, ..., N_L; $N_1 + ... + N_L = N$. Assuming for now that N_1, ..., N_L are known, we define $p_1 = N_1/N$, ..., $p_L = N_L/N$ as the proportion of subjects in each stratum, and let

$$\bar{v} = p_1\bar{v}_1 + ... + p_L\bar{v}_L$$

where $\bar{v}_1$, $\bar{v}_2$, ..., $\bar{v}_L$ are the average values of the characteristic v being measured in the 1st, 2nd, ..., L-th strata. Suppose that n_1, n_2, ..., n_L items are drawn without replacement from each stratum $\bar{X}_1$, $\bar{X}_2$, ..., $\bar{X}_L$ are the sample averages for each stratum. We simply are sampling from each stratum as a separate finite population, and the theory developed in the preceding section yields for i = 1, 2, ..., L:

$$E(\overline{X}_i) = \overline{v}_i$$

$$Var(\overline{X}_i) = \frac{N_i - n_i}{N_i - 1} \frac{\sigma_i^2}{n_i} ,$$

where σ_i^2 is the variance of a single observation sampled from the i-th stratum. The separate estimates $\overline{X}_i$ for $\overline{v}_i$ can be combined to produce an unbiased estimator X^* for $\overline{v}$, the population mean,

$$X^* = p_1 \overline{X}_1 + p_2 \overline{X}_2 + \ldots + p_L \overline{X}_L .$$

It is straightforward to show that X^* is unbiased

$$E(X^*) = \overline{v}$$

and

$$Var(X^*) = p_1^2 Var(\overline{X}_1) + \ldots + p_L^2 Var(\overline{X}_L)$$

$$= \sum_{i=1}^{L} p_i^2 \frac{N_i - n_i}{N_i - 1} \frac{\sigma_i^2}{n_i}$$

$$= \sum_{i=1}^{L} p_i^2 \frac{N_i}{N_i - 1} \frac{\sigma_i^2}{n_i} - \sum_{i=1}^{L} p_i^2 \frac{1}{N_i - 1} \sigma_i^2 .$$

Let us now assume that N_i is sufficiently large that we may approximate $N_i/(N_i-1)$ by 1. This assumption is unimportant, but it is made because the neater formula now holds:

$$Var(X^*) \doteq \sum_{i=1}^{L} p_i^2 \sigma_i^2 / n_i - \sum_{i=1}^{L} p_i^2 \sigma_i^2 / N_i$$

$$= \sum_{i=1}^{L} p_i^2 \sigma_i^2 / n_i - \frac{1}{N} \Sigma p_i \sigma_i^2 .$$

If the total sample size $n = n_1 + \ldots + n_L$ is fixed how should we allocate the sample among the strata? A natural procedure is to allocate the sample proportionally to the size of the strata. *Proportional sampling* uses

$$n_i \doteq n\, p_i \qquad i = 1, \ldots, L .$$

This results in

$$\text{Var}(X^*) \doteq \frac{p_1 \sigma_1^2 + \ldots + p_L \sigma_L^2}{n} \times \left(\frac{N-n}{N}\right) .$$

This variance is always less than or equal to

$$\text{Var}(\overline{X}) = \frac{\sigma^2}{n} \frac{N-n}{N}$$

(where we have again replaced N-1 in the denominator of the finite sampling correction factor $(N-n)/(N-1)$ for $\text{Var}(\overline{X})$). In fact, $\text{Var}(\overline{X}) = \text{Var}(X^*) + \left(\frac{N-n}{N}\right) \times \frac{1}{n} \Sigma_1^L\, p_i (\overline{v}_i - \overline{v})^2$ since $\sigma^2 = \Sigma\, p_i \sigma_i^2 + \Sigma\, p_i (\overline{v}_i - \overline{v})^2$, $\overline{v} = \Sigma\, p_i \overline{v}_i$ so that the improvement depends on choosing strata with different means $(\overline{v}_1, \ldots, \overline{v}_L)$ and the improvement increases as the $\{\overline{v}_i\}$ are more variable.

The *optimal stratified sampling allocation* for known variances for a sample of size n, given a set of strata, makes $n_1, \ldots, n_L$ proportional to $p_1 \sigma_1, \ldots, p_L \sigma_L$ so that

$$n_i = n\, \frac{p_i \sigma_i}{\Sigma_j p_j \sigma_j} \quad \text{for } i = 1, \ldots, L$$

and the variance of X^* for this design is

$$\text{Var}(X^*) = \frac{1}{n}(\Sigma p_i \sigma_i)^2 - \frac{1}{N} \Sigma p_i \sigma_i^2 .$$

This diminishes the variance obtained from proportional sampling by the amount $\Sigma p_i (\sigma_i - \overline{\sigma})^2$ where $\overline{\sigma} \equiv \Sigma p_i \sigma_i$ is the average standard deviation. Thus, there is no improvement relative to proportional sampling unless the standard deviations differ over strata, and the improvement then is proportional to the variance of the $\{\sigma_i\}$.

This scheme requires not only knowledge of strata sizes $N_1, \ldots, N_L$ but also strata standard deviations $\sigma_1, \ldots, \sigma_L$. Modifications such as double sampling and sequential sampling designs can be used in the presence of unknown N_i and σ_i by estimating their value for use in the second stage from the observed first stage values.

The preceding ignores potential differential costs for sampling the different strata. Now suppose the sampling costs are known to be c_1 for the first stratum, $\ldots, c_L$ for the last stratum. If $c_1 = \ldots = c_L$ then no change would be made from the preceding. Suppose a total budget $C = \Sigma c_i n_i$ is allowed. We are then to choose the $\{n_i\}$ to minimize $\text{Var}(X^*) = \Sigma p_i^2 \sigma_i^2 / n_i - \frac{1}{N} \Sigma p_i \sigma_i^2$. The solution is to take n_i proportional to $p_i \sigma_i / \sqrt{c_i}$. Thus

$$n_i = \left\{ \frac{C}{\Sigma \sqrt{c_i}\, p_i \sigma_i} \right\} \frac{p_i \sigma_i}{\sqrt{c_i}} .$$

Consequently, if the cost of a stratum goes up relative to the other strata, the sample size for that stratum is decreased by $1/\sqrt{c_i}$, but the expenditure $n_i c_i$ for that stratum is increased by $\sqrt{c_i}$. The variance is now reduced to

$$\text{Var}(X^*) = \frac{(\Sigma \sqrt{c_i}\, p_i \sigma_i)^2}{C}$$

which improves $\text{Var}(X^*)$ for the design using information about the $\{N_i, \sigma_i\}$ but not about the $\{c_i\}$, taking $n = C \Sigma p_i \sigma_i / \Sigma c_i p_i \sigma_i$ to make total costs of the two methods be equivalent. The amount of improvement is

327

$$\frac{1}{C}(\Sigma \ p_i\sigma_i)\Sigma(\sqrt{c_i} - \sqrt{c})^2 \ p_i\sigma_i, \quad \sqrt{c} \equiv \frac{\Sigma\sqrt{c_i}p_i\sigma_i}{\Sigma p_i\sigma_i} \ .$$

Thus the improvement is greater when $\sqrt{c_i}$ is more variable.

13.4. RATIO AND REGRESSION ESTIMATORS

So far we have looked only at stratifying the population before drawing the sample and then estimating the population mean μ as a simple or weighted average of the sampled X_i. There are many sampling situations where there exists some easily measured auxiliary variable S_i which is known to be highly correlated with X_i. By using the information contained in the auxiliary variable, the sampling variance can be reduced. Ratio and regression estimates are used for doing this.

For ratio estimation, the two variables (X_i, S_i) are observed in each sample and the population mean μ_S of S_i is known. The ratio estimate of μ, the population mean of X_i, assuming a sample of size n, is

$$\overline{X}_R = \left(\frac{\sum_{i=1}^{n} X_i}{\sum_{i=1}^{n} S_i}\right) \mu_S \ .$$

If the ratio X_i/S_i is constant $\overline{X}_R$ is unbiased. Otherwise it is biased, but for large samples this bias is negligible. The exact sampling distribution of this ratio estimate is complicated, but for large enough samples, the ratio is approximately normally distributed with mean μ and variance

$$\frac{(N-n)}{n(N-1)} \frac{1}{N} \sum_{i=1}^{N} (x_i - Rs_i)^2$$

328

where

$$R = \sum_{i=1}^{N} x_i / \sum_{i=1}^{N} s_i \; .$$

This variance can be estimated by

$$\frac{1}{n} \frac{(N-n)}{(N-1)} \frac{1}{(n-1)} \sum_{i=1}^{n} (x_i - \hat{R}s_i)^2$$

where

$$\hat{R} = \frac{\bar{x}}{\bar{s}} \; .$$

The exact relationship between X and S is not required to use the ratio estimate, but the variance of $\bar{X}_R$ is less than the variance of $\bar{X}$ if

$$\rho_{XS} > \frac{CV_S}{2CV_X}, \; \text{with } CV_S = \sigma_S/\mu_S \; \text{and} \; CV_X = \sigma_X/\mu_X$$

where ρ_{XS} is the correlation coefficient between X and S and CV_X and CV_S are the coefficients of variation of X and S. The coefficient of variation of a variable is the ratio of its standard deviation to its mean. Thus the variability of S is an important factor in deciding whether to use the ratio estimate. It is interesting to note that the ratio estimator can correct for bias under certain circumstances. For example, if each respondent underreported both his x_i and his s_i by the same factor, the ratio estimator still would be approximately unbiased while $\bar{X}$ would be biased by an amount depending on the factor.

If the relationship between X and S is linear, the *regression estimator* is more appropriate. Here the estimate of μ_X is

$$\hat{\mu}_X = \bar{x} + b(\mu_S - \bar{s})$$

where b is the usual least-squares estimate of the regression coefficient between X and S from the sample. The regression estimator may be biased if the relation between X and S is nonlinear. For large samples the regression estimate is more precise than the sample mean provided that the correlation between X and S is not zero.

Both the ratio estimator and the regression estimator can be generalized to include more than one ancillary variable S. Also, stratification can be combined with the use of regression or ratio estimates by making a separate regression estimate for each stratum. Of if we believe the regression coefficients to be similar across strata, we can use a combined stratified regression mean. The best choice depends on the joint relationships between X, S and the stratifying variables.

13.5. OTHER SAMPLING METHODS

Many sampling methods are designed to reduce sampling costs as well as sampling variance. Several of these are described below.

Quota Sampling is "nonrandom" stratified sampling. Frequently opinion or market research surveys choose n_i for each stratum from proportional stratification. The strata might be geographical area, age, sex, race, and economic level. Instead of filling each quota n_i at random, the interviewer is given some latitude in choosing his quota within each stratum. Since the sample within each stratum is not random, the sampling error formula cannot be applied with confidence. Experience indicates that the quota method tends to produce samples that are biased with respect to income, education, and occupation. Still, it often works well on questions of opinion and attitude.

In *cluster sampling*, the sampling unit consists of a group or *cluster* of smaller units. For example when sampling a city, the area of interest may be broken into city blocks. A number of these blocks are chosen at random. Then all (or a sample of) the families living in the selected blocks are included in the sample. This method often is preferred to simple random sampling because it reduces costs by reducing interviewer travel time. Estimates based on cluster samples are generally more variable than estimates based on a simple random sample of the same size because responses tend to be positively correlated. However, they frequently are less variable per unit cost because larger samples are possible at the same cost.

Another sampling procedure is *systematic sampling*, for example, every 20th name on a list. Usually the starting name or unit is chosen randomly. This method is used because it is easier, but one has to watch for the danger of hidden periodicities. For example, if we inspect every 20th part on an assembly line, and the machine tends to put blemishes on every 10th part, our sample would yield biased results.

Finally, we mention the possibility of *double sampling*. An initial sample is drawn and analyzed. Then the second sampling is designed using information from the first sample. Using double sampling to design stratified samples is common when strata sizes and strata standard deviations are not known.

Chapter 14

EXPERIMENTAL DESIGN: PROBLEMS IN PLANNING EXPERIMENTS
INVOLVING HUMAN POPULATIONS TO EVALUATE SOCIAL PROGRAMS

14.1. INTRODUCTION

An experiment is a deliberate and planned process to observe the
responses of experimental subjects to more than one treatment in order
to establish a causal link between the treatments and the responses,
and thereby to compare the treatments. Its main feature is that the
experimenter assigns the treatments to the subjects exogenously so
that inferences can be made about the treatment effects without fear
of systematic errors due to the assignments of the treatments.

Experimental design is a portion of experimental planning that
occurs before making observations. It determines which and how many
experimental subjects and treatments are needed and assigns the subjects to the treatments in order to efficiently attain the objectives
of the experiment. Other decisions for structuring the experiment
also are part of the experimental design. The design must consider
every aspect of the experiment including a proposed method of analysis,
which measurements will be made on the subjects, and how they are to
be obtained.

Experiments have been used to evaluate public policy programs
only recently. Important recent and current examples and the prime
contractors include experiments in negative income tax and income maintenance (Mathematica and Stanford Research Institute), housing demand
(Abt Associates), housing supply (Rand), health insurance (Rand); and
electricity rate setting (Rand and many others). These experiments

necessarily involve working with human subjects. Experiments with
nonhuman subjects have been performed in many fields and the literature
of the design of these experiments abounds. Experiments in social
policy include all the problems of nonhuman experiments and many new
complications in the political and social area. Literature on this
subject is growing rapidly. Reports on experiments conducted at Rand
include Newhouse (1974), Manning et al. (1976), Lowry (1974).

14.2. ALTERNATIVES TO SOCIAL EXPERIMENTATION

Social experiments tend to be extremely costly and very complicated. Useful information may take years to generate. They may intrude
in sensitive social and political areas, resulting in nonmonetary costs
and disruptions. Political and economic changes during the course of
the experiment can act to invalidate the results. For these reasons,
social experimentation should be considered the last resort in obtaining
information needed for improvement of public policy.

Alternatives to be considered before experimentation include the
use of expert opinion, the use of surveys, nonexperimental data, theoretical models, mathematical or simulation methods, and "natural experiments." These methods are generally much less costly than social
experiments, but they have their drawbacks, too. Expert opinion is
only as good as the expert. Surveys may generate reliable information
on subjects' responses to present and past social programs but may be
unreliable for obtaining behavioral responses to hypothetical future
changes in public policy. Nonexperimental data and observational
studies can generate information relevant to existing processes and
not to alternatives to present policies. Natural experiments occur

333

when variations in policies occur in different regions within the country
or in different countries. If the appropriate data are available and
if these experiments provide appropriate information for proposed social
programs, an evaluation of these experiments would be much less expensive and more timely than performing a social experiment. But such
experiments are generally unavailable.

14.3. THE OBJECTIVES OF AN EXPERIMENT

An experiment to evaluate social programs is warranted when insufficient knowledge exists to determine the best program, when other
methods for obtaining this knowledge are inadequate, and if the experiment can provide the needed knowledge. In a large experiment, many
objectives will be under consideration. The primary objectives will
compare responses to different social programs and predict future
responses to possible programs. The objective must be carefully defined.
It is not sufficient to say that programs are to be compared, but one
must say under what conditions the comparisons are to be made and for
what target population. Generally these comparisons must be made with
"everything else being equal," but there remains the question of "equal
to what?" Subsidiary research goals may measure the effect of treatments on certain subpopulations. Although the experiment should pursue
well defined goals, one should also watch for and expect unplanned dis-
coveries that yield other information concerning the operation and
administration of the programs.

After considering alternatives to the experiment, it may be determined that many policy questions could be answered if only a few gaps
in knowledge were filled. The objectives of an appropriate and less
ambitious experiment might be to collect the missing information only.

14.4. THE SIZE OF THE EXPERIMENT

An experiment will be attempted only when the expected long-run marginal benefit of the information it will provide exceeds its marginal cost. Although a social experiment may cost millions, social policies often involve billions of dollars, making the cost of an experiment seem trivial by comparison. The size of the experiment should increase with the number N of people who stand to benefit from improved public policy and with the amount (in dollars) of per capita improvement in policy expected from the experiment.

In one highly theoretical model, the optimal cost of experimentation C^* is determined to be equal to the total long-run social cost of using a suboptimal social policy based on imprecise knowledge available after the experiment. If less than C^* is spent, then each additional dollar allocated to the experiment would improve policy more than its cost, while experimental expenditures exceeding C^* would improve policy less than their cost. This optimal cost C^* is proportional to the square root of N, the number of subjects affected by the policy. Of course C^* is to be spent on the most informative experiment available for a cost of C^*.

Although this model is idealized, it clearly implies that a cost-benefit analysis is required when considering the size of an experiment, and benefits are derived only from improvement in social policy. Benefits therefore are greater in social programs whose value is less certain because experimentation can provide greatly increased understanding. Of course, to be useful, this new knowledge must be available in time to be used, and policymakers must use it for policy formulation.

14.5. EXPERIMENTAL TREATMENTS, RESPONSES, AND SUBJECTS

Treatments in social experiments are program variations. One or more null treatments act as "control." The experimenter assigns subjects to the treatments as an integral part of the design. Each subject is assigned to only one treatment, at least at any one time. The control group is observed, acting under an existing program, which permits comparisons between new and old programs. Other reasons for control groups are discussed later.

The number of treatments should depend on the range of policy interest, but otherwise should be minimized so sufficiently many subjects are assigned to each treatment in order to know with adequate precision the response to each treatment. Otherwise, a sample scattered over many treatments will inadequately determine the form of the response function, variation within and between treatments being confounded. In the first case, a regression functional form still can be used for interpolation and extrapolation to other interesting treatments not used directly.

The treatment effects are measured by the responses of the subjects assigned to them. The subjects therefore must be assigned to the treatments so the estimated effect of the response will be due to the treatment, and not to differences in subjects allocated to the treatments. Measurements will be multidimensional if a subject responds in several interesting ways to a treatment, and the experiment must collect all these measurements.

The characteristics of the experimental subjects should be determined primarily by the characteristics of the target population specified in the objectives. The New Jersey Negative Income Tax Experiment has been faulted for ignoring this because they mainly observed male-headed families, whereas most welfare families are female-headed. However, while the sample selected must provide a basis for determining the response of the target population to the treatments, it need not necessarily have the same distribution of characteristics as the target population. Instead, it may be more efficient to distort the sample and use models to adjust the estimated response to one expected of the target population. This practice can be risky, of course, because the adequacy of the final estimates will depend on using the proper model.

Social experiments usually are designed to obtain longitudinal responses of subjects to the treatments. These responses may be collected together with other "concomitant" characteristics peculiar to the subject. Knowing the concomitant characteristics in advance has several advantages. The concomitant or independent or explanatory variables may reduce the error variance in a regression model, and more importantly, subjects can be assigned to treatments to make the distribution of the concomitant variables as similar as possible for the group of subjects assigned to each treatment. This is important in that it reduces variances and some sources of potential bias.

14.6. RANDOMIZATION AND LATENT VARIABLES

If the subjects assigned to the treatments differ systematically from treatment to treatment, then provided the model is appropriate, an analysis of covariance can be applied to those variables measured to make appropriate adjustments and the estimates of treatment coefficients will be unbiased. The only loss in this case, provided the analysis of covariance model is appropriate, is a loss in precision.

If these variables differ from treatment to treatment but are never measured (in this case they are called "latent variables") then they will lead to biased estimates of treatment coefficients if they are correlated with the responses. Latent variables always exist, but two steps may be taken to minimize their potential for bias: as many important concomitant variables should be measured as is feasible, and randomization should be used to control bias by limiting latent differences to the amount that randomness allows.

Randomization removes biases due to latent variables by putting them into the error variance. Their values then average to zero and their variation can be measured. If many subjects are assigned randomly to each treatment, the latent variables will be distributed similarly for each treatment. For example, suppose n subjects are assigned randomly (probability 1/2) to two treatments. Let $u_i = \pm 1$ depending on which treatment was assigned to the i-th subject and x_i be the value of some latent variable for the subject. Then

$$E\, r^2(u, x) \doteq \frac{1}{n}$$

where $r(u,x)$ is the correlation coefficient between the treatment assignment and the latent variable. Therefore the correlation between the treatments and latent variables is small if n is large. Ideally, latent variables would be uncorrelated with all measured variables so that they could be ignored, their only effect being to increase the error variance. Under favorable circumstances randomization nearly accomplishes this.

However, randomization is no panacea if n is not large or if there are many independent latent variables, for then some latent variable

338

probably will interfere with estimation of the treatment effects. Flaws found in many experiments center around this issue: after the experiment a latent variable is identified that appears to differ systematically over treatments. Therefore it may have caused the treatment differences, and not the treatments themselves.

Insight into the problem of latent variables is provided by the smoking-lung cancer example. In naturally occurring data, lung cancer is positively correlated with smoking. This in itself does not prove smoking causes lung cancer, because both may be correlated with a latent variable, being uncorrelated if that variable is held constant. That is, the partial correlation between smoking and lung cancer may be zero, given the latent variable. This could happen if nervous people were more likely to smoke and also were more susceptible to lung cancer. Suppose now that there were zero correlation between smoking and lung cancer among nervous people, and zero correlation between these variables among people who are not nervous. This would produce positive correlation between smoking and lung cancer with no causal link. Were an experiment feasible, it would show no causal link. The experimenter ideally would assign the treatment (smoking) to subjects at random, so that the fraction of nervous people who smoke would be nearly the same as those not nervous. (Ethical considerations prevent doing this with humans, but this is done with laboratory experiments.) He then would observe the response (incidence of lung cancer). Because self-selection (endogenous choice) of treatments by subjects is not permitted, and instead smoking is determined by the experimenter's (exogenous) choice, the causal link can be estimated. Of course, "nervousness" could be measured and accounted for without an experiment, but one would still

339

have to worry about other possible latent variables, and one never can measure all variables. Only an experiment, with exogenous assignment of subjects to treatment, can counteract these objections effectively.

Collecting extra concomitant observations during an experiment is relatively cheap. Since one may never be able to obtain them after the experiment, it is best to make as many such measurements when one can, provided doing so does not overburden the experimental subjects. One hopes they are not partially correlated with the response, given the treatment. Even if so, they can be used in the analysis to correct biases.

14.7. THE PROPOSED ANALYSIS

A proposed analysis is needed to choose among alternative designs. Since the appropriate model usually is unknown before experimentation, one must guess about the analysis. Any statistical analyses may be used, but the general linear model, including regression, ANOVA and ANOCOVA, perhaps in their multivariate forms, are most commonly used. Simple models should be used in the planning stage, linear models being among the simplest. An inappropriate model leads to an inappropriate design, while highly specialized forms will not produce a "robust" design. If the designer is careless, linear models can lead to designs frustrating analysis if it later happens that nonlinear terms must be estimated.

The analysis actually used may differ from that assumed during the design. In fact, the actual analysis should include a search for unanticipated relationships by studying outliers and anomalies that could suggest further investigation.

14.8. EXPERIMENTAL DESIGN

Experimental design is an art aided greatly by knowledge. Non-experimental analyses can provide very useful data for a design, about the model and parameters. The design of a social experiment is constrained in many ways: legal, social, economic, and political. The object is to choose sample sizes and assign subjects to produce the best estimates for a given budget. Efficiency and robustness must be considered: efficiency relative to the stated models and objectives, and robustness to account for possible variation from these models and objectives.

Randomization traditionally is required in most designs to protect against bias (both conscious and unconscious), against latent variables, and to improve distributional assumptions. In fact, randomization can decrease efficiency relative to more controlled assignments through failure to balance concomitant variables across treatments. But randomization still should be practiced wherever it does not seriously inhibit efficiency, because it offers protection against nonrobust choices, and makes experimental results more widely acceptable.

The problem with randomization is that it may produce a very "unbalanced" sample (covariates may have very different distributions for some pairs of treatments). When this happens, the randomization should be redone, although this is not a true randomization. Suppose, for example, that five cities are chosen at random from the United States. If all chosen happen to be in one region, e.g. the South, it is folly to claim that results from an analysis based on these cities extend to other regions simply because cities in other regions were given an equal chance to be selected. "Representativeness" or "balance"

341

is required, not randomness. Randomness is likely to produce balance only with large samples.

Stratified random sampling increases balance of the concomitant variables and thereby reduces these concerns. But carrying stratification to its logical limit, each stratum containing only one experimental subject, ends with purposive (nonrandom) assignments to improve efficiencies.

"Blocking" provides another alternative to simple random sampling. In this case experimental subjects are grouped into homogeneous groups (blocks) and then each treatment is assigned one subject at random from each block. In effect, each block is a stratum, and each treatment samples proportionally to its size from that stratum. Any design method, such as blocking and stratified random sampling, must be accounted for in the analysis, as Chapters 7 and 13 show, for example.

14.9. USING LINEAR MODELS FOR DESIGN

To illustrate the use of a simple model, suppose k cities are to be chosen from the United States and an average response y_i is to be measured in the i-th city. The object is to estimate the average value μ of y for all cities. Suppose each city has concomitant characteristic x_i and m is the known national mean for x_i. If a linear relationship holds between responses y and characteristics x then

$$y = \mu + \beta(x - m) .$$

Let $\bar{x} = \Sigma x_i/k$. The national mean therefore is μ, but unless $\bar{x} = m$ (or $\beta = 0$) the mean of the sites $\bar{y} = \Sigma y_i/k$ is an unbiased estimate of $\mu + \beta(\bar{x} - m)$ and therefore is a biased estimate of μ. The best unbiased estimate of μ if $\bar{x} \neq m$ is

$$\hat{\mu} = \bar{y} - \hat{\beta}(\bar{x} - m)$$

with $\hat{\beta} \equiv \Sigma y_i (x_i - \bar{x})/ks_x^2$, $s_x^2 \equiv \Sigma(x_i - \bar{x})^2/k$. But the variance of this estimator of μ is $\left(1 + (\bar{x} - m)^2/s_x^2\right)\sigma^2/k$. This would be reduced by taking $\bar{x} = m$, and such a choice only could be guaranteed by making a purposive selection. Further, the variance of $\hat{\beta}$ is proportional to $1/s_x^2$. Consequently, a purposive design would attempt to choose cities so that $\bar{x} = m$ (approximately) and s_x is large. These values of $\bar{x}$ and s_x are left to chance in a random selection which therefore would be expected to be inefficient. Purposive selection methods can be extended to permit simultaneous adjustments for several concomitant characteristics by use of multiple regression methods.

Of course, if the true relationship is quadratic

$$y = \mu + \beta(x - m) + \gamma(x - m)^2$$

then the optimal assignment discussed in the preceding paragraph, having all cities with x_i far from m, will permit very inefficient estimates of γ. This illustrates the lack of robustness of a design chosen optimally for one model when used to analyze a different one. Simple random sampling would have produced a sample suitable for both models, although even better procedures are available.

In general (see Chapter 6), the variance of a regression estimate b_1 of β_1 in $y = \beta_0 + \beta_1 x_1 + \ldots + \beta_k x_k + \text{error}$ is

$$\text{Var}(b_1) = \frac{\sigma^2}{n} \frac{1}{s_1^2(1 - R_1^2)}$$

with σ^2 the residual variance, s_1^2 the variance of the sampled x_{1i} and R_1^2 the multiple correlation between x_1 and the other $(x_2, \ldots, x_k)$. The most efficient estimate of β_1 therefore makes s_1^2 large and

343

uncorrelates x_1 from $(x_2, \ldots, x_k)$. Were β_1 the only parameter of interest, and the assumed multiple regression model correct, then the experimental design would make $s_1^2(1 - R_1^2)$ as large as possible and make σ^2 small by measuring many concomitant variables $x_2, \ldots, x_k$. Uncorrelating the independent variables is required in most classical design theory.

When blocking is impossible, substantial effectiveness can be obtained by assigning subjects to treatments so that the covariates have the same means and variances for each treatment. This makes estimates of treatment differences in an ANOCOVA model, and estimates of the effects of concomitant characteristics be optimally efficient. This assignment is improved if the dispersion of the concomitant characteristics is large.

Classical symmetrical designs with their uncorrelated concomitant variables are not the best when costs and desired precision are asymmetric. A computerized model was developed for the Negative Income Tax Experiment to make optimal stratified allocations when a specified general linear model is used, assuming relative importances of estimates and costs are specified (Conlisk-Watts, 1969). The resulting designs have symmetry and balance only when these inputs are symmetrical.

14.10. SEQUENTIAL DESIGN

The many uncertainties in planning an experiment strongly suggest it be designed sequentially if possible. Early analysis of results may reveal problems requiring new measurements, new stratification, and reformulation of questionnaires or interviewing procedures. The ultimate precision will be revealed as the experiment unfolds,

344

requiring sample size adjustments. Consequently, decisions should be delayed when possible so that options are not closed earlier than necessary.

Adoption of a sequential design affects the analysis, if the dependent variable is used to determine when to stop, but the way in which it is affected is quite complicated. Analytical adjustments therefore would be considered informally, or even ignored. A sequential experiment has so many advantages over one with a fixed size that purist considerations must be secondary.

14.11. SAMPLE SIZE

After a model is tentatively accepted, the precision of the experiment depends only on sample size and the values of unknown parameters, especially the residual variance. The experimenter must estimate the unknown parameters and choose the sample size to give required precision. The best designed experiment will provide inadequate information if sample sizes are too small. As the experiment progresses, the unknown parameters may be estimated and adjustments made to the sample size, if possible.

The mathematics needed for choosing the appropriate sample size may be complicated, but many helpful formulas and tables exist. Simulation of the experiment is always an applicable method to estimate precision, although this approach can be costly.

There may be a choice between using many subjects in one experimental site or a few subjects in many sites. The best alternative depends on intersite variation and the cost of operating additional sites. Some theory has been generated on choosing the number of sites, and a mathematical solution may be derived under idealized conditions.

14.12. PROBLEMS PECULIAR TO SOCIAL EXPERIMENTS AND HUMAN SURVEYS

Using human subjects in experiments introduces many complications and potential sources of bias to experimentation. Some of these problems are outlined below.

Human subjects cannot be forced to participate or be interviewed. They can refuse, not respond, or drop out (attrite). Therefore, those who enroll may respond differently to treatments than those who do not participate. The absence of "uncooperative" subjects may frustrate generalization to a population which includes such people. Worse yet, treatment comparisons may result from variable refusal and attrition rates among the treatments. Of course attrition (quitting) also may represent a valid response to an ineffective treatment or simply be random. The former may provide useful information, while the latter does not cause bias. Follow-up procedures are required to determine which is the case.

"Hawthorne effects" or "reactive effects" refer to measurement problems caused by self-conscious subjects who respond differently to an experimental treatment than they would to the same treatment in a real situation. Participants may be enthusiastic during the temporary experiment, or try to impress interviewers by engaging in preferred behavior (the "halo effect"). Control groups help to estimate these biases if treatment and control groups are affected alike by such effects.

Transitory experimental behavior can be serious. An experiment may fail to reach "steady-state" for some time, while participants learn how to take advantage of the experimental programs. Near the end of experimentation, subjects may respond differently because they

346

know the experiment soon will end. If the experiment involves only a fraction of a community, then massive community responses that might occur with a real program as a result of community awareness may not occur. Saturation experiments may be required to observe such effects, or any effects on the supply system. For these reasons, long-term experiments, or at least long-term guarantees of benefits, may be needed for subjects to reach steady-state. The length of the experiment may be varied for subjects to provide a basis for determining the steady-state interval.

Unexpected experimental changes, such as drastic political, social, legislative, or economic changes during the experiment can greatly reduce its validity. A control group can be effective for some of these situations, because environment changes may have equal effect on treatments and controls, but have little effect on differences between them.

Individual responses depend on the subject's ability to receive and comprehend information. In fact, subjects respond to their perceptions of the treatments, and not to the treatments themselves. When this happens, the analyst encounters an "errors in variables" problem, biasing standard regression estimates. Alternative estimation methods are then required, and perhaps additional measurements.

There are many other potential problems in surveying human populations. The usefulness of questionnaire responses depends on the subject's knowledge and truthfulness, which cannot be known. Resentment by those people in and those not in the experiment may stem from unequal sharing of treatment benefits. Legal restrictions limit the design and require extensive rules of operation. Confidentiality of sensitive data must be protected. Enormous data storage is required and complicated data retrieval systems are needed to extract and organize data for analysis.

Policymakers understandably tend to press for early results. Public attention to large important experiments may be keen and create problems. Social experiments may be forced into the political arena even before they are complete. This could affect behavior of the subjects.

Few standard remedies exist to combat the biases inherent in social experiments. The designers and analysts must be sensitive to such possibilities and clever enough to devise ways to avoid, minimize, measure, or estimate them. A word of caution: bias is not bad in itself, and unbiased estimators can be worse then other biased ones (Morris, Newhouse, and Archibald, 1979). The real problems center around unknown biases when errors are computed as if no bias exists.

The experimental study of human populations has problems and advantages that do not exist in experiments on other subjects. Disadvantages are that refusal and nonresponse are a much greater threat, and that because humans are intelligent and conscious, they may not respond as they would in a real program. Offsetting this, each person monitors himself. Thus, the experimenter need not follow the subject and record each significant response, but instead can ask the subject himself to remember and report the response. Obviously this is possible only with human subjects.

The reader will find more on this topic in two volumes on experimentation in economics: Aigner and Morris (1979), *Experimental Design in Econometrics*, and Smith (1979), *Research in Experimental Economics*.

There are few absolute rules for designing an experiment. Knowledge, intuition, thoroughness, genius, and foresightedness are minimal requirements of those who plan large social experiments. An interdisciplinary team of planners helps greatly. The rest depends on good sense and doing one's best.

Table 1

FIVE DIGIT RANDOM NUMBERS

55034 81217	90564 81943	11241 84512	12288 89862	00760 76159
25521 99536	43233 48786	49221 06960	31564 21458	88199 06312
85421 72744	97242 66383	00132 05661	96442 37388	57671 27916
61219 48390	47344 30413	39392 91365	56203 79204	05330 31196
20230 03147	58854 11650	28415 12821	58931 30508	65989 26675
95776 83206	56144 55953	89787 64426	08448 45707	80364 60262
07603 17344	01148 83300	96955 65027	31713 89013	79557 49755
00645 17459	78742 39005	36027 98807	72666 54484	68262 38827
62950 83162	61504 31557	80590 47893	72360 72720	08396 33674
79350 10276	81933 26347	08068 67816	06659 87917	74166 85519
48339 69834	59047 82175	92010 58446	69591 56205	95700 86211
05842 08439	79836 50957	32059 32910	15842 13918	41365 80115
25855 02209	07307 59942	71389 76159	11263 38787	61541 22606
25272 16152	82323 70718	98081 38631	91956 49909	76253 33970
73003 29058	17605 49298	47675 90445	68919 05676	23823 84892
81310 94430	22663 06584	38142 00146	17496 51115	61458 65790
10024 44713	59832 80721	63711 67882	25100 45345	55743 67618
84671 52806	89124 37691	20897 82339	22627 06142	05773 03547
29296 58162	21858 33732	94056 88806	54603 00384	66340 69232
51771 94074	70630 41286	90583 87680	13961 55627	23670 35109
42166 56251	60770 51672	36031 77273	85218 14812	90758 23677
78355 67041	22492 51522	31164 30450	27600 44428	96380 26772
09552 51347	33864 89018	73418 81538	77399 30448	97740 18158
15771 63127	34847 05660	06156 48970	55699 61818	91763 20821
13231 99058	93754 36730	44286 44326	15729 37500	47269 13333
50583 03570	38472 73236	67613 72780	78174 18718	99092 64114
99485 57330	10634 74905	90671 19643	69903 60950	17968 37217
54676 39524	73785 48864	69835 62798	65205 69187	05572 74741
99343 71549	10248 76036	31702 76868	88909 69574	27642 00336
35492 40231	34868 55356	12847 68093	52643 32732	67016 46784
98170 25384	03841 23920	47954 10359	70114 11177	63298 99903
02670 86155	56860 02592	01646 42200	79950 37764	82341 71952
36934 42879	81637 79952	07066 41625	96804 92388	88860 68580
56851 12778	24309 73660	84264 24668	16686 02239	66022 64133
05464 28892	14271 23778	88599 17081	33884 88783	39015 57118
15025 20237	63386 71122	06620 07415	94982 32324	79427 70387
95610 08030	81469 91066	88857 56583	01224 28097	19726 71465
09026 40378	05731 55128	74298 49196	31669 42605	30368 96424
81431 99955	52462 67667	97322 69808	21240 65921	12629 92896
21431 59335	58627 94822	65484 09641	41018 85100	16110 32077
95832 76145	11636 80284	17787 97934	12822 73890	66009 27521
99813 44631	43746 99790	86823 12114	31706 05024	28156 04202
77210 31148	50543 11603	50934 02498	09184 95875	85840 71954
13268 02609	79833 66058	80277 08533	28676 37532	70535 82356
44285 71735	26620 54691	14909 52132	81110 74548	78853 31996
70526 45953	79637 57374	05053 31965	33376 13232	85666 86615
88386 11222	25080 71462	09818 46001	19065 68981	18310 74178
83161 73994	17209 79441	64091 49790	11936 44864	86978 34538
50214 71721	33851 45144	05696 29935	12823 01594	08453 52825
97689 29341	67747 80643	13620 23943	49396 83686	37302 95350

Reprinted, with permission, from The Rand Corporation, *A Million Random Digits with 100,000 Normal Deviates*, The Free Press, Publishers, Glencoe, Illinois, 1955.

Table 2

THE NORMAL CUMULATIVE DISTRIBUTION FUNCTION FOR $z \geq 0$

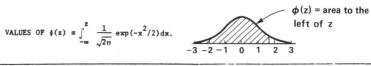

VALUES OF $\phi(z) \equiv \int_{-\infty}^{z} \frac{1}{\sqrt{2\pi}} \exp(-x^2/2)\,dx.$

$\phi(z)$ = area to the left of z

	.00	.01	.02	.03	.04	.05	.06	.07	.08	.09
.0	.50000	.50399	.50798	.51197	.51595	.51994	.52392	.52790	.53188	.53586
.1	.53983	.54380	.54776	.55172	.55567	.55962	.56356	.56749	.57142	.57535
.2	.57926	.58317	.58706	.59095	.59483	.59871	.60257	.60642	.61026	.61409
.3	.61791	.62172	.62552	.62930	.63307	.63683	.64058	.64431	.64803	.65173
.4	.65542	.65910	.66276	.66640	.67003	.67364	.67724	.68082	.68439	.68793
.5	.69146	.69497	.69847	.70194	.70540	.70884	.71226	.71566	.71904	.72240
.6	.72575	.72907	.73237	.73565	.73891	.74215	.74537	.74857	.75175	.75490
.7	.75804	.76115	.76424	.76730	.77035	.77337	.77637	.77935	.78230	.78524
.8	.78814	.79103	.79389	.79673	.79955	.80234	.80511	.80785	.81057	.81327
.9	.81594	.81859	.82121	.82381	.82639	.82894	.83147	.83398	.83646	.83891
1.0	.84134	.84375	.84614	.84849	.85083	.85314	.85543	.85769	.85993	.86214
1.1	.86433	.86650	.86864	.87076	.87286	.87493	.87698	.87900	.88100	.88298
1.2	.88493	.88686	.88877	.89065	.89251	.89435	.89617	.89796	.89973	.90147
1.3	.90320	.90490	.90658	.90824	.90988	.91149	.91308	.91466	.91621	.91774
1.4	.91924	.92073	.92220	.92364	.92507	.92647	.92785	.92922	.93056	.93189
1.5	.93319	.93448	.93574	.93699	.93822	.93943	.94062	.94179	.94295	.94408
1.6	.94520	.94630	.94738	.94845	.94950	.95053	.95154	.95254	.95352	.95449
1.7	.95543	.95637	.95728	.95818	.95907	.95994	.96080	.96164	.96246	.96327
1.8	.96407	.96485	.96562	.96638	.96712	.96784	.96856	.96926	.96995	.97062
1.9	.97128	.97193	.97257	.97320	.97381	.97441	.97500	.97558	.97615	.97670
2.0	.97725	.97778	.97831	.97882	.97932	.97982	.98030	.98077	.98124	.98169
2.1	.98214	.98257	.98300	.98341	.98382	.98422	.98461	.98500	.98537	.98574
2.2	.98610	.98645	.98679	.98713	.98745	.98778	.98809	.98840	.98870	.98899
2.3	.98928	.98956	.98983	.99010	.99036	.99061	.99086	.99111	.99134	.99158
2.4	.99180	.99202	.99224	.99245	.99266	.99286	.99305	.99324	.99343	.99361
2.5	.99379	.99396	.99413	.99430	.99446	.99461	.99477	.99492	.99506	.99520
2.6	.99534	.99547	.99560	.99573	.99585	.99598	.99609	.99621	.99632	.99643
2.7	.99653	.99664	.99674	.99683	.99693	.99702	.99711	.99720	.99728	.99736
2.8	.99744	.99752	.99760	.99767	.99774	.99781	.99788	.99795	.99801	.99807
2.9	.99813	.99819	.99825	.99831	.99836	.99841	.99846	.99851	.99856	.99861
3.0	.99865	.99869	.99874	.99878	.99882	.99886	.99889	.99893	.99896	.99900
3.1	.99903	.99906	.99910	.99913	.99916	.99918	.99921	.99924	.99926	.99929
3.2	.99931	.99934	.99936	.99938	.99940	.99942	.99944	.99946	.99948	.99950
3.3	.99952	.99953	.99955	.99957	.99958	.99960	.99961	.99962	.99964	.99965
3.4	.99966	.99968	.99969	.99970	.99971	.99972	.99973	.99974	.99975	.99976
3.5	.99977	.99978	.99978	.99979	.99980	.99981	.99981	.99982	.99983	.99983
3.6	.99984	.99985	.99985	.99986	.99986	.99987	.99987	.99988	.99988	.99989
3.7	.99989	.99990	.99990	.99990	.99991	.99991	.99992	.99992	.99992	.99992
3.8	.99993	.99993	.99993	.99994	.99994	.99994	.99994	.99995	.99995	.99995
3.9	.99995	.99995	.99996	.99996	.99996	.99996	.99996	.99996	.99997	.99997
4.0	.99997	.99997	.99997	.99997	.99997	.99997	.99998	.99998	.99998	.99998
4.1	.99998	.99998	.99998	.99998	.99998	.99998	.99998	.99998	.99999	.99999
4.2	.99999	.99999	.99999	.99999	.99999	.99999	.99999	.99999	.99999	.99999
4.3	.99999	.99999	.99999	.99999	.99999	.99999	.99999	.99999	.99999	.99999
4.4	.99999	.99999	1.00000	1.00000	1.00000	1.00000	1.00000	1.00000	1.00000	1.00000

$\phi(z)$	.90	.95	.975	.99	.995	.999	.9995	.99995
$2[1-\phi(z)]$	.20	.10	.05	.02	.01	.002	.001	.0001
z	1.2816	1.6449	1.9600	2.3263	2.5756	3.0903	3.2906	3.8907

For large z, giving the correct value to within a *factor* of one tenth of one percent (0.001) for all $z \geq 2.4$,

$$1 - \phi(z) \doteq \frac{1}{\sqrt{2\pi}\, z} \exp(-z^2/2)\left[1 - \frac{1}{z^2 + 2.5}\right].$$

352

Table 3

VALUES OF THE BINOMIAL CUMULATIVE DISTRIBUTION FUNCTION

$$F(x) = P(X \le x) = \sum_{i=0}^{x} \binom{n}{i} p^i (1-p)^{n-i} \text{ for } p \le 0.5,$$

X HAVING THE Bin(n,p) DISTRIBUTION. NOTE, FOR p > 0.5,
$P(X \le x) = 1 - P(Y \le n-x-1)$ with $Y = n - X \sim$ Bin(n, 1-p).

n	x	.05	.10	.15	.20	.25	.30	.35	.40	.45	.50
2	0	.9025	.8100	.7225	.6400	.5625	.4900	.4225	.3600	.3025	.2500
2	1	.9975	.9900	.9775	.9600	.9375	.9100	.8775	.8400	.7975	.7500
3	0	.8574	.7290	.6141	.5120	.4219	.3430	.2746	.2160	.1664	.1250
3	1	.9928	.9720	.9393	.8960	.8438	.7840	.7183	.6480	.5748	.5000
3	2	.9999	.9990	.9966	.9920	.9844	.9730	.9571	.9360	.9089	.8750
4	0	.8145	.6561	.5220	.4096	.3164	.2401	.1785	.1296	.0915	.0625
4	1	.9860	.9477	.8905	.8192	.7383	.6517	.5630	.4752	.3910	.3125
4	2	.9995	.9963	.9880	.9728	.9492	.9163	.8735	.8208	.7585	.6875
4	3	1.0000	.9999	.9995	.9984	.9961	.9919	.9850	.9744	.9590	.9375
5	0	.7738	.5905	.4437	.3277	.2373	.1681	.1160	.0778	.0503	.0313
5	1	.9774	.9185	.8352	.7373	.6328	.5282	.4284	.3370	.2562	.1875
5	2	.9988	.9914	.9734	.9421	.8965	.8369	.7648	.6826	.5931	.5000
5	3	1.0000	.9995	.9978	.9933	.9844	.9692	.9460	.9130	.8688	.8125
5	4	1.0000	1.0000	.9999	.9997	.9990	.9976	.9947	.9898	.9815	.9688
6	0	.7351	.5314	.3771	.2621	.1780	.1176	.0754	.0467	.0277	.0156
6	1	.9672	.8857	.7765	.6554	.5339	.4202	.3191	.2333	.1636	.1094
6	2	.9978	.9842	.9527	.9011	.8306	.7443	.6471	.5443	.4415	.3438
6	3	.9999	.9987	.9941	.9830	.9624	.9295	.8826	.8208	.7447	.6563
6	4	1.0000	.9999	.9996	.9984	.9954	.9891	.9777	.9590	.9308	.8906
6	5	1.0000	1.0000	1.0000	.9999	.9998	.9993	.9982	.9959	.9917	.9844
7	0	.6983	.4783	.3206	.2097	.1335	.0824	.0490	.0280	.0152	.0078
7	1	.9556	.8503	.7166	.5767	.4449	.3294	.2338	.1586	.1024	.0625
7	2	.9962	.9743	.9262	.8520	.7564	.6471	.5323	.4199	.3164	.2266
7	3	.9998	.9973	.9879	.9667	.9294	.8740	.8002	.7102	.6083	.5000
7	4	1.0000	.9998	.9988	.9953	.9871	.9712	.9444	.9037	.8471	.7734
7	5	1.0000	1.0000	.9999	.9996	.9987	.9962	.9910	.9812	.9643	.9375
7	6	1.0000	1.0000	1.0000	1.0000	.9999	.9998	.9994	.9984	.9963	.9922
8	0	.6634	.4305	.2725	.1678	.1001	.0576	.0319	.0168	.0084	.0039
8	1	.9428	.8131	.6572	.5033	.3671	.2553	.1691	.1064	.0632	.0352
8	2	.9942	.9619	.8948	.7969	.6785	.5518	.4278	.3154	.2201	.1445
8	3	.9996	.9950	.9786	.9437	.8862	.8059	.7064	.5941	.4770	.3633
8	4	1.0000	.9996	.9971	.9896	.9727	.9420	.8939	.8263	.7396	.6367
8	5	1.0000	1.0000	.9998	.9988	.9958	.9887	.9747	.9502	.9115	.8555
8	6	1.0000	1.0000	1.0000	.9999	.9996	.9987	.9964	.9915	.9819	.9648
8	7	1.0000	1.0000	1.0000	1.0000	1.0000	.9999	.9998	.9993	.9983	.9961
9	0	.6302	.3874	.2316	.1342	.0751	.0404	.0207	.0101	.0046	.0020
9	1	.9288	.7748	.5995	.4362	.3003	.1960	.1211	.0705	.0385	.0195
9	2	.9916	.9470	.8591	.7382	.6007	.4628	.3373	.2318	.1495	.0898
9	3	.9994	.9917	.9661	.9144	.8343	.7297	.6089	.4826	.3614	.2539
9	4	1.0000	.9991	.9944	.9804	.9511	.9012	.8283	.7334	.6214	.5000
9	5	1.0000	.9999	.9994	.9969	.9900	.9747	.9464	.9006	.8342	.7461
9	6	1.0000	1.0000	1.0000	.9997	.9987	.9957	.9888	.9750	.9502	.9102
9	7	1.0000	1.0000	1.0000	1.0000	.9999	.9996	.9986	.9962	.9909	.9805
9	8	1.0000	1.0000	1.0000	1.0000	1.0000	1.0000	.9999	.9997	.9992	.9980

Compiled by Carl Morris, The Rand Corporation, Santa Monica, California.

Table 3 (cont.)

VALUES OF THE BINOMIAL CUMULATIVE DISTRIBUTION FUNCTION

$$F(x) = P(X \le x) = \sum_{i=0}^{x} \binom{n}{i} p^i (1-p)^{n-i} \text{ for } p \le 0.5,$$

X HAVING THE Bin(n,p) DISTRIBUTION. NOTE, FOR p > 0.5,
$$P(X \le x) = 1 - P(Y \le n-x-1) \text{ with } Y = n-X \sim \text{Bin}(n, 1-p).$$

n	x	.05	.10	.15	.20	.25	.30	.35	.40	.45	.50
10	0	.5987	.3487	.1969	.1074	.0563	.0282	.0135	.0060	.0025	.0010
10	1	.9139	.7361	.5443	.3758	.2440	.1493	.0860	.0464	.0233	.0107
10	2	.9885	.9298	.8202	.6778	.5256	.3828	.2616	.1673	.0996	.0547
10	3	.9990	.9872	.9500	.8791	.7759	.6496	.5138	.3823	.2660	.1719
10	4	.9999	.9984	.9901	.9672	.9219	.8497	.7515	.6331	.5044	.3770
10	5	1.0000	.9999	.9986	.9936	.9803	.9527	.9051	.8338	.7384	.6230
10	6	1.0000	1.0000	.9999	.9991	.9965	.9894	.9740	.9452	.8980	.8281
10	7	1.0000	1.0000	1.0000	.9999	.9996	.9984	.9952	.9877	.9726	.9453
10	8	1.0000	1.0000	1.0000	1.0000	1.0000	.9999	.9995	.9983	.9955	.9893
10	9	1.0000	1.0000	1.0000	1.0000	1.0000	1.0000	1.0000	.9999	.9997	.9990
11	0	.5688	.3138	.1673	.0859	.0422	.0198	.0088	.0036	.0014	.0005
11	1	.8981	.6974	.4922	.3221	.1971	.1130	.0606	.0302	.0139	.0059
11	2	.9848	.9104	.7788	.6174	.4552	.3127	.2001	.1189	.0652	.0327
11	3	.9984	.9815	.9306	.8389	.7133	.5696	.4256	.2963	.1911	.1133
11	4	.9999	.9972	.9841	.9496	.8854	.7897	.6683	.5328	.3971	.2744
11	5	1.0000	.9997	.9973	.9883	.9657	.9218	.8513	.7535	.6331	.5000
11	6	1.0000	1.0000	.9997	.9980	.9924	.9784	.9499	.9006	.8262	.7256
11	7	1.0000	1.0000	1.0000	.9998	.9988	.9957	.9878	.9707	.9390	.8867
11	8	1.0000	1.0000	1.0000	1.0000	.9999	.9994	.9980	.9941	.9852	.9673
11	9	1.0000	1.0000	1.0000	1.0000	1.0000	1.0000	.9998	.9993	.9978	.9941
11	10	1.0000	1.0000	1.0000	1.0000	1.0000	1.0000	1.0000	1.0000	.9998	.9995
12	0	.5404	.2824	.1422	.0687	.0317	.0138	.0057	.0022	.0008	.0002
12	1	.8816	.6590	.4435	.2749	.1584	.0850	.0424	.0196	.0083	.0032
12	2	.9804	.8891	.7358	.5583	.3907	.2528	.1513	.0834	.0421	.0193
12	3	.9978	.9744	.9078	.7946	.6488	.4925	.3467	.2253	.1345	.0730
12	4	.9998	.9957	.9761	.9274	.8424	.7237	.5833	.4382	.3044	.1938
12	5	1.0000	.9995	.9954	.9806	.9456	.8822	.7873	.6652	.5269	.3872
12	6	1.0000	.9999	.9993	.9961	.9857	.9614	.9154	.8418	.7393	.6128
12	7	1.0000	1.0000	.9999	.9994	.9972	.9905	.9745	.9427	.8883	.8062
12	8	1.0000	1.0000	1.0000	.9999	.9996	.9983	.9944	.9847	.9644	.9270
12	9	1.0000	1.0000	1.0000	1.0000	1.0000	.9998	.9992	.9972	.9921	.9807
12	10	1.0000	1.0000	1.0000	1.0000	1.0000	1.0000	.9999	.9997	.9989	.9968
12	11	1.0000	1.0000	1.0000	1.0000	1.0000	1.0000	1.0000	1.0000	.9999	.9998
13	0	.5133	.2542	.1209	.0550	.0238	.0097	.0037	.0013	.0004	.0001
13	1	.8646	.6213	.3983	.2336	.1267	.0637	.0296	.0126	.0049	.0017
13	2	.9755	.8661	.6920	.5017	.3326	.2025	.1132	.0579	.0269	.0112
13	3	.9969	.9658	.8820	.7473	.5843	.4206	.2783	.1686	.0929	.0461
13	4	.9997	.9935	.9658	.9009	.7940	.6543	.5005	.3530	.2279	.1334
13	5	1.0000	.9991	.9925	.9700	.9198	.8346	.7159	.5744	.4268	.2905
13	6	1.0000	.9999	.9987	.9930	.9757	.9376	.8705	.7712	.6437	.5000
13	7	1.0000	1.0000	.9998	.9988	.9944	.9818	.9538	.9023	.8212	.7095
13	8	1.0000	1.0000	1.0000	.9998	.9990	.9960	.9874	.9679	.9302	.8666
13	9	1.0000	1.0000	1.0000	1.0000	.9999	.9993	.9975	.9922	.9797	.9539
13	10	1.0000	1.0000	1.0000	1.0000	1.0000	.9999	.9997	.9987	.9959	.9888
13	11	1.0000	1.0000	1.0000	1.0000	1.0000	1.0000	1.0000	.9999	.9995	.9983
13	12	1.0000	1.0000	1.0000	1.0000	1.0000	1.0000	1.0000	1.0000	1.0000	.9999

Table 3 (cont.)

VALUES OF THE BINOMIAL CUMULATIVE DISTRIBUTION FUNCTION

$$F(x) = P(X \le x) = \sum_{i=0}^{x} \binom{n}{i} p^i (1-p)^{n-i} \text{ for } p \le 0.5,$$

X HAVING THE Bin(n,p) DISTRIBUTION. NOTE, FOR $p > 0.5$,
$P(X \le x) = 1 - P(Y \le n-x-1)$ with $Y = n-X \sim \text{Bin}(n, 1-p)$.

n	x	.05	.10	.15	.20	.25	.30	.35	.40	.45	.50
14	0	.4877	.2288	.1028	.0440	.0178	.0068	.0024	.0008	.0002	.0001
14	1	.8470	.5846	.3567	.1979	.1010	.0475	.0205	.0081	.0029	.0009
14	2	.9699	.8416	.6479	.4481	.2811	.1608	.0839	.0398	.0170	.0065
14	3	.9958	.9559	.8535	.6982	.5213	.3552	.2205	.1243	.0632	.0287
14	4	.9996	.9908	.9533	.8702	.7415	.5842	.4227	.2793	.1672	.0898
14	5	1.0000	.9985	.9885	.9561	.8883	.7805	.6405	.4859	.3373	.2120
14	6	1.0000	.9998	.9978	.9884	.9617	.9067	.8164	.6925	.5461	.3953
14	7	1.0000	1.0000	.9997	.9976	.9897	.9685	.9247	.8499	.7414	.6047
14	8	1.0000	1.0000	1.0000	.9996	.9978	.9917	.9757	.9417	.8811	.7880
14	9	1.0000	1.0000	1.0000	1.0000	.9997	.9983	.9940	.9825	.9574	.9102
14	10	1.0000	1.0000	1.0000	1.0000	1.0000	.9998	.9989	.9961	.9886	.9713
14	11	1.0000	1.0000	1.0000	1.0000	1.0000	1.0000	.9999	.9994	.9978	.9935
14	12	1.0000	1.0000	1.0000	1.0000	1.0000	1.0000	1.0000	.9999	.9997	.9991
14	13	1.0000	1.0000	1.0000	1.0000	1.0000	1.0000	1.0000	1.0000	1.0000	.9999
15	0	.4633	.2059	.0874	.0352	.0134	.0047	.0016	.0005	.0001	.0000
15	1	.8290	.5490	.3186	.1671	.0802	.0353	.0142	.0052	.0017	.0005
15	2	.9638	.8159	.6042	.3980	.2361	.1268	.0617	.0271	.0107	.0037
15	3	.9945	.9444	.8227	.6482	.4613	.2969	.1727	.0905	.0424	.0176
15	4	.9994	.9873	.9383	.8358	.6865	.5155	.3519	.2173	.1204	.0592
15	5	.9999	.9978	.9832	.9389	.8516	.7216	.5643	.4032	.2608	.1509
15	6	1.0000	.9997	.9964	.9819	.9434	.8689	.7548	.6098	.4522	.3036
15	7	1.0000	1.0000	.9994	.9958	.9827	.9500	.8868	.7869	.6535	.5000
15	8	1.0000	1.0000	.9999	.9992	.9958	.9848	.9578	.9050	.8182	.6964
15	9	1.0000	1.0000	1.0000	.9999	.9992	.9963	.9876	.9662	.9231	.8491
15	10	1.0000	1.0000	1.0000	1.0000	.9999	.9993	.9972	.9907	.9745	.9408
15	11	1.0000	1.0000	1.0000	1.0000	1.0000	.9999	.9995	.9981	.9937	.9824
15	12	1.0000	1.0000	1.0000	1.0000	1.0000	1.0000	.9999	.9997	.9989	.9963
15	13	1.0000	1.0000	1.0000	1.0000	1.0000	1.0000	1.0000	1.0000	.9999	.9995
15	14	1.0000	1.0000	1.0000	1.0000	1.0000	1.0000	1.0000	1.0000	1.0000	1.0000
16	0	.4401	.1853	.0743	.0281	.0100	.0033	.0010	.0003	.0001	.0000
16	1	.8108	.5147	.2839	.1407	.0635	.0261	.0098	.0033	.0010	.0003
16	2	.9571	.7892	.5614	.3518	.1971	.0994	.0451	.0183	.0066	.0021
16	3	.9930	.9316	.7899	.5981	.4050	.2459	.1339	.0651	.0281	.0106
16	4	.9991	.9830	.9209	.7982	.6302	.4499	.2892	.1666	.0853	.0384
16	5	.9999	.9967	.9765	.9183	.8103	.6598	.4900	.3288	.1976	.1051
16	6	1.0000	.9995	.9944	.9733	.9204	.8247	.6881	.5272	.3660	.2272
16	7	1.0000	.9999	.9989	.9930	.9729	.9256	.8406	.7161	.5629	.4018
16	8	1.0000	1.0000	.9998	.9985	.9925	.9743	.9329	.8577	.7441	.5982
16	9	1.0000	1.0000	1.0000	.9998	.9984	.9929	.9771	.9417	.8759	.7728
16	10	1.0000	1.0000	1.0000	1.0000	.9997	.9984	.9938	.9809	.9514	.8949
16	11	1.0000	1.0000	1.0000	1.0000	1.0000	.9997	.9987	.9951	.9851	.9616
16	12	1.0000	1.0000	1.0000	1.0000	1.0000	1.0000	.9998	.9991	.9965	.9894
16	13	1.0000	1.0000	1.0000	1.0000	1.0000	1.0000	1.0000	.9999	.9994	.9979
16	14	1.0000	1.0000	1.0000	1.0000	1.0000	1.0000	1.0000	1.0000	.9999	.9997
16	15	1.0000	1.0000	1.0000	1.0000	1.0000	1.0000	1.0000	1.0000	1.0000	1.0000

Table 3 (cont.)

VALUES OF THE BINOMIAL CUMULATIVE DISTRIBUTION FUNCTION

$$F(x) = P(X \leq x) = \sum_{i=0}^{x} \binom{n}{i} p^i (1-p)^{n-i} \text{ for } p \leq 0.5,$$

X HAVING THE Bin(n,p) DISTRIBUTION. NOTE, FOR $p > 0.5$,
$P(X \leq x) = 1 - P(Y \leq n-x-1)$ with $Y = n-X \sim \text{Bin}(n, 1-p)$.

n	x	.05	.10	.15	.20	.25	.30	.35	.40	.45	.50
17	0	.4181	.1668	.0631	.0225	.0075	.0023	.0007	.0002	.0000	.0000
17	1	.7922	.4818	.2525	.1182	.0501	.0193	.0067	.0021	.0006	.0001
17	2	.9497	.7618	.5198	.3096	.1637	.0774	.0327	.0123	.0041	.0012
17	3	.9912	.9174	.7556	.5489	.3530	.2019	.1028	.0464	.0184	.0064
17	4	.9988	.9779	.9013	.7582	.5739	.3887	.2348	.1260	.0596	.0245
17	5	.9999	.9953	.9681	.8943	.7653	.5968	.4197	.2639	.1471	.0717
17	6	1.0000	.9992	.9917	.9623	.8929	.7752	.6188	.4478	.2902	.1662
17	7	1.0000	.9999	.9983	.9891	.9598	.8954	.7872	.6405	.4743	.3145
17	8	1.0000	1.0000	.9997	.9974	.9876	.9597	.9006	.8011	.6626	.5000
17	9	1.0000	1.0000	1.0000	.9995	.9969	.9873	.9617	.9081	.8166	.6855
17	10	1.0000	1.0000	1.0000	.9999	.9994	.9968	.9880	.9652	.9174	.8338
17	11	1.0000	1.0000	1.0000	1.0000	.9999	.9993	.9970	.9894	.9699	.9283
17	12	1.0000	1.0000	1.0000	1.0000	1.0000	.9999	.9994	.9975	.9914	.9755
17	13	1.0000	1.0000	1.0000	1.0000	1.0000	1.0000	.9999	.9995	.9981	.9936
17	14	1.0000	1.0000	1.0000	1.0000	1.0000	1.0000	1.0000	.9999	.9997	.9988
17	15	1.0000	1.0000	1.0000	1.0000	1.0000	1.0000	1.0000	1.0000	1.0000	.9999
17	16	1.0000	1.0000	1.0000	1.0000	1.0000	1.0000	1.0000	1.0000	1.0000	1.0000
18	0	.3972	.1501	.0536	.0180	.0056	.0016	.0004	.0001	.0000	.0000
18	1	.7735	.4503	.2241	.0991	.0395	.0142	.0046	.0013	.0003	.0001
18	2	.9419	.7338	.4797	.2713	.1353	.0600	.0236	.0082	.0025	.0007
18	3	.9891	.9018	.7202	.5010	.3057	.1646	.0783	.0328	.0120	.0038
18	4	.9985	.9718	.8794	.7164	.5187	.3327	.1886	.0942	.0411	.0154
18	5	.9998	.9936	.9581	.8671	.7175	.5344	.3550	.2088	.1077	.0481
18	6	1.0000	.9988	.9882	.9487	.8610	.7217	.5491	.3743	.2258	.1189
18	7	1.0000	.9998	.9973	.9837	.9431	.8593	.7283	.5634	.3915	.2403
18	8	1.0000	1.0000	.9995	.9957	.9807	.9404	.8609	.7368	.5778	.4073
18	9	1.0000	1.0000	.9999	.9991	.9946	.9790	.9403	.8653	.7473	.5927
18	10	1.0000	1.0000	1.0000	.9998	.9988	.9939	.9788	.9424	.8720	.7597
18	11	1.0000	1.0000	1.0000	1.0000	.9998	.9986	.9938	.9797	.9463	.8811
18	12	1.0000	1.0000	1.0000	1.0000	1.0000	.9997	.9986	.9942	.9817	.9519
18	13	1.0000	1.0000	1.0000	1.0000	1.0000	1.0000	.9997	.9987	.9951	.9846
18	14	1.0000	1.0000	1.0000	1.0000	1.0000	1.0000	1.0000	.9998	.9990	.9962
18	15	1.0000	1.0000	1.0000	1.0000	1.0000	1.0000	1.0000	1.0000	.9999	.9993
18	16	1.0000	1.0000	1.0000	1.0000	1.0000	1.0000	1.0000	1.0000	1.0000	.9999
18	17	1.0000	1.0000	1.0000	1.0000	1.0000	1.0000	1.0000	1.0000	1.0000	1.0000

Table 3 (cont.)

VALUES OF THE BINOMIAL CUMULATIVE DISTRIBUTION FUNCTION

$$F(x) = P(X \le x) = \sum_{i=0}^{x} \binom{n}{i} p^i (1-p)^{n-i} \text{ for } p \le 0.5,$$

X HAVING THE Bin(n,p) DISTRIBUTION. NOTE, FOR p > 0.5,
$P(X \le x) = 1-P(Y \le n-x-1)$ with $Y = n-X \sim Bin(n, 1-p)$.

n	x	.05	.10	.15	.20	.25	.30	.35	.40	.45	.50
19	0	.3774	.1351	.0456	.0144	.0042	.0011	.0003	.0001	.0000	.0000
19	1	.7547	.4203	.1985	.0829	.0310	.0104	.0031	.0008	.0002	.0000
19	2	.9335	.7054	.4413	.2369	.1113	.0462	.0170	.0055	.0015	.0004
19	3	.9868	.8850	.6841	.4551	.2631	.1332	.0591	.0230	.0077	.0022
19	4	.9980	.9648	.8556	.6733	.4654	.2822	.1500	.0696	.0280	.0096
19	5	.9998	.9914	.9463	.8369	.6678	.4739	.2968	.1629	.0777	.0318
19	6	1.0000	.9983	.9837	.9324	.8251	.6655	.4812	.3081	.1727	.0835
19	7	1.0000	.9997	.9959	.9767	.9225	.8180	.6656	.4878	.3169	.1796
19	8	1.0000	1.0000	.9992	.9933	.9713	.9161	.8145	.6675	.4940	.3238
19	9	1.0000	1.0000	.9999	.9984	.9911	.9674	.9125	.8139	.6710	.5000
19	10	1.0000	1.0000	1.0000	.9997	.9977	.9895	.9653	.9115	.8159	.6762
19	11	1.0000	1.0000	1.0000	1.0000	.9995	.9972	.9886	.9648	.9129	.8204
19	12	1.0000	1.0000	1.0000	1.0000	.9999	.9994	.9969	.9884	.9658	.9165
19	13	1.0000	1.0000	1.0000	1.0000	1.0000	.9999	.9993	.9969	.9891	.9682
19	14	1.0000	1.0000	1.0000	1.0000	1.0000	1.0000	.9999	.9994	.9972	.9904
19	15	1.0000	1.0000	1.0000	1.0000	1.0000	1.0000	1.0000	.9999	.9995	.9978
19	16	1.0000	1.0000	1.0000	1.0000	1.0000	1.0000	1.0000	1.0000	.9999	.9996
19	17	1.0000	1.0000	1.0000	1.0000	1.0000	1.0000	1.0000	1.0000	1.0000	1.0000
19	18	1.0000	1.0000	1.0000	1.0000	1.0000	1.0000	1.0000	1.0000	1.0000	1.0000
20	0	.3585	.1216	.0388	.0115	.0032	.0008	.0002	.0000	.0000	.0000
20	1	.7358	.3917	.1756	.0692	.0243	.0076	.0021	.0005	.0001	.0000
20	2	.9245	.6769	.4049	.2061	.0913	.0355	.0121	.0036	.0009	.0002
20	3	.9841	.8670	.6477	.4114	.2252	.1071	.0444	.0160	.0049	.0013
20	4	.9974	.9568	.8298	.6296	.4148	.2375	.1182	.0510	.0189	.0059
20	5	.9997	.9887	.9327	.8042	.6172	.4164	.2454	.1256	.0553	.0207
20	6	1.0000	.9976	.9781	.9133	.7858	.6080	.4166	.2500	.1299	.0577
20	7	1.0000	.9996	.9941	.9679	.8982	.7723	.6010	.4159	.2520	.1316
20	8	1.0000	.9999	.9987	.9900	.9591	.8867	.7624	.5956	.4143	.2517
20	9	1.0000	1.0000	.9998	.9974	.9861	.9520	.8782	.7553	.5914	.4119
20	10	1.0000	1.0000	1.0000	.9994	.9961	.9829	.9468	.8725	.7507	.5881
20	11	1.0000	1.0000	1.0000	.9999	.9991	.9949	.9804	.9435	.8692	.7483
20	12	1.0000	1.0000	1.0000	1.0000	.9998	.9987	.9940	.9790	.9420	.8684
20	13	1.0000	1.0000	1.0000	1.0000	1.0000	.9997	.9985	.9935	.9786	.9423
20	14	1.0000	1.0000	1.0000	1.0000	1.0000	1.0000	.9997	.9984	.9936	.9793
20	15	1.0000	1.0000	1.0000	1.0000	1.0000	1.0000	1.0000	.9997	.9985	.9941
20	16	1.0000	1.0000	1.0000	1.0000	1.0000	1.0000	1.0000	1.0000	.9997	.9987
20	17	1.0000	1.0000	1.0000	1.0000	1.0000	1.0000	1.0000	1.0000	1.0000	.9998
20	18	1.0000	1.0000	1.0000	1.0000	1.0000	1.0000	1.0000	1.0000	1.0000	1.0000
20	19	1.0000	1.0000	1.0000	1.0000	1.0000	1.0000	1.0000	1.0000	1.0000	1.0000

Table 3 (cont.)

VALUES OF THE BINOMIAL CUMULATIVE DISTRIBUTION FUNCTION

$$F(x) = P(X \leq x) = \sum_{i=0}^{x} \binom{n}{i} p^i (1-p)^{n-i} \text{ for } p \leq 0.5,$$

X HAVING THE Bin(n,p) DISTRIBUTION. NOTE, FOR $p > 0.5$,
$P(X \leq x) = 1 - P(Y \leq n-x-1)$ with $Y = n-X \sim Bin(n, 1-p)$.

n	x	.05	.10	.15	.20	.25	.30	.35	.40	.45	.50
21	0	.3406	.1094	.0329	.0092	.0024	.0006	.0001	.0000	.0000	.0000
21	1	.7170	.3647	.1550	.0576	.0190	.0056	.0014	.0003	.0001	.0000
21	2	.9151	.6484	.3705	.1787	.0745	.0271	.0086	.0024	.0006	.0001
21	3	.9811	.8480	.6113	.3704	.1917	.0856	.0331	.0110	.0031	.0007
21	4	.9968	.9478	.8025	.5860	.3674	.1984	.0924	.0370	.0126	.0036
21	5	.9996	.9856	.9173	.7693	.5666	.3627	.2009	.0957	.0389	.0133
21	6	1.0000	.9967	.9713	.8915	.7436	.5505	.3567	.2002	.0964	.0392
21	7	1.0000	.9994	.9917	.9569	.8701	.7230	.5365	.3495	.1971	.0946
21	8	1.0000	.9999	.9980	.9856	.9439	.8523	.7059	.5237	.3413	.1917
21	9	1.0000	1.0000	.9996	.9959	.9794	.9324	.8377	.6914	.5117	.3318
21	10	1.0000	1.0000	.9999	.9990	.9936	.9736	.9228	.8256	.6790	.5000
21	11	1.0000	1.0000	1.0000	.9998	.9983	.9913	.9687	.9151	.8159	.6682
21	12	1.0000	1.0000	1.0000	1.0000	.9996	.9976	.9892	.9648	.9092	.8083
21	13	1.0000	1.0000	1.0000	1.0000	.9999	.9994	.9969	.9877	.9621	.9054
21	14	1.0000	1.0000	1.0000	1.0000	1.0000	.9999	.9993	.9964	.9868	.9608
21	15	1.0000	1.0000	1.0000	1.0000	1.0000	1.0000	.9999	.9992	.9963	.9867
21	16	1.0000	1.0000	1.0000	1.0000	1.0000	1.0000	1.0000	.9998	.9992	.9964
21	17	1.0000	1.0000	1.0000	1.0000	1.0000	1.0000	1.0000	1.0000	.9999	.9993
21	18	1.0000	1.0000	1.0000	1.0000	1.0000	1.0000	1.0000	1.0000	1.0000	.9999
21	19	1.0000	1.0000	1.0000	1.0000	1.0000	1.0000	1.0000	1.0000	1.0000	1.0000
21	20	1.0000	1.0000	1.0000	1.0000	1.0000	1.0000	1.0000	1.0000	1.0000	1.0000
22	0	.3235	.0985	.0280	.0074	.0018	.0004	.0001	.0000	.0000	.0000
22	1	.6982	.3392	.1367	.0480	.0149	.0041	.0010	.0002	.0000	.0000
22	2	.9052	.6200	.3382	.1545	.0606	.0207	.0061	.0016	.0003	.0001
22	3	.9778	.8281	.5752	.3320	.1624	.0681	.0245	.0076	.0020	.0004
22	4	.9960	.9379	.7738	.5429	.3235	.1645	.0716	.0266	.0083	.0022
22	5	.9994	.9818	.9001	.7326	.5168	.3134	.1629	.0722	.0271	.0085
22	6	.9999	.9956	.9632	.8670	.6994	.4942	.3022	.1584	.0705	.0262
22	7	1.0000	.9991	.9886	.9439	.8385	.6713	.4736	.2898	.1518	.0669
22	8	1.0000	.9999	.9970	.9799	.9254	.8135	.6466	.4540	.2764	.1431
22	9	1.0000	1.0000	.9993	.9939	.9705	.9084	.7916	.6244	.4350	.2617
22	10	1.0000	1.0000	.9999	.9984	.9900	.9613	.8930	.7720	.6037	.4159
22	11	1.0000	1.0000	1.0000	.9997	.9971	.9860	.9526	.8793	.7543	.5841
22	12	1.0000	1.0000	1.0000	.9999	.9993	.9957	.9820	.9449	.8672	.7383
22	13	1.0000	1.0000	1.0000	1.0000	.9999	.9989	.9942	.9785	.9383	.8569
22	14	1.0000	1.0000	1.0000	1.0000	1.0000	.9998	.9984	.9930	.9757	.9331
22	15	1.0000	1.0000	1.0000	1.0000	1.0000	1.0000	.9997	.9981	.9920	.9738
22	16	1.0000	1.0000	1.0000	1.0000	1.0000	1.0000	.9999	.9996	.9979	.9915
22	17	1.0000	1.0000	1.0000	1.0000	1.0000	1.0000	1.0000	.9999	.9995	.9978
22	18	1.0000	1.0000	1.0000	1.0000	1.0000	1.0000	1.0000	1.0000	.9999	.9996
22	19	1.0000	1.0000	1.0000	1.0000	1.0000	1.0000	1.0000	1.0000	1.0000	.9999
22	20	1.0000	1.0000	1.0000	1.0000	1.0000	1.0000	1.0000	1.0000	1.0000	1.0000
22	21	1.0000	1.0000	1.0000	1.0000	1.0000	1.0000	1.0000	1.0000	1.0000	1.0000

Table 3 (cont.)

VALUES OF THE BINOMIAL CUMULATIVE DISTRIBUTION FUNCTION

$$F(x) = P(X \le x) = \sum_{i=0}^{x} \binom{n}{i} p^i (1-p)^{n-i} \text{ for } p \le 0.5,$$

X HAVING THE Bin(n,p) DISTRIBUTION. NOTE, FOR $p > 0.5$,
$P(X \le x) = 1 - P(Y \le n-x-1)$ with $Y = n-X \sim Bin(n, 1-p)$.

n	x	.05	.10	.15	.20	.25	.30	.35	.40	.45	.50
23	0	.3074	.0886	.0238	.0059	.0013	.0003	.0000	.0000	.0000	.0000
23	1	.6794	.3151	.1204	.0398	.0116	.0030	.0007	.0001	.0000	.0000
23	2	.8948	.5920	.3080	.1332	.0492	.0157	.0043	.0010	.0002	.0000
23	3	.9742	.8073	.5396	.2965	.1370	.0538	.0181	.0052	.0012	.0002
23	4	.9951	.9269	.7440	.5007	.2832	.1356	.0551	.0190	.0055	.0013
23	5	.9992	.9774	.8811	.6947	.4685	.2688	.1309	.0540	.0186	.0053
23	6	.9999	.9942	.9537	.8402	.6537	.4399	.2534	.1240	.0510	.0173
23	7	1.0000	.9988	.9848	.9285	.8037	.6181	.4136	.2373	.1152	.0466
23	8	1.0000	.9998	.9958	.9727	.9037	.7709	.5860	.3884	.2203	.1050
23	9	1.0000	1.0000	.9990	.9911	.9592	.8799	.7408	.5562	.3636	.2024
23	10	1.0000	1.0000	.9998	.9975	.9851	.9454	.8575	.7129	.5278	.3388
23	11	1.0000	1.0000	1.0000	.9994	.9954	.9786	.9318	.8364	.6865	.5000
23	12	1.0000	1.0000	1.0000	.9999	.9988	.9928	.9717	.9187	.8164	.6612
23	13	1.0000	1.0000	1.0000	1.0000	.9997	.9979	.9900	.9651	.9063	.7976
23	14	1.0000	1.0000	1.0000	1.0000	.9999	.9995	.9970	.9872	.9589	.8950
23	15	1.0000	1.0000	1.0000	1.0000	1.0000	.9999	.9992	.9960	.9847	.9534
23	16	1.0000	1.0000	1.0000	1.0000	1.0000	1.0000	.9998	.9990	.9952	.9827
23	17	1.0000	1.0000	1.0000	1.0000	1.0000	1.0000	1.0000	.9998	.9988	.9947
23	18	1.0000	1.0000	1.0000	1.0000	1.0000	1.0000	1.0000	1.0000	.9998	.9987
23	19	1.0000	1.0000	1.0000	1.0000	1.0000	1.0000	1.0000	1.0000	1.0000	.9998
23	20	1.0000	1.0000	1.0000	1.0000	1.0000	1.0000	1.0000	1.0000	1.0000	1.0000
23	21	1.0000	1.0000	1.0000	1.0000	1.0000	1.0000	1.0000	1.0000	1.0000	1.0000
23	22	1.0000	1.0000	1.0000	1.0000	1.0000	1.0000	1.0000	1.0000	1.0000	1.0000
24	0	.2920	.0798	.0202	.0047	.0010	.0002	.0000	.0000	.0000	.0000
24	1	.6608	.2925	.1059	.0331	.0090	.0022	.0005	.0001	.0000	.0000
24	2	.8841	.5643	.2798	.1145	.0398	.0119	.0030	.0007	.0001	.0000
24	3	.9702	.7857	.5049	.2639	.1150	.0424	.0133	.0035	.0008	.0001
24	4	.9940	.9149	.7134	.4599	.2466	.1111	.0422	.0134	.0036	.0008
24	5	.9990	.9723	.8606	.6559	.4222	.2288	.1044	.0400	.0127	.0033
24	6	.9999	.9925	.9428	.8111	.6074	.3886	.2106	.0960	.0364	.0113
24	7	1.0000	.9983	.9801	.9108	.7662	.5647	.3575	.1919	.0863	.0320
24	8	1.0000	.9997	.9941	.9638	.8787	.7250	.5257	.3279	.1730	.0758
24	9	1.0000	.9999	.9985	.9874	.9453	.8472	.6866	.4891	.2991	.1537
24	10	1.0000	1.0000	.9997	.9962	.9787	.9258	.8167	.6502	.4539	.2706
24	11	1.0000	1.0000	.9999	.9990	.9928	.9686	.9058	.7870	.6151	.4194
24	12	1.0000	1.0000	1.0000	.9998	.9979	.9885	.9577	.8857	.7580	.5806
24	13	1.0000	1.0000	1.0000	1.0000	.9995	.9964	.9836	.9465	.8659	.7294
24	14	1.0000	1.0000	1.0000	1.0000	.9999	.9990	.9945	.9783	.9352	.8463
24	15	1.0000	1.0000	1.0000	1.0000	1.0000	.9998	.9984	.9925	.9731	.9242
24	16	1.0000	1.0000	1.0000	1.0000	1.0000	1.0000	.9996	.9978	.9905	.9680
24	17	1.0000	1.0000	1.0000	1.0000	1.0000	1.0000	.9999	.9995	.9972	.9887
24	18	1.0000	1.0000	1.0000	1.0000	1.0000	1.0000	1.0000	.9999	.9993	.9967
24	19	1.0000	1.0000	1.0000	1.0000	1.0000	1.0000	1.0000	1.0000	.9999	.9992
24	20	1.0000	1.0000	1.0000	1.0000	1.0000	1.0000	1.0000	1.0000	1.0000	.9999
24	21	1.0000	1.0000	1.0000	1.0000	1.0000	1.0000	1.0000	1.0000	1.0000	1.0000
24	22	1.0000	1.0000	1.0000	1.0000	1.0000	1.0000	1.0000	1.0000	1.0000	1.0000
24	23	1.0000	1.0000	1.0000	1.0000	1.0000	1.0000	1.0000	1.0000	1.0000	1.0000

Table 3 (cont.)

VALUES OF THE BINOMIAL CUMULATIVE DISTRIBUTION FUNCTION

$$F(x) = P(X \le x) = \sum_{i=0}^{x} \binom{n}{i} p^i (1-p)^{n-i} \text{ for } p \le 0.5,$$

X HAVING THE Bin(n,p) DISTRIBUTION. NOTE, FOR $p > 0.5$,
$P(X \le x) = 1 - P(Y \le n-x-1)$ with $Y = n-X \sim$ Bin(n, 1-p).

n	x	.05	.10	.15	.20	.25	.30	.35	.40	.45	.50
25	0	.2774	.0718	.0172	.0038	.0008	.0001	.0000	.0000	.0000	.0000
25	1	.6424	.2712	.0931	.0274	.0070	.0016	.0003	.0001	.0000	.0000
25	2	.8729	.5371	.2537	.0982	.0321	.0090	.0021	.0004	.0001	.0000
25	3	.9659	.7636	.4711	.2340	.0962	.0332	.0097	.0024	.0005	.0001
25	4	.9928	.9020	.6821	.4207	.2137	.0905	.0320	.0095	.0023	.0005
25	5	.9988	.9666	.8385	.6167	.3783	.1935	.0826	.0294	.0086	.0020
25	6	.9998	.9905	.9305	.7800	.5611	.3407	.1734	.0736	.0258	.0073
25	7	1.0000	.9977	.9745	.8909	.7265	.5118	.3061	.1536	.0639	.0216
25	8	1.0000	.9995	.9920	.9532	.8506	.6769	.4668	.2735	.1340	.0539
25	9	1.0000	.9999	.9979	.9827	.9287	.8106	.6303	.4246	.2424	.1148
25	10	1.0000	1.0000	.9995	.9944	.9703	.9022	.7712	.5858	.3843	.2122
25	11	1.0000	1.0000	.9999	.9985	.9893	.9558	.8746	.7323	.5426	.3450
25	12	1.0000	1.0000	1.0000	.9996	.9966	.9825	.9396	.8462	.6937	.5000
25	13	1.0000	1.0000	1.0000	.9999	.9991	.9940	.9745	.9222	.8173	.6550
25	14	1.0000	1.0000	1.0000	1.0000	.9998	.9982	.9907	.9656	.9040	.7878
25	15	1.0000	1.0000	1.0000	1.0000	1.0000	.9995	.9971	.9868	.9560	.8852
25	16	1.0000	1.0000	1.0000	1.0000	1.0000	.9999	.9992	.9957	.9826	.9461
25	17	1.0000	1.0000	1.0000	1.0000	1.0000	1.0000	.9998	.9988	.9942	.9784
25	18	1.0000	1.0000	1.0000	1.0000	1.0000	1.0000	1.0000	.9997	.9984	.9927
25	19	1.0000	1.0000	1.0000	1.0000	1.0000	1.0000	1.0000	.9999	.9996	.9980
25	20	1.0000	1.0000	1.0000	1.0000	1.0000	1.0000	1.0000	1.0000	.9999	.9995
25	21	1.0000	1.0000	1.0000	1.0000	1.0000	1.0000	1.0000	1.0000	1.0000	.9999
25	22	1.0000	1.0000	1.0000	1.0000	1.0000	1.0000	1.0000	1.0000	1.0000	1.0000
25	23	1.0000	1.0000	1.0000	1.0000	1.0000	1.0000	1.0000	1.0000	1.0000	1.0000
25	24	1.0000	1.0000	1.0000	1.0000	1.0000	1.0000	1.0000	1.0000	1.0000	1.0000

Table 4

PERCENTAGE POINTS OF THE t-DISTRIBUTION
(Values of t in Terms of A and ν)

ν \ A	0.2	0.5	0.8	0.9	0.95	0.98	0.99	0.995	0.998	0.999	0.9999	0.99999	0.999999
1	0.325	1.000	3.078	6.314	12.706	31.821	63.657	127.321	318.309	636.619	6366.198	63661.977	636619.772
2	0.289	0.816	1.886	2.920	4.303	6.965	9.925	14.089	22.327	31.598	99.992	316.225	999.999
3	0.277	0.765	1.638	2.353	3.182	4.541	5.841	7.453	10.214	12.924	28.000	60.397	130.155
4	0.271	0.741	1.533	2.132	2.776	3.747	4.604	5.598	7.173	8.610	15.544	27.771	49.459
5	0.267	0.727	1.476	2.015	2.571	3.365	4.032	4.773	5.893	6.869	11.178	17.897	28.477
6	0.265	0.718	1.440	1.943	2.447	3.143	3.707	4.317	5.208	5.959	9.082	13.555	20.047
7	0.263	0.711	1.415	1.895	2.365	2.998	3.499	4.029	4.785	5.408	7.885	11.215	15.764
8	0.262	0.706	1.397	1.860	2.306	2.896	3.355	3.833	4.501	5.041	7.120	9.782	13.257
9	0.261	0.703	1.383	1.833	2.262	2.821	3.250	3.690	4.297	4.781	6.594	8.827	11.637
10	0.260	0.700	1.372	1.812	2.228	2.764	3.169	3.581	4.144	4.587	6.211	8.150	10.516
11	0.260	0.697	1.363	1.796	2.201	2.718	3.106	3.497	4.025	4.437	5.921	7.648	9.702
12	0.259	0.695	1.356	1.782	2.179	2.681	3.055	3.428	3.930	4.318	5.694	7.261	9.085
13	0.259	0.694	1.350	1.771	2.160	2.650	3.012	3.372	3.852	4.221	5.513	6.955	8.604
14	0.258	0.692	1.345	1.761	2.145	2.624	2.977	3.326	3.787	4.140	5.363	6.706	8.218
15	0.258	0.691	1.341	1.753	2.131	2.602	2.947	3.286	3.733	4.073	5.239	6.502	7.903
16	0.258	0.690	1.337	1.746	2.120	2.583	2.921	3.252	3.686	4.015	5.134	6.330	7.642
17	0.257	0.689	1.333	1.740	2.110	2.567	2.898	3.223	3.646	3.965	5.044	6.184	7.421
18	0.257	0.688	1.330	1.734	2.101	2.552	2.878	3.197	3.610	3.922	4.966	6.059	7.232
19	0.257	0.688	1.328	1.729	2.093	2.539	2.861	3.174	3.579	3.883	4.897	5.949	7.069
20	0.257	0.687	1.325	1.725	2.086	2.528	2.845	3.153	3.552	3.850	4.837	5.854	6.927
21	0.257	0.686	1.323	1.721	2.080	2.518	2.831	3.135	3.527	3.819	4.784	5.769	6.802
22	0.256	0.686	1.321	1.717	2.074	2.508	2.819	3.119	3.505	3.792	4.736	5.694	6.692
23	0.256	0.685	1.319	1.714	2.069	2.500	2.807	3.104	3.485	3.768	4.693	5.627	6.593
24	0.256	0.685	1.318	1.711	2.064	2.492	2.797	3.090	3.467	3.745	4.654	5.566	6.504
25	0.256	0.684	1.316	1.708	2.060	2.485	2.787	3.078	3.450	3.725	4.619	5.511	6.424
26	0.256	0.684	1.315	1.706	2.056	2.479	2.779	3.067	3.435	3.707	4.587	5.461	6.352
27	0.256	0.684	1.314	1.703	2.052	2.473	2.771	3.057	3.421	3.690	4.558	5.415	6.286
28	0.256	0.683	1.313	1.701	2.048	2.467	2.763	3.047	3.408	3.674	4.530	5.373	6.225
29	0.256	0.683	1.311	1.699	2.045	2.462	2.756	3.038	3.396	3.659	4.506	5.335	6.170
30	0.256	0.683	1.310	1.697	2.042	2.457	2.750	3.030	3.385	3.646	4.482	5.299	6.119
40	0.255	0.681	1.303	1.684	2.021	2.423	2.704	2.971	3.307	3.551	4.321	5.053	5.768
60	0.254	0.679	1.296	1.671	2.000	2.390	2.660	2.915	3.232	3.460	4.169	4.825	5.449
120	0.254	0.677	1.289	1.658	1.980	2.358	2.617	2.860	3.160	3.373	3.997	4.573	5.102
∞	0.253	0.674	1.282	1.645	1.960	2.326	2.576	2.807	3.090	3.291	3.891	4.417	4.892

$$A = A(t\nu) = \left[\sqrt{\nu}B\left(\frac{1}{2},\frac{\nu}{2}\right) \right]^{-1} \int_{-1}^{t} \left(1+\frac{x^2}{\nu}\right)^{-\left(\frac{\nu+1}{2}\right)} dx$$

From E. S. Pearson and H. O. Hartley (editors), Biometrika tables for statisticians, vol. I. Cambridge Univ. Press, Cambridge, England, 1954, for A < 0.999, from E. T. Federighi, Extended tables of the percentage points of Student's t-distribution, *J. Amer. Statist. Assoc.* 54, 683-688 (1959) for A < 0.999 (with permission).

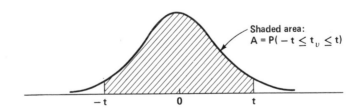

Shaded area:
$A = P(-t \leq t_\nu \leq t)$

−t 0 t

Table 5

PERCENTAGE POINTS OF THE χ^2-DISTRIBUTION
(Values of χ^2 in Terms of Q and ν)

ν \ Q	0.995	0.99	0.975	0.95	0.9	0.75	0.5	0.25
1	(−5)3.92704	(−4)1.57088	(−4)9.82069	(−3)3.93214	0.0157908	0.101531	0.454937	1.32330
2	(−2)1.00251	(−2)2.01007	(−2)5.06356	0.102587	0.210720	0.575364	1.38629	2.77259
3	(−2)7.17212	0.114832	0.215795	0.351846	0.584375	1.212534	2.36597	4.10835
4	0.206990	0.297110	0.484419	0.710721	1.063623	1.92255	3.35670	5.38527
5	0.411740	0.554300	0.831211	1.145476	1.61031	2.67460	4.35146	6.62568
6	0.675727	0.872085	1.237347	1.63539	2.20413	3.45460	5.34812	7.84080
7	0.989265	1.239043	1.68987	2.16735	2.83311	4.25485	6.34581	9.03715
8	1.344419	1.646482	2.17973	2.73264	3.48954	5.07064	7.34412	10.2188
9	1.734926	2.087912	2.70039	3.32511	4.16816	5.89883	8.34283	11.3887
10	2.15585	2.55821	3.24697	3.94030	4.86518	6.73720	9.34182	12.5489
11	2.60321	3.05347	3.81575	4.57481	5.57779	7.58412	10.3410	13.7007
12	3.07382	3.57056	4.40379	5.22603	6.30380	8.43842	11.3403	14.8454
13	3.56503	4.10691	5.00874	5.89186	7.04150	9.29906	12.3398	15.9839
14	4.07468	4.66043	5.62872	6.57063	7.78953	10.1653	13.3393	17.1170
15	4.60094	5.22935	6.26214	7.26094	8.54675	11.0365	14.3389	18.2451
16	5.14224	5.81221	6.90766	7.96164	9.31223	11.9122	15.3385	19.3688
17	5.69724	6.40776	7.56418	8.67176	10.0852	12.7919	16.3381	20.4887
18	6.26481	7.01491	8.23075	9.39046	10.8649	13.6753	17.3379	21.6049
19	6.84398	7.63273	8.90655	10.1170	11.6509	14.5620	18.3376	22.7178
20	7.43386	8.26040	9.59083	10.8508	12.4426	15.4518	19.3374	23.8277
21	8.03366	8.89720	10.28293	11.5913	13.2396	16.3444	20.3372	24.9348
22	8.64272	9.54249	10.9823	12.3380	14.0415	17.2396	21.3370	26.0393
23	9.26042	10.19567	11.6885	13.0905	14.8479	18.1373	22.3369	27.1413
24	9.88623	10.8564	12.4011	13.8484	15.6587	19.0372	23.3367	28.2412
25	10.5197	11.5240	13.1197	14.6114	16.4734	19.9393	24.3366	29.3389
26	11.1603	12.1981	13.8439	15.3791	17.2919	20.8434	25.3364	30.4345
27	11.8076	12.8786	14.5733	16.1513	18.1138	21.7494	26.3363	31.5284
28	12.4613	13.5648	15.3079	16.9279	18.9392	22.6572	27.3363	32.6205
29	13.1211	14.2565	16.0471	17.7083	19.7677	23.5666	28.3362	33.7109
30	13.7867	14.9535	16.7908	18.4926	20.5992	24.4776	29.3360	34.7998
40	20.7065	22.1643	24.4331	26.5093	29.0505	33.6603	39.3354	45.6160
50	27.9907	29.7067	32.3574	34.7642	37.6886	42.9421	49.3349	56.3336
60	35.5346	37.4848	40.4817	43.1879	46.4589	52.2938	59.3347	66.9814
70	43.2752	45.4418	48.7576	51.7393	55.3290	61.6983	69.3344	77.5766
80	51.1720	53.5400	57.1532	60.3915	64.2778	71.1445	79.3343	88.1303
90	59.1963	61.7541	65.6466	69.1260	73.2912	80.6247	89.3342	98.6499
100	67.3276	70.0648	74.2219	77.9295	82.3581	90.1332	99.3341	109.141
X	−2.5758	−2.3263	−1.9600	−1.6449	−1.2816	−0.6745	0.0000	0.6745

$$Q(\chi^2 \mid \nu) = \left[2^{\frac{\nu}{2}} \Gamma \left(\frac{\nu}{2} \right) \right]^{-1} \int_{\chi^2}^{\infty} e^{-\frac{t}{2}} t^{\frac{\nu}{2}-1} \, dt$$

From E. S. Pearson and H. O. Hartley (editors), Biometrika tables for statisticians, vol. I. Cambridge Univ. Press, Cambridge, England, 1954 (with permission) for Q > 0.0005.

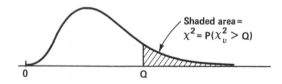

Shaded area =
$\chi^2 = P(\chi^2_{\nu} > Q)$

Table 5 (cont.)

PERCENTAGE POINTS OF THE χ^2 DISTRIBUTION
(Values of χ^2 in Terms of Q and ν)

$\nu \backslash Q$	0.1	0.05	0.025	0.01	0.005	0.001	0.0005	0.0001
1	2.70554	3.84146	5.02389	6.63490	7.87944	10.828	12.116	15.137
2	4.60517	5.99147	7.37776	9.21034	10.5966	13.816	15.202	18.421
3	6.25139	7.81473	9.34840	11.3449	12.8381	16.266	17.730	21.108
4	7.77944	9.48773	11.1433	13.2767	14.8602	18.467	19.997	23.513
5	9.23635	11.0705	12.8325	15.0863	16.7496	20.515	22.105	25.745
6	10.6446	12.5916	14.4494	16.8119	18.5476	22.458	24.103	27.856
7	12.0170	14.0671	16.0128	18.4753	20.2777	24.322	26.018	29.877
8	13.3616	15.5073	17.5346	20.0902	21.9550	26.125	27.868	31.828
9	14.6837	16.9190	19.0228	21.6660	23.5893	27.877	29.666	33.720
10	15.9871	18.3070	20.4831	23.2093	25.1882	29.588	31.420	35.564
11	17.2750	19.6751	21.9200	24.7250	26.7569	31.264	33.137	37.367
12	18.5494	21.0261	23.3367	26.2170	28.2995	32.909	34.821	39.134
13	19.8119	22.3621	24.7356	27.6883	29.8194	34.528	36.478	40.871
14	21.0642	23.6848	26.1190	29.1413	31.3193	36.123	38.109	42.579
15	22.3072	24.9958	27.4884	30.5779	32.8013	37.697	39.719	44.263
16	23.5418	26.2962	28.8454	31.9999	34.2672	39.252	41.308	45.925
17	24.7690	27.5871	30.1910	33.4087	35.7185	40.790	42.879	47.566
18	25.9894	28.8693	31.5264	34.8053	37.1564	42.312	44.434	49.189
19	27.2036	30.1435	32.8523	36.1908	38.5822	43.820	45.973	50.796
20	28.4120	31.4104	34.1696	37.5662	39.9968	45.315	47.498	52.386
21	29.6151	32.6705	35.4789	38.9321	41.4010	46.797	49.011	53.962
22	30.8133	33.9244	36.7807	40.2894	42.7956	48.268	50.511	55.525
23	32.0069	35.1725	38.0757	41.6384	44.1813	49.728	52.000	57.075
24	33.1963	36.4151	39.3641	42.9798	45.5585	51.179	53.479	58.613
25	34.3816	37.6525	40.6465	44.3141	46.9278	52.620	54.947	60.140
26	35.5631	38.8852	41.9232	45.6417	48.2899	54.052	56.407	61.657
27	36.7412	40.1133	43.1944	46.9630	49.6449	55.476	57.858	63.164
28	37.9159	41.3372	44.4607	48.2782	50.9933	56.892	59.300	64.662
29	39.0875	42.5569	45.7222	49.5879	52.3356	58.302	60.735	66.152
30	40.2560	43.7729	46.9792	50.8922	53.6720	59.703	62.162	67.633
40	51.8050	55.7585	59.3417	63.6907	66.7659	73.402	76.095	82.062
50	63.1671	67.5048	71.4202	76.1539	79.4900	86.661	89.560	95.969
60	74.3970	79.0819	83.2976	88.3794	91.9517	99.607	102.695	109.503
70	85.5271	90.5312	95.0231	100.425	104.215	112.317	115.578	122.755
80	96.5782	101.879	106.629	112.329	116.321	124.839	128.261	135.783
90	107.565	113.145	118.136	124.116	128.299	137.208	140.782	148.627
100	118.498	124.342	129.561	135.807	140.169	149.449	153.167	161.319
X	1.2816	1.6449	1.9600	2.3263	2.5758	3.0902	3.2905	3.7190

$$Q(\chi^2|\nu) = \left[2^{\frac{\nu}{2}} \Gamma\left(\frac{\nu}{2}\right) \right]^{-1} \int_{\chi^2}^{\infty} e^{-\frac{t}{2}} t^{\frac{\nu}{2}-1} \, dt$$

Table 6

PERCENTAGE POINTS OF THE F-DISTRIBUTION
(Values of F in Terms of Q, ν_1, ν_2)

$$Q(F|\nu_1,\nu_2) = 0.5$$

$\nu_2 . \nu_1$	1	2	3	4	5	6	8	12	15	20	30	60	∞
1	1.00	1.50	1.71	1.82	1.89	1.94	2.00	2.07	2.09	2.12	2.15	2.17	2.20
2	0.667	1.00	1.13	1.21	1.25	1.28	1.32	1.36	1.38	1.39	1.41	1.43	1.44
3	0.585	0.881	1.00	1.06	1.10	1.13	1.16	1.20	1.21	1.23	1.24	1.25	1.27
4	0.549	0.828	0.941	1.00	1.04	1.06	1.09	1.13	1.14	1.15	1.16	1.18	1.19
5	0.528	0.799	0.907	0.965	1.00	1.02	1.05	1.09	1.10	1.11	1.12	1.14	1.15
6	0.515	0.780	0.886	0.942	0.977	1.00	1.03	1.06	1.07	1.08	1.10	1.11	1.12
7	0.506	0.767	0.871	0.926	0.960	0.983	1.01	1.04	1.05	1.07	1.08	1.09	1.10
8	0.499	0.757	0.860	0.915	0.948	0.971	1.00	1.03	1.04	1.05	1.07	1.08	1.09
9	0.494	0.749	0.852	0.906	0.939	0.962	0.990	1.02	1.03	1.04	1.05	1.07	1.08
10	0.490	0.743	0.845	0.899	0.932	0.954	0.983	1.01	1.02	1.03	1.05	1.06	1.07
11	0.486	0.739	0.840	0.893	0.926	0.948	0.977	1.01	1.02	1.03	1.04	1.05	1.06
12	0.484	0.735	0.835	0.888	0.921	0.943	0.972	1.00	1.01	1.02	1.03	1.05	1.06
13	0.481	0.731	0.832	0.885	0.917	0.939	0.967	0.996	1.01	1.02	1.03	1.04	1.05
14	0.479	0.729	0.828	0.881	0.914	0.936	0.964	0.992	1.00	1.01	1.03	1.04	1.05
15	0.478	0.726	0.826	0.878	0.911	0.933	0.960	0.989	1.00	1.01	1.02	1.03	1.05
16	0.476	0.724	0.823	0.876	0.908	0.930	0.958	0.986	0.997	1.01	1.02	1.03	1.04
17	0.475	0.722	0.821	0.874	0.906	0.928	0.955	0.983	0.995	1.01	1.02	1.03	1.04
18	0.474	0.721	0.819	0.872	0.904	0.926	0.953	0.981	0.992	1.00	1.02	1.03	1.04
19	0.473	0.719	0.818	0.870	0.902	0.924	0.951	0.979	0.990	1.00	1.01	1.02	1.04
20	0.472	0.718	0.816	0.868	0.900	0.922	0.950	0.977	0.989	1.00	1.01	1.02	1.03
21	0.471	0.716	0.815	0.867	0.899	0.921	0.948	0.976	0.987	0.998	1.01	1.02	1.03
22	0.470	0.715	0.814	0.866	0.898	0.919	0.947	0.974	0.986	0.997	1.01	1.02	1.03
23	0.470	0.714	0.813	0.864	0.896	0.918	0.945	0.973	0.984	0.996	1.01	1.02	1.03
24	0.469	0.714	0.812	0.863	0.895	0.917	0.944	0.972	0.983	0.994	1.01	1.02	1.03
25	0.468	0.713	0.811	0.862	0.894	0.916	0.943	0.971	0.982	0.993	1.00	1.02	1.03
26	0.468	0.712	0.810	0.861	0.893	0.915	0.942	0.970	0.981	0.992	1.00	1.01	1.03
27	0.467	0.711	0.809	0.861	0.892	0.914	0.941	0.969	0.980	0.991	1.00	1.01	1.03
28	0.467	0.711	0.808	0.860	0.892	0.913	0.940	0.968	0.979	0.990	1.00	1.01	1.02
29	0.466	0.710	0.808	0.859	0.891	0.912	0.940	0.967	0.978	0.990	1.00	1.01	1.02
30	0.466	0.709	0.807	0.858	0.890	0.912	0.939	0.966	0.978	0.989	1.00	1.01	1.02
40	0.463	0.705	0.802	0.854	0.885	0.907	0.934	0.961	0.972	0.983	0.994	1.01	1.02
60	0.461	0.701	0.798	0.849	0.880	0.901	0.928	0.956	0.967	0.978	0.989	1.00	1.01
120	0.458	0.697	0.793	0.844	0.875	0.896	0.923	0.950	0.961	0.972	0.983	0.994	1.01
∞	0.455	0.693	0.789	0.839	0.870	0.891	0.918	0.945	0.956	0.967	0.978	0.989	1.00

$$Q(F|\nu_1,\nu_2) = 0.25$$

$\nu_2 . \nu_1$	1	2	3	4	5	6	8	12	15	20	30	60	∞
1	5.83	7.50	8.20	8.58	8.82	8.98	9.19	9.41	9.49	9.58	9.67	9.76	9.85
2	2.57	3.00	3.15	3.23	3.28	3.31	3.35	3.39	3.41	3.43	3.44	3.46	3.48
3	2.02	2.28	2.36	2.39	2.41	2.42	2.44	2.45	2.46	2.46	2.47	2.47	2.47
4	1.81	2.00	2.05	2.06	2.07	2.08	2.08	2.08	2.08	2.08	2.08	2.08	2.08
5	1.69	1.85	1.88	1.89	1.89	1.89	1.89	1.89	1.89	1.88	1.88	1.87	1.87
6	1.62	1.76	1.78	1.79	1.79	1.78	1.78	1.77	1.76	1.76	1.75	1.74	1.74
7	1.57	1.70	1.72	1.72	1.71	1.71	1.70	1.68	1.68	1.67	1.66	1.65	1.65
8	1.54	1.66	1.67	1.66	1.66	1.65	1.64	1.62	1.62	1.61	1.60	1.59	1.58
9	1.51	1.62	1.63	1.63	1.62	1.61	1.60	1.58	1.57	1.56	1.55	1.54	1.53
10	1.49	1.60	1.60	1.59	1.59	1.58	1.56	1.54	1.53	1.52	1.51	1.50	1.48
11	1.47	1.58	1.58	1.57	1.56	1.55	1.53	1.51	1.50	1.49	1.48	1.47	1.45
12	1.46	1.56	1.56	1.55	1.54	1.53	1.51	1.49	1.48	1.47	1.45	1.44	1.42
13	1.45	1.55	1.55	1.53	1.52	1.51	1.49	1.47	1.46	1.45	1.43	1.42	1.40
14	1.44	1.53	1.53	1.52	1.51	1.50	1.48	1.45	1.44	1.43	1.41	1.40	1.38
15	1.43	1.52	1.52	1.51	1.49	1.48	1.46	1.44	1.43	1.41	1.40	1.38	1.36
16	1.42	1.51	1.51	1.50	1.48	1.47	1.45	1.43	1.41	1.40	1.38	1.36	1.34
17	1.42	1.51	1.50	1.49	1.47	1.46	1.44	1.41	1.40	1.39	1.37	1.35	1.33
18	1.41	1.50	1.49	1.48	1.46	1.45	1.43	1.40	1.39	1.38	1.36	1.34	1.32
19	1.41	1.49	1.49	1.47	1.46	1.44	1.42	1.40	1.38	1.37	1.35	1.33	1.30
20	1.40	1.49	1.48	1.47	1.45	1.44	1.42	1.39	1.37	1.36	1.34	1.32	1.29
21	1.40	1.48	1.48	1.46	1.44	1.43	1.41	1.38	1.37	1.35	1.33	1.31	1.28
22	1.40	1.48	1.47	1.45	1.44	1.42	1.40	1.37	1.36	1.34	1.32	1.30	1.28
23	1.39	1.47	1.47	1.45	1.43	1.42	1.40	1.37	1.35	1.34	1.32	1.30	1.27
24	1.39	1.47	1.46	1.44	1.43	1.41	1.39	1.36	1.35	1.33	1.31	1.29	1.26
25	1.39	1.47	1.46	1.44	1.42	1.41	1.39	1.36	1.34	1.33	1.31	1.28	1.25
26	1.38	1.46	1.45	1.44	1.42	1.41	1.38	1.35	1.34	1.32	1.30	1.28	1.25
27	1.38	1.46	1.45	1.43	1.42	1.40	1.38	1.35	1.33	1.32	1.30	1.27	1.24
28	1.38	1.46	1.45	1.43	1.41	1.40	1.38	1.34	1.33	1.31	1.29	1.27	1.24
29	1.38	1.45	1.45	1.43	1.41	1.40	1.37	1.34	1.32	1.31	1.29	1.26	1.23
30	1.38	1.45	1.44	1.42	1.41	1.39	1.37	1.34	1.32	1.30	1.28	1.26	1.23
40	1.36	1.44	1.42	1.40	1.39	1.37	1.35	1.31	1.30	1.28	1.25	1.22	1.19
60	1.35	1.42	1.41	1.38	1.37	1.35	1.32	1.29	1.27	1.25	1.22	1.19	1.15
120	1.34	1.40	1.39	1.37	1.35	1.33	1.30	1.26	1.24	1.22	1.19	1.16	1.10
∞	1.32	1.39	1.37	1.35	1.33	1.31	1.28	1.24	1.22	1.19	1.16	1.12	1.00

Compiled from E. S. Pearson and H. O. Hartley (editors), Biometrika tables for statisticians, vol. I. Cambridge Univ. Press, Cambridge, England, 1954 (with permission).

Table 6 (cont.)

PERCENTAGE POINTS OF THE F-DISTRIBUTION
(Values of F in Terms of Q, ν_1, ν_2)

$$Q(F|\nu_1,\nu_2)=0.1$$

$\nu_2\backslash\nu_1$	1	2	3	4	5	6	8	12	15	20	30	60	∞
1	39.86	49.50	53.59	55.83	57.24	58.20	59.44	60.71	61.22	61.74	62.26	62.79	63.33
2	8.53	9.00	9.16	9.24	9.29	9.33	9.37	9.41	9.42	9.44	9.46	9.47	9.49
3	5.54	5.46	5.39	5.34	5.31	5.28	5.25	5.22	5.20	5.18	5.17	5.15	5.13
4	4.54	4.32	4.19	4.11	4.05	4.01	3.95	3.90	3.87	3.84	3.82	3.79	3.76
5	4.06	3.78	3.62	3.52	3.45	3.40	3.34	3.27	3.24	3.21	3.17	3.14	3.10
6	3.78	3.46	3.29	3.18	3.11	3.05	2.98	2.90	2.87	2.84	2.80	2.76	2.72
7	3.59	3.26	3.07	2.96	2.88	2.83	2.75	2.67	2.63	2.59	2.56	2.51	2.47
8	3.46	3.11	2.92	2.81	2.73	2.67	2.59	2.50	2.46	2.42	2.38	2.34	2.29
9	3.36	3.01	2.81	2.69	2.61	2.55	2.47	2.38	2.34	2.30	2.25	2.21	2.16
10	3.29	2.92	2.73	2.61	2.52	2.46	2.38	2.28	2.24	2.20	2.16	2.11	2.06
11	3.23	2.86	2.66	2.54	2.45	2.39	2.30	2.21	2.17	2.12	2.08	2.03	1.97
12	3.18	2.81	2.61	2.48	2.39	2.33	2.24	2.15	2.10	2.06	2.01	1.96	1.90
13	3.14	2.76	2.56	2.43	2.35	2.28	2.20	2.10	2.05	2.01	1.96	1.90	1.85
14	3.10	2.73	2.52	2.39	2.31	2.24	2.15	2.05	2.01	1.96	1.91	1.86	1.80
15	3.07	2.70	2.49	2.36	2.27	2.21	2.12	2.02	1.97	1.92	1.87	1.82	1.76
16	3.05	2.67	2.46	2.33	2.24	2.18	2.09	1.99	1.94	1.89	1.84	1.78	1.72
17	3.03	2.64	2.44	2.31	2.22	2.15	2.06	1.96	1.91	1.86	1.81	1.75	1.69
18	3.01	2.62	2.42	2.29	2.20	2.13	2.04	1.93	1.89	1.84	1.78	1.72	1.66
19	2.99	2.61	2.40	2.27	2.18	2.11	2.02	1.91	1.86	1.81	1.76	1.70	1.63
20	2.97	2.59	2.38	2.25	2.16	2.09	2.00	1.89	1.84	1.79	1.74	1.68	1.61
21	2.96	2.57	2.36	2.23	2.14	2.08	1.98	1.87	1.83	1.78	1.72	1.66	1.59
22	2.95	2.56	2.35	2.22	2.13	2.06	1.97	1.86	1.81	1.76	1.70	1.64	1.57
23	2.94	2.55	2.34	2.21	2.11	2.05	1.95	1.84	1.80	1.74	1.69	1.62	1.55
24	2.93	2.54	2.33	2.19	2.10	2.04	1.94	1.83	1.78	1.73	1.67	1.61	1.53
25	2.92	2.53	2.32	2.18	2.09	2.02	1.93	1.82	1.77	1.72	1.66	1.59	1.52
26	2.91	2.52	2.31	2.17	2.08	2.01	1.92	1.81	1.76	1.71	1.65	1.58	1.50
27	2.90	2.51	2.30	2.17	2.07	2.00	1.91	1.80	1.75	1.70	1.64	1.57	1.49
28	2.89	2.50	2.29	2.16	2.06	2.00	1.90	1.79	1.74	1.69	1.63	1.56	1.48
29	2.89	2.50	2.28	2.15	2.06	1.99	1.89	1.78	1.73	1.68	1.62	1.55	1.47
30	2.88	2.49	2.28	2.14	2.05	1.98	1.88	1.77	1.72	1.67	1.61	1.54	1.46
40	2.84	2.44	2.23	2.09	2.00	1.93	1.83	1.71	1.66	1.61	1.54	1.47	1.38
60	2.79	2.39	2.18	2.04	1.95	1.87	1.77	1.66	1.60	1.54	1.48	1.40	1.29
120	2.75	2.35	2.13	1.99	1.90	1.82	1.72	1.60	1.55	1.48	1.41	1.32	1.19
∞	2.71	2.30	2.08	1.94	1.85	1.77	1.67	1.55	1.49	1.42	1.34	1.24	1.00

$$Q(F|\nu_1,\nu_2)=0.05$$

$\nu_2\backslash\nu_1$	1	2	3	4	5	6	8	12	15	20	30	60	∞
1	161.4	199.5	215.7	224.6	230.2	234.0	238.9	243.9	245.9	248.0	250.1	252.2	254.3
2	18.51	19.00	19.16	19.25	19.30	19.33	19.37	19.41	19.43	19.45	19.46	19.48	19.50
3	10.13	9.55	9.28	9.12	9.01	8.94	8.85	8.74	8.70	8.66	8.62	8.57	8.53
4	7.71	6.94	6.59	6.39	6.26	6.16	6.04	5.91	5.86	5.80	5.75	5.69	5.63
5	6.61	5.79	5.41	5.19	5.05	4.95	4.82	4.68	4.62	4.56	4.50	4.43	4.36
6	5.99	5.14	4.76	4.53	4.39	4.28	4.15	4.00	3.94	3.87	3.81	3.74	3.67
7	5.59	4.74	4.35	4.12	3.97	3.87	3.73	3.57	3.51	3.44	3.38	3.30	3.23
8	5.32	4.46	4.07	3.84	3.69	3.58	3.44	3.28	3.22	3.15	3.08	3.01	2.93
9	5.12	4.26	3.86	3.63	3.48	3.37	3.23	3.07	3.01	2.94	2.86	2.79	2.71
10	4.96	4.10	3.71	3.48	3.33	3.22	3.07	2.91	2.85	2.77	2.70	2.62	2.54
11	4.84	3.98	3.59	3.36	3.20	3.09	2.95	2.79	2.72	2.65	2.57	2.49	2.40
12	4.75	3.89	3.49	3.26	3.11	3.00	2.85	2.69	2.62	2.54	2.47	2.38	2.30
13	4.67	3.81	3.41	3.18	3.03	2.92	2.77	2.60	2.53	2.46	2.38	2.30	2.21
14	4.60	3.74	3.34	3.11	2.96	2.85	2.70	2.53	2.46	2.39	2.31	2.22	2.13
15	4.54	3.68	3.29	3.06	2.90	2.79	2.64	2.48	2.40	2.33	2.25	2.16	2.07
16	4.49	3.63	3.24	3.01	2.85	2.74	2.59	2.42	2.35	2.28	2.19	2.11	2.01
17	4.45	3.59	3.20	2.96	2.81	2.70	2.55	2.38	2.31	2.23	2.15	2.06	1.96
18	4.41	3.55	3.16	2.93	2.77	2.66	2.51	2.34	2.27	2.19	2.11	2.02	1.92
19	4.38	3.52	3.13	2.90	2.74	2.63	2.48	2.31	2.23	2.16	2.07	1.98	1.88
20	4.35	3.49	3.10	2.87	2.71	2.60	2.45	2.28	2.20	2.12	2.04	1.95	1.84
21	4.32	3.47	3.07	2.84	2.68	2.57	2.42	2.25	2.18	2.10	2.01	1.92	1.81
22	4.30	3.44	3.05	2.82	2.66	2.55	2.40	2.23	2.15	2.07	1.98	1.89	1.78
23	4.28	3.42	3.03	2.80	2.64	2.53	2.37	2.20	2.13	2.05	1.96	1.86	1.76
24	4.26	3.40	3.01	2.78	2.62	2.51	2.36	2.18	2.11	2.03	1.94	1.84	1.73
25	4.24	3.39	2.99	2.76	2.60	2.49	2.34	2.16	2.09	2.01	1.92	1.82	1.71
26	4.23	3.37	2.98	2.74	2.59	2.47	2.32	2.15	2.07	1.99	1.90	1.80	1.69
27	4.21	3.35	2.96	2.73	2.57	2.46	2.31	2.13	2.06	1.97	1.88	1.79	1.67
28	4.20	3.34	2.95	2.71	2.56	2.45	2.29	2.12	2.04	1.96	1.87	1.77	1.65
29	4.18	3.33	2.93	2.70	2.55	2.43	2.28	2.10	2.03	1.94	1.85	1.75	1.64
30	4.17	3.32	2.92	2.69	2.53	2.42	2.27	2.09	2.01	1.93	1.84	1.74	1.62
40	4.08	3.23	2.84	2.61	2.45	2.34	2.18	2.00	1.92	1.84	1.74	1.64	1.51
60	4.00	3.15	2.76	2.53	2.37	2.25	2.10	1.92	1.84	1.75	1.65	1.53	1.39
120	3.92	3.07	2.68	2.45	2.29	2.17	2.02	1.83	1.75	1.66	1.55	1.43	1.25
∞	3.84	3.00	2.60	2.37	2.21	2.10	1.94	1.75	1.67	1.57	1.46	1.32	1.00

Table 6 (cont.)

PERCENTAGE POINTS OF THE F-DISTRIBUTION
(Values of F in Terms of Q, ν_1, ν_2)

$Q(F|\nu_1,\nu_2)=0.025$

$\nu_2\backslash\nu_1$	1	2	3	4	5	6	8	12	15	20	30	60	∞
1	647.8	799.5	864.2	899.6	921.8	937.1	956.7	976.7	984.9	993.1	1001	1010	1018
2	38.51	39.00	39.17	39.25	39.30	39.33	39.37	39.41	39.43	39.45	39.46	39.48	39.50
3	17.44	16.04	15.44	15.10	14.88	14.73	14.54	14.34	14.25	14.17	14.08	13.99	13.90
4	12.22	10.65	9.98	9.60	9.36	9.20	8.98	8.75	8.66	8.56	8.46	8.36	8.26
5	10.01	8.43	7.76	7.39	7.15	6.98	6.76	6.52	6.43	6.33	6.23	6.12	6.02
6	8.81	7.26	6.60	6.23	5.99	5.82	5.60	5.37	5.27	5.17	5.07	4.96	4.85
7	8.07	6.54	5.89	5.52	5.29	5.12	4.90	4.67	4.57	4.47	4.36	4.25	4.14
8	7.57	6.06	5.42	5.05	4.82	4.65	4.43	4.20	4.10	4.00	3.89	3.78	3.67
9	7.21	5.71	5.08	4.72	4.48	4.32	4.10	3.87	3.77	3.67	3.56	3.45	3.33
10	6.94	5.46	4.83	4.47	4.24	4.07	3.85	3.62	3.52	3.42	3.31	3.20	3.08
11	6.72	5.26	4.63	4.28	4.04	3.88	3.66	3.43	3.33	3.23	3.12	3.00	2.88
12	6.55	5.10	4.47	4.12	3.89	3.73	3.51	3.28	3.18	3.07	2.96	2.85	2.72
13	6.41	4.97	4.35	4.00	3.77	3.60	3.39	3.15	3.05	2.95	2.84	2.72	2.60
14	6.30	4.86	4.24	3.89	3.66	3.50	3.29	3.05	2.95	2.84	2.73	2.61	2.49
15	6.20	4.77	4.15	3.80	3.58	3.41	3.20	2.96	2.86	2.76	2.64	2.52	2.40
16	6.12	4.69	4.08	3.73	3.50	3.34	3.12	2.89	2.79	2.68	2.57	2.45	2.32
17	6.04	4.62	4.01	3.66	3.44	3.28	3.06	2.82	2.72	2.62	2.50	2.38	2.25
18	5.98	4.56	3.95	3.61	3.38	3.22	3.01	2.77	2.67	2.56	2.44	2.32	2.19
19	5.92	4.51	3.90	3.56	3.33	3.17	2.96	2.72	2.62	2.51	2.39	2.27	2.13
20	5.87	4.46	3.86	3.51	3.29	3.13	2.91	2.68	2.57	2.46	2.35	2.22	2.09
21	5.83	4.42	3.82	3.48	3.25	3.09	2.87	2.64	2.53	2.42	2.31	2.18	2.04
22	5.79	4.38	3.78	3.44	3.22	3.05	2.84	2.60	2.50	2.39	2.27	2.14	2.00
23	5.75	4.35	3.75	3.41	3.18	3.02	2.81	2.57	2.47	2.36	2.24	2.11	1.97
24	5.72	4.32	3.72	3.38	3.15	2.99	2.78	2.54	2.44	2.33	2.21	2.08	1.94
25	5.69	4.29	3.69	3.35	3.13	2.97	2.75	2.51	2.41	2.30	2.18	2.05	1.91
26	5.66	4.27	3.67	3.33	3.10	2.94	2.73	2.49	2.39	2.28	2.16	2.03	1.88
27	5.63	4.24	3.65	3.31	3.08	2.92	2.71	2.47	2.36	2.25	2.13	2.00	1.85
28	5.61	4.22	3.63	3.29	3.06	2.90	2.69	2.45	2.34	2.23	2.11	1.98	1.83
29	5.59	4.20	3.61	3.27	3.04	2.88	2.67	2.43	2.32	2.21	2.09	1.96	1.81
30	5.57	4.18	3.59	3.25	3.03	2.87	2.65	2.41	2.31	2.20	2.07	1.94	1.79
40	5.42	4.05	3.46	3.13	2.90	2.74	2.53	2.29	2.18	2.07	1.94	1.80	1.64
60	5.29	3.93	3.34	3.01	2.79	2.63	2.41	2.17	2.06	1.94	1.82	1.67	1.48
120	5.15	3.80	3.23	2.89	2.67	2.52	2.30	2.05	1.94	1.82	1.69	1.53	1.31
∞	5.02	3.69	3.12	2.79	2.57	2.41	2.19	1.94	1.83	1.71	1.57	1.39	1.00

$Q(F|\nu_1,\nu_2)=0.01$

$\nu_2\backslash\nu_1$	1	2	3	4	5	6	8	12	15	20	30	60	∞
1	4052	4999.5	5403	5625	5764	5859	5982	6106	6157	6209	6261	6313	6366
2	98.50	99.00	99.17	99.25	99.30	99.33	99.37	99.42	99.43	99.45	99.47	99.48	99.50
3	34.12	30.82	29.46	28.71	28.24	27.91	27.49	27.05	26.87	26.69	26.50	26.32	26.13
4	21.20	18.00	16.69	15.98	15.52	15.21	14.80	14.37	14.20	14.02	13.84	13.65	13.46
5	16.26	13.27	12.06	11.39	10.97	10.67	10.29	9.89	9.72	9.55	9.38	9.20	9.02
6	13.75	10.92	9.78	9.15	8.75	8.47	8.10	7.72	7.56	7.40	7.23	7.06	6.88
7	12.25	9.55	8.45	7.85	7.46	7.19	6.84	6.47	6.31	6.16	5.99	5.82	5.65
8	11.26	8.65	7.59	7.01	6.63	6.37	6.03	5.67	5.52	5.36	5.20	5.03	4.86
9	10.56	8.02	6.99	6.42	6.06	5.80	5.47	5.11	4.96	4.81	4.65	4.48	4.31
10	10.04	7.56	6.55	5.99	5.64	5.39	5.06	4.71	4.56	4.41	4.25	4.08	3.91
11	9.65	7.21	6.22	5.67	5.32	5.07	4.74	4.40	4.25	4.10	3.94	3.78	3.60
12	9.33	6.93	5.95	5.41	5.06	4.82	4.50	4.16	4.01	3.86	3.70	3.54	3.36
13	9.07	6.70	5.74	5.21	4.86	4.62	4.30	3.96	3.82	3.66	3.51	3.34	3.17
14	8.86	6.51	5.56	5.04	4.69	4.46	4.14	3.80	3.66	3.51	3.35	3.18	3.00
15	8.68	6.36	5.42	4.89	4.56	4.32	4.00	3.67	3.52	3.37	3.21	3.05	2.87
16	8.53	6.23	5.29	4.77	4.44	4.20	3.89	3.55	3.41	3.26	3.10	2.93	2.75
17	8.40	6.11	5.18	4.67	4.34	4.10	3.79	3.46	3.31	3.16	3.00	2.83	2.65
18	8.29	6.01	5.09	4.58	4.25	4.01	3.71	3.37	3.23	3.08	2.92	2.75	2.57
19	8.18	5.93	5.01	4.50	4.17	3.94	3.63	3.30	3.15	3.00	2.84	2.67	2.49
20	8.10	5.85	4.94	4.43	4.10	3.87	3.56	3.23	3.09	2.94	2.78	2.61	2.42
21	8.02	5.78	4.87	4.37	4.04	3.81	3.51	3.17	3.03	2.88	2.72	2.55	2.36
22	7.95	5.72	4.82	4.31	3.99	3.76	3.45	3.12	2.98	2.83	2.67	2.50	2.31
23	7.88	5.66	4.76	4.26	3.94	3.71	3.41	3.07	2.93	2.78	2.62	2.45	2.26
24	7.82	5.61	4.72	4.22	3.90	3.67	3.36	3.03	2.89	2.74	2.58	2.40	2.21
25	7.77	5.57	4.68	4.18	3.85	3.63	3.32	2.99	2.85	2.70	2.54	2.36	2.17
26	7.72	5.53	4.64	4.14	3.82	3.59	3.29	2.96	2.81	2.66	2.50	2.33	2.13
27	7.68	5.49	4.60	4.11	3.78	3.56	3.26	2.93	2.78	2.63	2.47	2.29	2.10
28	7.64	5.45	4.57	4.07	3.75	3.53	3.23	2.90	2.75	2.60	2.44	2.26	2.06
29	7.60	5.42	4.54	4.04	3.73	3.50	3.20	2.87	2.73	2.57	2.41	2.23	2.03
30	7.56	5.39	4.51	4.02	3.70	3.47	3.17	2.84	2.70	2.55	2.39	2.21	2.01
40	7.31	5.18	4.31	3.83	3.51	3.29	2.99	2.66	2.52	2.37	2.20	2.02	1.80
60	7.08	4.98	4.13	3.65	3.34	3.12	2.82	2.50	2.35	2.20	2.03	1.84	1.60
120	6.85	4.79	3.95	3.48	3.17	2.96	2.66	2.34	2.19	2.03	1.86	1.66	1.38
∞	6.63	4.61	3.78	3.32	3.02	2.80	2.51	2.18	2.04	1.88	1.70	1.47	1.00

Table 6 (cont.)

PERCENTAGE POINTS OF THE F-DISTRIBUTION
(Values of F in Terms of Q, ν_1, ν_2)

$$Q(F|\nu_1,\nu_2)=0.005$$

ν_2\\ν_1	1	2	3	4	5	6	8	12	15	20	30	60	∞
1	16211	20000	21615	22500	23056	23437	23925	24426	24630	24836	25044	25253	25465
2	198.5	199.0	199.2	199.2	199.3	199.3	199.4	199.4	199.4	199.4	199.5	199.5	199.5
3	55.55	49.80	47.47	46.19	45.39	44.84	44.13	43.39	43.08	42.78	42.47	42.15	41.83
4	31.33	26.28	24.26	23.15	22.46	21.97	21.35	20.70	20.44	20.17	19.89	19.61	19.32
5	22.78	18.31	16.53	15.56	14.94	14.51	13.96	13.38	13.15	12.90	12.66	12.40	12.14
6	18.63	14.54	12.92	12.03	11.46	11.07	10.57	10.03	9.81	9.59	9.36	9.12	8.88
7	16.24	12.40	10.88	10.05	9.52	9.16	8.68	8.18	7.97	7.75	7.53	7.31	7.08
8	14.69	11.04	9.60	8.81	8.30	7.95	7.50	7.01	6.81	6.61	6.40	6.18	5.95
9	13.61	10.11	8.72	7.96	7.47	7.13	6.69	6.23	6.03	5.83	5.62	5.41	5.19
10	12.83	9.43	8.08	7.34	6.87	6.54	6.12	5.66	5.47	5.27	5.07	4.86	4.64
11	12.23	8.91	7.60	6.88	6.42	6.10	5.68	5.24	5.05	4.86	4.65	4.44	4.23
12	11.75	8.51	7.23	6.52	6.07	5.76	5.35	4.91	4.72	4.53	4.33	4.12	3.90
13	11.37	8.19	6.93	6.23	5.79	5.48	5.08	4.64	4.46	4.27	4.07	3.87	3.65
14	11.06	7.92	6.68	6.00	5.56	5.26	4.86	4.43	4.25	4.06	3.86	3.66	3.44
15	10.80	7.70	6.48	5.80	5.37	5.07	4.67	4.25	4.07	3.88	3.69	3.48	3.26
16	10.58	7.51	6.30	5.64	5.21	4.91	4.52	4.10	3.92	3.73	3.54	3.33	3.11
17	10.38	7.35	6.16	5.50	5.07	4.78	4.39	3.97	3.79	3.61	3.41	3.21	2.98
18	10.22	7.21	6.03	5.37	4.96	4.66	4.28	3.86	3.68	3.50	3.30	3.10	2.87
19	10.07	7.09	5.92	5.27	4.85	4.56	4.18	3.76	3.59	3.40	3.21	3.00	2.78
20	9.94	6.99	5.82	5.17	4.76	4.47	4.09	3.68	3.50	3.32	3.12	2.92	2.69
21	9.83	6.89	5.73	5.09	4.68	4.39	4.01	3.60	3.43	3.24	3.05	2.84	2.61
22	9.73	6.81	5.65	5.02	4.61	4.32	3.94	3.54	3.36	3.18	2.98	2.77	2.55
23	9.63	6.73	5.58	4.95	4.54	4.26	3.88	3.47	3.30	3.12	2.92	2.71	2.48
24	9.55	6.66	5.52	4.89	4.49	4.20	3.83	3.42	3.25	3.06	2.87	2.66	2.43
25	9.48	6.60	5.46	4.84	4.43	4.15	3.78	3.37	3.20	3.01	2.82	2.61	2.38
26	9.41	6.54	5.41	4.79	4.38	4.10	3.73	3.33	3.15	2.97	2.77	2.56	2.33
27	9.34	6.49	5.36	4.74	4.34	4.06	3.69	3.28	3.11	2.93	2.73	2.52	2.29
28	9.28	6.44	5.32	4.70	4.30	4.02	3.65	3.25	3.07	2.89	2.69	2.48	2.25
29	9.23	6.40	5.28	4.66	4.26	3.98	3.61	3.21	3.04	2.86	2.66	2.45	2.21
30	9.18	6.35	5.24	4.62	4.23	3.95	3.58	3.18	3.01	2.82	2.63	2.42	2.18
40	8.83	6.07	4.98	4.37	3.99	3.71	3.35	2.95	2.78	2.60	2.40	2.18	1.93
60	8.49	5.79	4.73	4.14	3.76	3.49	3.13	2.74	2.57	2.39	2.19	1.96	1.69
120	8.18	5.54	4.50	3.92	3.55	3.28	2.93	2.54	2.37	2.19	1.98	1.75	1.43
∞	7.88	5.30	4.28	3.72	3.35	3.09	2.74	2.36	2.19	2.00	1.79	1.53	1.00

$$Q(F|\nu_1,\nu_2)=0.001$$

ν_2\\ν_1	1	2	3	4	5	6	8	12	15	20	30	60	∞
1	(5)4.053	(5)5.000	(5)5.404	(5)5.625	(5)5.764	(5)5.859	(5)5.981	(5)6.107	(5)6.158	(5)6.209	(5)6.261	(5)6.313	(5)6.366
2	998.5	999.0	999.2	999.2	999.3	999.3	999.4	999.4	999.4	999.4	999.5	999.5	999.5
3	167.0	148.5	141.1	137.1	134.6	132.8	130.6	128.3	127.4	126.4	125.4	124.5	123.5
4	74.14	61.25	56.18	53.44	51.71	50.53	49.00	47.41	46.76	46.10	45.43	44.75	44.05
5	47.18	37.12	33.20	31.09	29.75	28.84	27.64	26.42	25.91	25.39	24.87	24.33	23.79
6	35.51	27.00	23.70	21.92	20.81	20.03	19.03	17.99	17.56	17.12	16.67	16.21	15.75
7	29.25	21.69	18.77	17.19	16.21	15.52	14.63	13.71	13.32	12.93	12.53	12.12	11.70
8	25.42	18.49	15.83	14.39	13.49	12.86	12.04	11.19	10.84	10.48	10.11	9.73	9.33
9	22.86	16.39	13.90	12.56	11.71	11.13	10.37	9.57	9.24	8.90	8.55	8.19	7.81
10	21.04	14.91	12.55	11.28	10.48	9.92	9.20	8.45	8.13	7.80	7.47	7.12	6.76
11	19.69	13.81	11.56	10.35	9.58	9.05	8.35	7.63	7.32	7.01	6.68	6.35	6.00
12	18.64	12.97	10.80	9.63	8.89	8.38	7.71	7.00	6.71	6.40	6.09	5.76	5.42
13	17.81	12.31	10.21	9.07	8.35	7.86	7.21	6.52	6.23	5.93	5.63	5.30	4.97
14	17.14	11.78	9.73	8.62	7.92	7.43	6.80	6.13	5.85	5.56	5.25	4.94	4.60
15	16.59	11.34	9.34	8.25	7.57	7.09	6.47	5.81	5.54	5.25	4.95	4.64	4.31
16	16.12	10.97	9.00	7.94	7.27	6.81	6.19	5.55	5.27	4.99	4.70	4.39	4.06
17	15.72	10.66	8.73	7.68	7.02	6.56	5.96	5.32	5.05	4.78	4.48	4.18	3.85
18	15.38	10.39	8.49	7.46	6.81	6.35	5.76	5.13	4.87	4.59	4.30	4.00	3.67
19	15.08	10.16	8.28	7.26	6.62	6.18	5.59	4.97	4.70	4.43	4.14	3.84	3.51
20	14.82	9.95	8.10	7.10	6.46	6.02	5.44	4.82	4.56	4.29	4.00	3.70	3.38
21	14.59	9.77	7.94	6.95	6.32	5.88	5.31	4.70	4.44	4.17	3.88	3.58	3.26
22	14.38	9.61	7.80	6.81	6.19	5.76	5.19	4.58	4.33	4.06	3.78	3.48	3.15
23	14.19	9.47	7.67	6.69	6.08	5.65	5.09	4.48	4.23	3.96	3.68	3.38	3.05
24	14.03	9.34	7.55	6.59	5.98	5.55	4.99	4.39	4.14	3.87	3.59	3.29	2.97
25	13.88	9.22	7.45	6.49	5.88	5.46	4.91	4.31	4.06	3.79	3.52	3.22	2.89
26	13.74	9.12	7.36	6.41	5.80	5.38	4.83	4.24	3.99	3.72	3.44	3.15	2.82
27	13.61	9.02	7.27	6.33	5.73	5.31	4.76	4.17	3.92	3.66	3.38	3.08	2.75
28	13.50	8.93	7.19	6.25	5.66	5.24	4.69	4.11	3.86	3.60	3.32	3.02	2.69
29	13.39	8.85	7.12	6.19	5.59	5.18	4.64	4.05	3.80	3.54	3.27	2.97	2.64
30	13.29	8.77	7.05	6.12	5.53	5.12	4.58	4.00	3.75	3.49	3.22	2.92	2.59
40	12.61	8.25	6.60	5.70	5.13	4.73	4.21	3.64	3.40	3.15	2.87	2.57	2.23
60	11.97	7.76	6.17	5.31	4.76	4.37	3.87	3.31	3.08	2.83	2.55	2.25	1.89
120	11.38	7.32	5.79	4.95	4.42	4.04	3.55	3.02	2.78	2.53	2.26	1.95	1.54
∞	10.83	6.91	5.42	4.62	4.10	3.74	3.27	2.74	2.51	2.27	1.99	1.66	1.00

Table 7

TWO THOUSAND NORMAL (N(0,1)) DEVIATES

0000	1.276-	1.218-	.453-	.350-	.723	.676	1.099-	.314-	.394-	.633-
0001	.318-	.799-	1.664-	1.391	.382	.733	.653	.219	.681-	1.129
0002	1.377-	1.257-	.495	.139-	.854-	.428	1.322-	.315-	.732-	1.348-
0003	2.334	.337-	1.955-	.636-	1.318-	.433-	.545	.428	.297-	.276
0004	1.136-	.642	3.436	1.667-	.847	1.173-	.355-	.035	.359	.930
0005	.414	.011-	.666	1.132-	.410-	1.077-	.734	1.484	.340-	.789
0006	.494-	.364	1.237-	.044-	.111-	.210-	.931	.616	.377-	.433-
0007	1.048	.037	.759	.609	2.043-	.290-	.404	.543-	.486	.869
0008	.347	2.816	.464-	.632-	1.614-	.372	.074-	.916-	1.314	.038-
0009	.637	.563	.107-	.131	1.808-	1.126-	.379	.610	.364-	2.626-
0010	2.176	.393	.924-	1.911	1.040-	1.168-	.485	.076	.769-	1.607
0011	1.185-	.944-	1.604-	.185	.258-	.300-	.591-	.545-	.018	.485-
0012	.972	1.710	2.682	2.813	1.531-	.490-	2.071	1.444	1.092-	.478
0013	1.210	.294	.248-	.719	1.103	1.090	.212	1.185-	.338-	1.134-
0014	2.647	.777	.450	2.247	1.151	1.676-	.384	1.133	1.393	.814
0015	.398	.318	.928-	2.416	.936-	1.036	.024	.560-	.203	.871-
0016	.846	.699-	.368-	.344	.926-	.797-	1.404-	1.472-	.118-	1.456
0017	.654	.955-	2.907	1.688	.752	.434-	.746	.149	.170-	.479-
0018	.522	.231	.619-	.265-	.419	.558	.549-	.192	.334-	1.373
0019	1.288-	.539-	.824-	.244	1.070-	.010	.482	.469-	.090-	1.171
0020	1.372	1.769	1.057-	1.646	.481	.600-	.592-	.610	.096-	1.375-
0021	.854	.535-	1.607	.428	.615-	.331	.336-	1.152-	.533	.833-
0022	.148-	1.144-	.913	.684	1.043	.554	.051-	.944-	.440-	.212-
0023	1.148-	1.056-	.635	.328-	1.221-	.118	2.045-	1.977-	1.133-	.338
0024	.348	.970	.017-	1.217	.974-	1.291-	.399-	1.209-	.248-	.480
0025	.284	.458	1.307	1.625-	.629-	.504-	.056-	.131-	.048	1.879
0026	1.016-	.360	.119-	2.331	1.672	1.053-	.840	.246-	.237	1.312-
0027	1.603	.952-	.566-	1.600	.465	1.951	.110	.251	.116	.957-
0028	.190-	1.479	.986-	1.249	1.934	.070	1.358-	1.246-	.959-	1.297-
0029	.722-	.925	.783	.402-	.619	1.826	1.272	.945-	.494	.050
0030	1.696-	1.879	.063	.132	.682	.544	.417-	.666-	.104-	.253-
0031	2.543-	1.333-	1.987	.668	.360	1.927	1.183	1.211	1.765	.035
0032	.359-	.193	1.023-	.222-	.616-	.060-	1.319-	.785	.430-	.298-
0033	.248	.088-	1.379-	.295	.115-	.621-	.618-	.209	.979	.906
0034	.099-	1.376-	1.047	.872-	2.200-	1.384-	1.425	.812-	.748	1.093-
0035	.463-	1.281-	2.514-	.675	1.145	1.083	.667-	.223-	1.592-	1.278-
0036	.503	1.434	.290	.397	.837-	.973-	.120-	1.594-	.996-	1.244-
0037	.857-	.371-	.216-	.148	2.106-	1.453-	.686	.075-	.243-	.170-
0038	.122-	1.107	1.039-	.636-	.860-	.895-	1.458-	.539-	.159-	.420-
0039	1.632	.586	.468-	.386-	.354-	.203	1.234-	2.381	.388-	.063-
0040	2.072	1.445-	.680-	.224	.120-	1.753	.571-	1.223	.126-	.034
0041	.435-	.375-	.985-	.585-	.203-	.556-	.024	.126	1.250	.615-
0042	.876	1.227-	2.647-	.745-	1.797	1.231-	.547	.634-	.836-	.719-
0043	.833	1.289	.022-	.431-	.582	.766	.574-	1.153-	.520	1.018-
0044	.891-	.332	.453-	1.127-	2.085	.722-	1.508-	.489	.496-	.025-
0045	.644	.233-	.153-	1.098	.757	.039-	.460-	.393	2.012	1.356
0046	.105	.171-	.110-	1.145-	.878	.909-	.328-	1.021	1.613-	1.560
0047	1.192-	1.770	.003-	.369	.052	.647	1.029	1.526	.237	1.328-
0048	.042-	.553	.770	.324	.489-	.367-	.378	.601	1.996-	.738-
0049	.498	1.072	1.567	.302	1.157	.720-	1.403-	.698	.370-	.551-

Table 7 (cont.)

TWO THOUSAND NORMAL (N(0,1)) DEVIATES

0050	1.329-	.238-	.838-	.988-	.445-	.964	.266-	.322-	1.726-	2.252
0051	1.284	.229-	1.058	.090	.050	.523	.016	.277	1.639	.554
0052	.619	.628	.005	.973	.058-	.150	.635-	.917-	.313	1.203-
0053	.699	.269-	.722	.994-	.807-	1.203-	1.163	1.244	1.306	1.210-
0054	.101	.202	.150-	.731	.420	.116	.496-	.037-	2.466-	.794
0055	1.381-	.301	.522	.233	.791	1.017-	.182-	.926	1.096-	1.001
0056	.574-	1.366	1.843-	.746	.890	.824	1.249-	.806-	.240-	.217
0057	.096	.210	1.091	.990	.900	.837-	1.097-	1.238-	.030	.311-
0058	1.389	.236-	.094	3.282	.295	.416-	.313	.720	.007	.354
0059	1.249	.706	1.453	.366	2.654-	1.400-	.212	.307	1.145-	.639
0060	.756	.397-	1.772-	.257-	1.120	1.188	.527-	.709	.479	.317
0061	.860-	.412	.327-	.178	.524	.672-	.831-	.758	.131	.771
0062	.778-	.979-	.236	1.033-	1.497	.661-	.906	1.169	1.582-	1.303
0063	.037	.062	.426	1.220	.471	.784	.719-	.465	1.559	1.326-
0064	2.619	.440-	.477	1.063	.320	1.406	.701-	.128-	.518	.676-
0065	.420-	.287-	.050-	.481-	1.521	1.367-	.609	.292	.048	.592
0066	1.048	.220	1.121	1.789-	1.211-	.871-	.740-	.513	.558-	.395-
0067	1.000	.638-	1.261	.510	.150-	.034	.054	.055-	.639	.825-
0068	.170	1.131-	.985-	.102	.939-	1.457-	1.766	1.087	1.275-	2.362
0069	.389	.435-	.171	.891	1.158	1.041	1.048	.324-	.404-	1.060
0070	.305-	.838	2.019-	.540-	.905	1.195	1.190-	.106	.571	.298
0071	.321-	.039-	1.799	1.032-	2.225-	.148-	.758	.862-	.158	.726-
0072	1.900	1.572	.244-	1.721-	1.130	.495	.484-	.014	.778-	1.483-
0073	.778-	.288-	.224-	1.324-	.072-	.890	.410-	.752	.376	.224-
0074	.617	1.718-	.183-	.100-	1.719	.696	1.339-	.614-	1.071	.386-
0075	1.430-	.953-	.770	.007-	1.872-	1.075	.913-	1.168-	1.775	.238
0076	.267	.048-	.972	.734	1.408-	1.955-	.848-	2.002	.232	1.273-
0077	.978	.520-	.368-	1.690	1.479-	.985	1.475	.098-	1.633-	2.399
0078	1.235-	1.168-	.325	1.421	2.652	.486-	1.253-	.270	1.103-	.118
0079	.258-	.638	2.309	.741	.161-	.679-	.336	1.973	.370	2.277-
0080	.243	.629	1.516-	.157-	.693	1.710	.800	.265-	1.218	.655
0081	.292-	1.455-	1.451-	1.492	.713-	.821	.031-	.780-	1.330	.977
0082	.505-	.389	.544	.042-	1.615	1.440-	.989-	.580-	.156	.052
0083	.397	.287-	1.712	.289	.904-	.259	.600-	1.635-	.009-	.799-
0084	.605-	.470-	.007	.721	1.117-	.635	.592	1.362-	1.441-	.672
0085	1.360	.182	1.476-	.599-	.875-	.292	.700-	.058	.340-	.639-
0086	.480	.699-	1.615	.225-	1.014	1.370-	1.097-	.294	.309	1.389-
0087	.027-	.487-	1.000-	.015-	.119	1.990-	.687-	1.964-	.366-	1.759
0088	1.482-	.815-	.121-	1.884	.185-	.601	.793	.430	1.181-	.426
0089	1.256-	.567-	.994-	1.011	1.071-	.623-	.420-	.309-	1.362	.863
0090	1.132-	2.039	1.934	.222-	.386	1.100	.284	1.597	1.718-	.560-
0091	.780-	.239-	.497-	.434-	.284-	.241-	.333-	1.348	.478-	.169-
0092	.859-	.215-	.241	1.471	.389	.952-	.245	.781	1.093	.240-
0093	.447	1.479	.067	.426	.370-	.675-	.972-	.225-	.815	.389
0094	.269	.735	.066-	.271-	1.439-	1.036	.306-	1.439-	.122-	.336-
0095	.097	1.883-	.218-	.202	.357-	.019	1.631	1.400	.223	.793-
0096	.686-	1.596	.286-	.722	.655	.275-	1.245	1.504-	.066	1.280-
0097	.957	.057	1.153-	.701	.280-	1.747	.745-	1.338	1.421-	.386
0098	.976-	1.789-	.696-	1.799-	.354-	.071	2.355	.135	.598-	1.883
0099	.274	.226	.909-	.572-	.181	1.115	.406	.453	1.218-	.115-

Table 7 (cont.)

TWO THOUSAND NORMAL (N(0,1)) DEVIATES

```
0100   1.752-   .329- 1.256-   .318  1.531   .349   .958-   .059-   .415  1.084-
0101    .291-   .085  1.701  1.087-  .443-   .292-  .248    .539- 1.382-   .318
0102    .933-   .130   .634   .899  1.409   .883-   .095-   .229   .129   .367
0103    .450-   .244-  .072  1.028  1.730   .056- 1.488-   .078- 2.361-   .992-
0104    .512    .882-  .490  1.304-  .266-  .757   .361-   .194  1.078-   .529

0105    .702-   .472   .429   .664-  .592- 1.443  1.515- 1.209- 1.043-   .278
0106    .284    .039   .518- 1.351  1.473   .889   .300    .339   .206- 1.392
0107    .509-  1.420   .782-   .429- 1.266-  .627 1.165-   .819   .261-   .409
0108   1.776-  1.033- 1.977   .014   .702   .435-   .816- 1.131   .656   .061
0109    .044-  1.807   .342  2.510- 1.071  1.220-  .060-   .764-   .079   .964-

0110    .263    .578- 1.612   .148-  .383- 1.007-  .414-   .638   .186-   .507
0111    .986    .439   .192-   .132-  .167   .883   .400- 1.440-   .385- 1.414-
0112    .441-   .852- 1.446-   .605-  .348- 1.018   .963    .004- 2.504    .847-
0113    .866-   .489   .097   .379   .192   .842-   .065  1.420   .426  1.191-
0114   1.215-   .675  1.621   .394  1.447- 2.199   .321-   .540-   .037-   .185

0115    .475-  1.210-  .183   .526   .495  1.297  1.613- 1.241  1.016-   .090-
0116   1.200    .131  2.502   .344  1.060-  .909- 1.695-   .666-   .838-   .866-
0117    .498-  1.202-  .057- 1.354- 1.441- 1.590-  .987    .441   .637  1.116-
0118    .743-   .894   .028- 1.119   .598-  .279  2.241    .830   .267    .156-
0119    .779    .780-  .954-   .705   .361-  .734- 1.365  1.297   .142- 1.387-

0120    .206-   .195- 1.017  1.167-  .079-  .452-  .058  1.068-   .394-   .406-
0121    .092-   .927-  .439-   .256   .503   .338  1.511    .465-   .118-   .454-
0122   1.222-  1.582- 1.786   .517- 1.080-  .409-  .474- 1.890-   .247    .575
0123    .068    .075  1.383-   .084-  .159  1.276  1.141    .186   .973-   .266-
0124    .183   1.600   .335- 1.553   .889   .896   .035-   .461   .486  1.246

0125    .811-  2.904-  .618   .588   .533   .803   .696-   .690   .820   .557
0126   1.010-  1.149  1.033   .336  1.306   .835  1.523    .296   .426-   .004
0127   1.453   1.210   .043-   .220   .256- 1.161- 2.030-   .046-   .243  1.082
0128    .759    .838-  .877-   .177- 1.183   .218- 3.154-   .963-   .822- 1.114-
0129    .287-   .278   .454-   .897   .122-  .013   .346    .921   .238    .586-

0130    .669-   .035  2.077- 1.077   .525   .154- 1.036-   .015   .220-   .882
0131    .392    .106  1.430-   .204-  .326-  .825   .432-   .094- 1.566-   .679-
0132    .337-   .199   .160-   .625   .891- 1.464-  .318-  1.297   .932    .032-
0133    .369   1.990- 1.190-   .666  1.614-  .082   .922    .139-   .833-   .091
0134   1.694-   .710   .655-   .546- 1.654   .134   .466    .033   .039-   .838

0135    .985    .340   .276   .911   .170-  .551- 1.000    .838-   .275    .304-
0136   1.063-   .594- 1.526-   .787-  .873   .405- 1.324-   .162   .163- 2.716-
0137    .033   1.527- 1.422   .308   .845   .151-  .741    .064  1.212    .823
0138    .597    .362  3.760- 1.159   .874   .794-  .915- 1.215  1.627  1.248-
0139   1.601-   .570-  .133   .660- 1.485   .682   .898-   .686   .658    .346

0140    .266-  1.309-  .597   .989   .934  1.079   .656-   .999-   .036-   .537-
0141    .901   1.531   .889- 1.019-  .084  1.531   .144- 1.920-   .678    .402-
0142   1.433-  1.008-  .990-   .090   .940   .207   .745-   .638  1.469  1.214
0143   1.327    .763  1.724-   .709- 1.100- 1.346-  .946-   .157-   .522  1.264-
0144    .248-   .788   .577-   .122   .536-  .293  1.207  2.243- 1.642  1.353

0145    .401-   .679-  .921   .476  1.121   .864-  .128    .551-   .872- 1.511
0146    .344    .324-  .686  1.487-  .126-  .803   .961-   .183   .358-   .184-
0147    .441    .372- 1.336-   .062  1.506   .315-  .112-   .452- 1.594    .264-
0148    .824    .040  1.734-   .251   .054   .379- 1.298    .126-   .104    .529-
0149   1.385   1.320   .509-   .381- 1.671-  .524-  .805- 1.348   .676    .799
```

Table 7 (cont.)

TWO THOUSAND NORMAL (N(0,1)) DEVIATES

0150	1.556	.119	.078-	.164	.455-	.077	.043-	.299-	.249	.182-
0151	.647	1.029	1.186	.887	1.204	.657-	.644	.410-	.652-	.165-
0152	.329	.407	1.169	2.072-	1.661	.891	.233	1.628-	.762-	.717-
0153	1.188-	1.171	1.170-	.291-	.863	.045-	.205-	.574	.926-	1.407
0154	.917-	.616-	1.589-	1.184	.266	.559	1.833-	.572-	.648-	1.090-
0155	.414	.469	.182-	.397	1.649	1.198	.067	1.526-	.081-	.192-
0156	.107	.187-	1.343	.472	.112-	1.182	.548	2.748	.249	.154
0157	.497-	1.907	.191	.136	.475-	.458	.183	1.640-	.058-	1.278
0158	.501	.083	.321-	1.133	1.126	.299-	1.299	1.617	1.581	2.455-
0159	1.382-	.738-	1.225	1.564	.363-	.548-	1.070	.390	1.398-	.524
0160	.590-	.699	.162-	.011-	1.049	.689-	1.225	.339	.539-	.445-
0161	1.125-	1.111	1.065-	.534	.102	.425	1.026-	.695	.057-	.795
0162	.849	.169	.351-	.584	2.177	.009	.696-	.426-	.692-	1.638-
0163	1.233-	.585-	.306	.773	1.304	1.304-	.282	1.705-	.187	.880-
0164	.104	.468-	.185	.498	.624-	.322-	.875-	1.478	.691-	.281-
0165	.261	1.883-	.181-	1.675	.324-	1.029-	.185-	.004	.101-	1.187-
0166	.007-	1.280	.568	1.270-	1.405	1.731	2.072	1.686	.728	.417-
0167	.794	.111-	.040	.536-	.976-	2.192	1.609	.190-	.279-	1.611-
0168	.431	2.300-	1.081-	1.370-	2.943	.653	2.523-	.756	.886	.983-
0169	.149-	1.294	.580-	.482	1.449-	1.067-	1.996	.274-	.721	.490
0170	.216-	1.647-	1.043	.481	.011-	.587-	.916-	1.016-	1.040-	1.117-
0171	1.604	.851-	.317-	.686-	.008-	1.939	.078	.465-	.533	.652
0172	.212-	.005	.535	.837	.362	1.103	.219	.488	1.332	.200-
0173	.007	.076-	1.484	.455	.207-	.554-	1.120	.913	.681-	1.751
0174	.217-	.937	.860	.323	1.321	.492-	1.386-	.003-	.230-	.539
0175	.649-	.300	.698-	.900	.569	.842	.804	1.025	.603	1.546-
0176	1.541-	.193	2.047	.552-	1.190	.087-	2.062	2.173-	.791-	.520-
0177	.274	.530-	.112	.385	.656	.436	.882	.312	2.265-	.218-
0178	.876	1.498-	.128-	.387-	1.259-	.856-	.353-	.714	.863	1.169
0179	.859-	1.083-	1.288	.078-	.081-	.210	.572	1.194	1.118-	1.543-
0180	.015-	.567-	.113	2.127	.719-	3.256	.721-	.663-	.779-	.930-
0181	1.529-	.231-	1.223	.300	.995-	.651-	.505	.138	.064-	1.341
0182	.278	.058-	2.740-	.296-	1.180-	.574	1.452	.846	.243-	1.208-
0183	1.428	.322	2.302	.852-	.782	1.322-	.092-	.546-	.560	1.430-
0184	.770	1.874-	.347	.994	.485-	1.179-	.048	1.324-	1.061	.449
0185	.303-	.629-	.764	.013	1.192-	.475-	1.085-	.880-	1.738	1.225-
0186	.263-	2.105-	.509	.645-	1.362	.504	.755-	1.274	1.448	.604
0187	.997	1.187-	.242-	.121	2.510	1.935-	.350	.073	.458	.446-
0188	.063-	.475-	1.802-	.476-	.193	1.199-	.339	.364	.684-	1.353
0189	.168-	1.904	.485-	.032-	.554-	.056	.710-	.778-	.722	.024-
0190	.366	.491-	.301	.008-	.894-	.945-	.384	1.748-	1.118-	.394
0191	.436	.464-	.539	.942	.458-	.445	1.883-	1.228	1.113	.218-
0192	.597	1.471-	.434-	.705	.788-	.575	.086	.504	1.445	.513-
0193	.805-	.624-	1.344	.649	1.124-	.680	.986-	1.845	1.152-	.393-
0194	1.681	1.910-	.440	.067	1.502-	.755-	.989-	.054-	2.320-	.474
0195	.007-	.459-	1.940	.220	1.259-	1.729-	.137	.520-	.412-	2.847
0196	.209	.633-	.299	.174	1.975	.271-	.119	.199-	.007	2.315-
0197	1.254	1.672	1.186-	1.310-	.474	.878	.725-	.191-	.642	1.212-
0198	1.016-	.697-	.017	.263-	.047-	1.294-	.339-	2.257	.078-	.049-
0199	1.169-	.355-	1.086	.199-	.031	.396	.143-	1.572	.276	.027

BIBLIOGRAPHY AND REFERENCES

Aigner, D. J., and C. N. Morris (eds.), "Experimental Design in Econometrics," *Journal of Econometrics*, Vol. 11, No. 1, September 1979, North-Holland.

Andrews, D., et al., *Robust Estimates of Location: Survey and Advances*, Princeton University Press, Princeton, New Jersey, 1972.

Ashton, W. D., *The Logit Transformation* (Griffin's Statistical Monographs and Courses, A. Stuart (ed.)), Hafner Publishing Company, New York, 1972.

Belsey, D. A., E. Kuh, and R. E. Welch, *Regression Diagnostics: Identifying Influential Data and Sources of Colinearity*, John Wiley & Sons, Inc., New York, 1980.

Blackwell, D., *Basic Statistics*, McGraw-Hill, New York, 1969, Chapters 6-8.

Blalock, H., *Social Statistics*, McGraw-Hill, New York, 1960, Chapters 17-19.

Box, G.E.P., and D. R. Cox, "An Analysis of Transformations," *Journal of the Royal Statistical Society*, Series B, No. 2, 1964, pp. 211-252.

------------, and N. Draper, "A Basis for the Selection of a Response Surface Design," *Journal of the American Statistical Association*, Vol. 54, No. 287, 1959, pp. 622-654.

Bradley, J., *Distribution-Free Statistical Tests*, Prentice-Hall, Englewood Cliffs, New Jersey, 1969, Chapters 1-3, 5, 7, 13.

Brelsford, W. A. and D. A. Relles, *STATLIB, A Statistical Computing Library* (Prentice-Hall, 1981).

Carter, G. M., and J. E. Rolph, *New York City Fire Alarm Prediction Models: 1. Box-reported Serious Fires*, The Rand Corporation, R-1214-NYC, 1973.

------------, "Empirical Bayes Methods Applied to Estimating Fire Alarm Probabilities," *Journal of the American Statistical Association*, Vol. 69, No. 348, December 1974, pp. 880-885.

Cochran, W. G., "The χ^2 Test of Goodness of Fit," *Annals of Mathematical Statistics*, Vol. 23, 1952, pp. 315-345.

------------, *Sampling Techniques*, 2d Ed., John Wiley & Sons, Inc., New York, 1953.

------------, and G. M. Cox, *Experimental Designs*, 2d Ed., John Wiley & Sons, Inc., New York, 1957.

Cox, D. R., *Planning of Experiments*, John Wiley & Sons, Inc., New York, 1958.

------------, *Analysis of Binary Data*, Methuen and Co., Ltd., London, 1970.

------------, and P.A.W. Lewis, *The Statistical Analysis of Series of Events*, Methuen and Co., Ltd., London, 1966, Chapter 3.

Cramer, H., *Mathematical Methods in Statistics*, Princeton University Press, Princeton, New Jersey, 1945.

Daniel, C., and F. S. Wood, *Fitting Equations to Data*, Wiley-Interscience, New York, 1971, Chapters 3, 4, 5, 6, 7.

Dixon, W. J., and J. W. Tukey, "Approximation Behavior of the Distribution of the Winsorized t (Trimming/Winsorization 2)," *Technometrics*, Vol. 10, No. 1, 1968, pp. 83-98.

Draper, N. R., and H. Smith, *Applied Regression Analysis*, John Wiley & Sons, Inc., New York, 1966, Chapters 3-4.

Efron, B., and C. Morris, "Stein's Estimation Rule and Its Competitors--An Empirical Bayes Approach," *Journal of the American Statistical Association*, Vol. 68, No. 541, 1973, pp. 117-130.

------------, "Stein's Paradox in Statistics," *Scientific American*, May 1977, pp. 119-127.

------------, *Data Analysis Using Stein's Estimator and Its Generalizations*, The Rand Corporation, R-1394-OEO, March 1974; also *Journal of the American Statistical Association*, Vol. 70, No. 350, June 1975, pp. 311-319.

Fay, R., and R. Herriott, "Estimates of Income for Small Places: An Application of James-Stein Procedures to Census Data," *Journal of the American Statistical Association*, Vol. 74, No. 366, June 1979, pp. 269-277.

Freund, J., *Modern Elementary Statistics*, 3d Ed., Prentice-Hall, Englewood Cliffs, New Jersey, 1967, Chapter 15.

Graybill, F., *An Introduction to Linear Statistical Models*, McGraw-Hill, New York, 1961.

Haggstrom, G. W., *The Foundations of Probability and Mathematical Statistics*, The Rand Corporation, P-5394, March 1975.

Hodges, J. L., Jr., and E. L. Lehmann, *Basic Concepts of Probability and Statistics*, Holden-Day, San Francisco, 1964, Chapter 12.

Hogg, R. V., and A. T. Craig, *Introduction to Mathematical Statistics*, 4th Ed., The Macmillan Company, New York, 1978.

Keeler, E. B., and W. H. Rogers, *A Classification of Large American Urban Areas*, The Rand Corporation, R-1246-NSF, May 1973.

Kendall, M. G., and A. Stuart, *The Advanced Theory of Statistics*, 2d Ed., Vol. 2, Hafner Publishing Company, New York, 1961, pp. 337-342.

Kish, L., *Survey Sampling*, John Wiley & Sons, Inc., New York, 1967.

Klitgaard, R., and G. Hall, *A Statistical Search for Unusually Effective Schools*, The Rand Corporation, R-1210-CC/RC, March 1973.

Kruskal, J. B., "Multidimensional Scaling by Optimizing Goodness-of-Fit to a Non-Metric Hypothesis," *Psychometrika*, Vol. 29, March 1964, pp. 1-27.

Lehmann, E. L., *Nonparametrics: Statistical Methods Based on Ranks*, Holden-Day, San Francisco, 1975.

Lohnes, P. R., and W. W. Cooley, *Introduction to Statistical Procedures: With Computer Exercises*, John Wiley & Sons, Inc., New York, 1968.

Lowry, I. S., et al., *First Annual Report of the Housing Assistance Supply Experiment*, The Rand Corporation, R-1659-HUD, October 1974.

Manning, W. G., B. M. Mitchell, and J. P. Acton, *Design of the Los Angeles Peak-Load Pricing Experiment for Electricity*, The Rand Corporation, R-1955-DWP, November 1976.

Maxwell, A. E., *Analyzing Qualitative Data*, Methuen and Co., Ltd., London, 1961.

Mood, A., F. Graybill, and D. C. Boes, *Introduction to the Theory of Statistics*, McGraw-Hill, New York, 1974.

Morris, C., *Considerations for a Sequential Housing Allowance Demand Experiment*, The Rand Corporation, P-4895, June 1972.

------------, J. P. Newhouse, and R. Archibald, "On the Theory and Practice of Obtaining Unbiased and Efficient Samples in Social Surveys and Experiments," in *Research in Experimental Economics*, Vernon L. Smith (ed.), Vol. 1, JAI Press, 1979.

Mosteller, F., et al. (eds.), *Statistics by Example: Detecting Patterns; Statistics by Example: Weighing Chances; Statistics by Example: Finding Models; Statistics by Example: Exploring Data*, Addison-Wesley, Menlo Park, California, 1973.

------------, and J. W. Tukey, *Data Analysis and Regression*, Addison-Wesley, Reading, Massachusetts, 1977.

Neter, J., and W. Wasserman, *Applied Linear Statistical Models*, R. D. Irwin, Inc., Homewood, Illinois, 1974, Chapters 6 and 7.

Newhouse, J. P., *The Health Insurance Study--A Summary*, The Rand Corporation, R-965-1-OEO, March 1974.

Novick, M. R., and P. H. Jackson, *Statistical Methods for Educational and Psychological Research*, McGraw-Hill, New York, 1974.

Owen, D., *Handbook of Statistical Tables*, Addison-Wesley, Reading, Massachusetts, 1962.

Prais, S. J., and J. Aitchison, "The Grouping of Observations in Regression Analysis," *Revue de L'Institut International de Statistique*, Vol. 22 (1954), pp. 1-22.

Press, S. J., *Applied Multivariate Analysis*, Holt, Rinehart, and Winston, New York, 1972, Chapter 15.

Puri, M. L., and P. K. Sen, *Nonparametric Methods in Multivariate Analysis*, John Wiley & Sons, Inc., New York, 1971.

Rao, C. R., *Linear Statistical Inference and Its Applications*, 2d Ed., John Wiley & Sons, Inc., New York, 1965.

Sampson, A. R., "A Tale of Two Regressions," *Journal of the American Statistical Association*, Vol. 69, No. 347, 1974, pp. 682-689.

Savage, I. R., *Statistics: Uncertainty and Behavior*, Houghton-Mifflin, Boston, Massachusetts, 1968.

Scheffe, H., *The Analysis of Variance*, John Wiley & Sons, Inc., New York, 1959.

Schmitt, S. A., *Measuring Uncertainty, An Elementary Introduction to Bayesian Statistics*, Addison-Wesley, Reading, Massachusetts, 1969.

Searle, S. R., *Linear Models*, John Wiley & Sons, Inc., New York, 1971.

Shapiro, S. S., and M. B. Wilk, "An Analysis of Variance Test for Normality (Complete Samples)," *Biometrika*, Vol. 52, 1965, pp. 591-611.

Siegel, S., *Nonparametric Statistics for the Behavioral Sciences*, McGraw-Hill, New York, 1956, pp. 42-47, 104-111, 175-179.

Smith, Vernon L. (ed.), *Research in Experimental Economics*, JAI Press, Vol. 1, 1979.

Spiegel, M. R., *Schaum's Outline of Theory and Problems in Statistics*, Schaum Publishing Co., New York, 1961, Chapters 7-15.

Stigler, S. M., "Do Robust Estimators Work with *Real* Data?" with discussion, *The Annals of Statistics*, Vol. 5, No. 6, 1977, pp. 1055-1098.

Tanur, J. M., F. Mosteller, W. Kruskal, et al., *Statistics: A Guide to the Unknown*, Houghton-Mifflin, Boston, Massachusetts, 1968.

Theil, H., *Principles of Econometrics*, John Wiley & Sons, Inc., New York, 1971.

Tufte, E. R. (ed.), *The Quantitative Analysis of Social Problems*, Addison-Wesley, Reading, Massachusetts, 1970.

------------, *Data Analysis for Politics and Policy*, Prentice-Hall, Inc., Englewood Cliffs, New Jersey, 1974.

Tukey, J. W., *Exploratory Data Analysis*, Addison-Wesley, Reading, Massachusetts, 1977.

------------. "Discussion, Emphasizing the Connection Between Analysis of Variance and Spectrum Analysis," *Technometrics*, Vol. 3, No. 2, May 1961, pp. 191-220.

------------, and D. H. McLaughlin, "Less Vulnerable Confidence and Significance Procedures for Location Based on a Single Sample (Trimming/Winsorization 1)," *Sankhya*, Series A, Vol. 25, No. 3, 1963, pp. 331-352.

Walker, W. E., et al. (eds.), *Fire Department Deployment Analysis*, Elsevier-North Holland, New York, 1979.

Wallis, W. A., and H. V. Roberts, *Statistics: A New Approach*, The Free Press, Glencoe, Illinois, 1956.

Watts, H. W., and J. Conlisk, "A Model for Optimizing Experimental Designs for Estimating Response Surfaces," *1969 Proceedings of the Social Statistics Section*, American Statistical Association, 1969, pp. 150-156.

Weisberg, S., *Applied Linear Regression*, John Wiley & Sons, Inc., New York, 1980.

Yarnold, J. K., "The Minimum Expectation in χ^2 Goodness-of-Fit Tests and the Accuracy of Approximation for the Null Distribution," *Journal of the American Statistical Association*, Vol. 65, 1970, pp. 864-886.

Yuen, K. K., and V. K. Murthy, "Percentage Points of the Distribution of the t Statistic When the Parent is Student's t," *Technometrics*, Vol. 16, No. 4, November 1974, pp. 495-497.

Rand Graduate Institute Published Books

deLeon, Peter. *Development and Diffusion of the Nuclear Power Reactor: A Comparative Analysis.* Cambridge, Massachusetts: Ballinger Publishing Company, 1979. A Rand Graduate Institute Book.

Goldhamer, Herbert. *The Adviser.* New York: Elsevier-North Holland, Inc., 1978. A Rand Graduate Institute Book.

Morris, Carl N. and John E. Rolph. *Introduction to Data Analysis and Statistical Inference.* Englewood Cliffs, New Jersey: Prentice-Hall, Inc., 1980. A Rand Graduate Institute Book.

Selected Rand Books on Policy Analysis

Armor, David J., J. Michael Polich, and Harriet B. Stambul. *Alcoholism and Treatment.* New York: John Wiley and Sons, Inc., 1978.

Brewer, Garry D., and Martin Shubik. *The War Game: A Critique of Military Problem Solving.* Cambridge, Massachusetts: Harvard University Press, 1979.

Fishman, George S. *Spectral Methods in Econometrics.* Cambridge, Massachusetts: Harvard University Press, 1969.

Jorgenson, D. W., J. J. McCall, and R. Radner. *Optimal Replacement Policy.* Amsterdam, The Netherlands: North-Holland Publishing Company, 1967.

Mitchell, Bridger, Willard G. Manning, Jr., and Jan P. Acton. *Peak-Load Pricing: European Lessons for U.S. Energy Policy.* Cambridge, Massachusetts: Ballinger Publishing Company, 1978.

Park, Rolla Edward. *The Role of Analysis in Regulatory Decision-Making.* Lexington, Massachusetts: D. C. Heath and Company, 1973.

Quade, E. S. *Analysis for Public Decisions.* New York: American Elsevier Publishing Company, 1975.

Rolph, Elizabeth S. *Nuclear Power and the Public Safety: A Study in Regulation.* Lexington, Massachusetts: D. C. Heath & Company, 1979.

Sackman, Harold. *Delphi Critique: Expert Opinion, Forecasting, and Group Process.* Lexington, Massachusetts: D. C. Heath & Company, 1975.

Smith, James P. *Female Labor Supply.* Princeton, New Jersey: Princeton University Press, 1980.

Walker, Warren E., Jan M. Chaiken, and Edward J. Ignall, eds. *Fire Department Deployment Analysis: A Public Policy Analysis Case Study.* New York: Elsevier North-Holland, Inc., 1980.

INDEX

384

Variance (*Contd.*):
  analysis of (*see* Analysis of variance)
  finite sampling correction factor for, 320-24
  heterogeneity of, weighted least squares and, 249-60
  in logit regression, 196-97, 202-203
  matrix notation of, 89
  in multiple regression, 66, 73-74, 76, 78, 83-84, 89, 91-92
  in probability plotting, 220-21, 224
  in random sampling, 112-13, 118-19, 125-26
  in simple regression, 38-52
  in statistical inference, 142-45, 152-53, 155-56, 261-64

W

Walker, W. E., 226, 252, 256, 377, 379
Wallis, W. A., 377
Wasserman, W., 89, 375
Watts, H. W., 344, 377
Weighted least squares:
  defined, 250

Weighted least squares (*Contd.*):
  heterogeneity of variance and, 249-60
  logit regression for grouped data using, 194-201
Weisberg, S., 236, 377
Welch, R. E., 236, 373
Wilcoxon one-sample test, 272-76, 281
Wilcoxon two-sample test, 280-81
Wilk, M. B., 224, 376
Winsorized mean, 262-68
Winsorized t-test, 276
Wood, F. S., 374

Y

Yarnold, J. K., 291, 377
Yuen, K., 266, 377

Z

Zero correlation, 117